Qualitative Research Methods in Air Transport Management

Yeşim Kurt
School of Edremit Civil Aviation, Department of Aviation Management, Balıkesir University, Turkey

IGI Global
Scientific Publishing
Publishing Tomorrow's Research Today

Vice President of Editorial	Melissa Wagner
Managing Editor of Acquisitions	Mikaela Felty
Managing Editor of Book Development	Jocelynn Hessler
Production Manager	Mike Brehm
Cover Design	Phillip Shickler

Published in the United States of America by
IGI Global Scientific Publishing
701 East Chocolate Avenue
Hershey, PA, 17033, USA
Tel: 717-533-8845
Fax: 717-533-8661
Website: https://www.igi-global.com E-mail: cust@igi-global.com

Library of Congress Cataloging-in-Publication Data

Names: Kurt, Yeşim, 1985- editor.
Title: Qualitative research methods in air transport management / edited by, Yeşim Kurt.
Description: Hershey, PA : IGI Global Scientific Publishing, [2025] | Includes bibliographical references and index. | Summary: "This book clearly establishes the relationship between air transport and qualitative research methods. It discusses the need for qualitative research to understand the unique nature of the industry"-- Provided by publisher.
Identifiers: LCCN 2024055937 (print) | LCCN 2024055938 (ebook) | ISBN 9798369374030 (hardcover) | ISBN 9798369374047 (paperback) | ISBN 9798369374054 (ebook)
Subjects: LCSH: Airlines--Management. | Qualitative research--Methodology. | Aeronautics, Commercial--Management. | Airports--Management.
Classification: LCC HE9780 .Q83 2025 (print) | LCC HE9780 (ebook) | DDC 387.7068--dc23/eng/20250203
LC record available at https://lccn.loc.gov/2024055937
LC ebook record available at https://lccn.loc.gov/2024055938

British Cataloguing in Publication Data
A Cataloguing in Publication record for this book is available from the British Library.

Table of Contents

Detailed Table of Contents

in addressing the evolving needs of the aviation sector and ensuring the highest standards of communication and safety in aviation operations.

Preface

The aviation industry stands at the crossroads of rapid innovation and complex challenges. To navigate these dynamics, it is crucial to integrate diverse research methods that allow for a deeper understanding of the multifaceted world of air transport management. This book, *Qualitative Research Methods in Air Transport Management*, brings together cutting-edge insights from scholars and experts to explore the value of qualitative research in this vital sector. It examines how qualitative methodologies offer unique contributions to the study of human behaviors, organizational dynamics, operational challenges, and sustainability in aviation.

In recent years, the aviation industry has recognized the importance of qualitative research methods in understanding the nuances of passenger experiences, organizational cultures, and management practices. While quantitative data provides essential metrics, qualitative research illuminates the human factors that shape decision-making, innovation, and service quality. This book aims to bridge the gap by offering a comprehensive exploration of qualitative research techniques and their application in various areas of air transport management.

CHAPTER OVERVIEW

Chapter 1: The Importance of Qualitative Research in the Aviation Industry

By Satya Subrahmanyam, this chapter underscores the significance of qualitative research in uncovering insights into operational challenges and the human dimensions of aviation. It demonstrates how qualitative approaches provide a deeper understanding of issues such as pilot fatigue, passenger experiences, and organizational culture, ultimately fostering innovation and informed decision-making.

Chapter 2: Qualitative Research Methods - Harnessing Interviews, Focus Groups, Observations, and Document Analysis

Nusrat Khan, Fehmina Khalique, and Kartikay Saini delve into the diverse data collection methods central to qualitative research. They explore interviews, focus groups, participant observations, and document analysis, offering guidance on selecting the most appropriate methods based on research design and objectives.

Chapter 3: Evaluating Qualitative Data Collection Methods in Aviation Management - Challenges and Opportunities

Mustafa Uzgör evaluates the challenges faced when employing qualitative data collection methods in aviation management. The chapter addresses issues such as data complexity, ethical considerations, and the rapid changes in the aviation environment, while offering insights into how qualitative data can guide effective decision-making.

Chapter 4: Focus Group Research and Applicability of Focus Groups in the Air Transport Studies

Ozlem Colak introduces the focus group methodology and its applicability in air transport research. By exploring the dynamic and interactive nature of focus groups, the chapter highlights their usefulness in understanding key areas like management, marketing, organizational culture, and operations.

Chapter 5: Case Study as a Research Design in Air Transport Management

Hilal Lapçin outlines the case study methodology and its application in air transport management. This chapter provides guidance on selecting cases, gathering data, and ensuring validity, offering an in-depth exploration of how this approach contributes to understanding complex aviation phenomena.

Chapter 6: Trustworthiness in Qualitative Research in the Aviation Industry

Satya Subrahmanyam revisits the theme of trustworthiness, emphasizing its critical role in maintaining the credibility of qualitative research findings. The chapter discusses strategies such as triangulation, member checking, and ethical considerations, which are essential in ensuring the validity of research outcomes in aviation.

Chapter 7: Qualitative Research Methods in Sustainability Aviation Research

Gülaçti Sen examines the role of qualitative methods in studying sustainability within the aviation industry. The chapter explores how qualitative approaches help address the sector's environmental and operational challenges, with a focus on human factors and decision-making processes in sustainability initiatives.

Chapter 8: Sustainable Operations in the Aviation Industry: A Qualitative Study on Behavioural Conformity and Contradictions

Armagan Macit and Deniz Macit present a qualitative study investigating sustainability practices within aviation. Through interviews with aviation interns, they reveal insights into sustainable and unsustainable behaviors, offering valuable lessons for enhancing sustainability efforts in the industry.

Chapter 9: Challenges and Solution Approaches for Unmanned Aerial Vehicles in Air Traffic Control, a Case Study

Arif Tuncal explores the challenges of integrating Unmanned Aerial Vehicles (UAVs) into air traffic control systems. This case study, based on interviews with air traffic controllers, identifies key issues such as training, technology, and legal frameworks, providing recommendations for future UAV management.

Chapter 10: Every Air Travel Begins with a Customer Journey: Applying Service Design Principles in Airport Planning and Operations

Elad Harison and Yael Lahav focus on improving the passenger experience by applying service design principles to airport operations. By mapping the entire passenger journey, they offer a framework for enhancing airport services, from addressing pain points to leveraging emerging technologies.

Chapter 11: Understanding Aviation English Course Design and Delivery Through Qualitative Research

Nicole Falzon and Daniel Xerri highlight the importance of Aviation English (AE) in ensuring safety and communication standards in the aviation sector. This chapter explores the role of qualitative research in enhancing AE course design and delivery, demonstrating its critical contribution to training personnel in the aviation industry.

CONCLUSION

The chapters within this book provide a comprehensive and multifaceted look at how qualitative research methods can be applied to diverse aspects of air transport management. From sustainability challenges and operational efficiencies to the human elements that shape passenger experiences and organizational behaviors, qualitative research offers a powerful toolkit for understanding the complexities of aviation. We hope this book inspires further exploration and innovation, promoting the broader adoption of qualitative methods in aviation studies and practices.

Acknowledgments

This book has been brought to life through the invaluable contributions of many individuals who shared their expertise and precious time.

First and foremost, I extend my deepest gratitude to the academics and researchers who authored the chapters of this book. Their knowledge and experience have significantly enriched the qualitative depth of this work.

I am also profoundly grateful to the reviewers and critics who helped maintain the high standards of quality throughout the review process. Their constructive feedback ensured that the book reached a more comprehensive and balanced content.

Finally, I would like to thank the IGI Global team for their unwavering support and guidance throughout the publication process. Their encouragement and assistance have been pivotal in completing this work on time and to the highest standards.

I hope this book contributes to advancing the understanding and application of qualitative research methods in the aviation sector, supporting processes such as data collection, analysis, and establishing trustworthiness in this critical field.

Associate Professor Yesim KURT

Balıkesir University, School of Edremit Civil Aviation, Department of Aviation Management. Turkey

Section 1
Theoretical Background

Theoretical background to the relationships between qualitative research methods and air transport management

Chapter 1
The Importance of Qualitative Research in the Aviation Industry

Satya Subrahmanyam
https://orcid.org/0000-0003-0441-2742
Holy Spirit University of Kaslik, Lebanon

ABSTRACT

This chapter emphasizes the indispensable role of qualitative research in advancing the aviation industry. It begins by outlining the necessity and significant contributions of qualitative methods, which delve into complex aviation dynamics often overlooked by quantitative approaches. The discussion explores how qualitative research uncovers nuanced insights into operational challenges, passenger experiences, and organizational cultures within aviation. Examples illustrate its impact, from improving safety protocols based on pilot fatigue studies to enhancing customer service through insights into passenger preferences. The chapter advocates for broader recognition and integration of qualitative research in aviation education and practice, aiming to foster innovation and informed decision-making across the industry.

1. INTRODUCTION

Qualitative research serves as a cornerstone of academic inquiry and practical exploration, offering deep insights into human experiences, behaviors, and perceptions. Unlike quantitative research, which focuses on numerical data and statistical analysis, qualitative research delves into the subjective and contextual dimensions of phenomena (Creswell & Poth, 2018). This approach allows researchers to capture nuanced understandings that are often overlooked by traditional quantitative

DOI: 10.4018/979-8-3693-7403-0.ch001

methods. By prioritizing the exploration of meaning, context, and lived experiences, qualitative research contributes to generating rich, descriptive data that can inform decision-making, policy development, and innovation across various fields (Denzin & Lincoln, 2017).

In the aviation industry, qualitative research has a unique and indispensable role. The aviation sector is inherently complex, involving intricate systems, high stakes, and interactions among diverse stakeholders, including passengers, crew members, regulators, and service providers. These complexities often go beyond what quantitative metrics can capture. For instance, understanding passenger satisfaction, pilot decision-making under stress, or the organizational culture of airlines requires an exploration of perceptions, emotions, and behaviors. Qualitative research methodologies such as interviews, focus groups, case studies, and ethnography enable a comprehensive examination of these aspects, offering actionable insights for industry leaders and policymakers.

This chapter aims to explore the importance of qualitative research in the aviation industry by highlighting its value in addressing key challenges and opportunities within the sector. Specifically, it will examine how qualitative methods contribute to understanding passenger experiences, improving safety protocols, enhancing employee engagement, and fostering innovation. By doing so, the chapter underscores the indispensable role of qualitative research in navigating the complexities of modern aviation.

1.1 Significance of Qualitative Research in Academic and Practical Contexts

Qualitative research has gained prominence for its ability to provide an in-depth understanding of phenomena that are difficult to quantify. In academic contexts, it enriches theoretical frameworks by uncovering the underlying mechanisms and meanings of observed behaviors and interactions (Bryman, 2016). Researchers can develop new theories or refine existing ones through grounded insights derived from qualitative data. For example, in organizational behavior studies, qualitative research has been instrumental in revealing the dynamics of leadership, communication, and team collaboration (Yin, 2018).

In practical applications, qualitative research serves as a tool for exploring real-world problems and generating context-specific solutions. Industries that rely heavily on human factors, such as healthcare, education, and aviation, benefit significantly from qualitative approaches. These methods allow practitioners to understand stakeholder needs, identify pain points, and co-create strategies that address unique challenges. For instance, in customer service, qualitative research provides insights into consumer preferences and expectations, enabling companies to tailor their offerings accordingly (Patton, 2015).

1.2 Relevance of Qualitative Research to the Aviation Industry

The aviation industry represents a complex intersection of technology, human interaction, and regulatory frameworks, requiring a multifaceted approach to understanding its dynamics. Qualitative research offers a critical lens to capture both the technical and human dimensions of the sector, making it an invaluable tool for addressing industry challenges. By focusing on subjective experiences, behaviors, and contextual insights, qualitative methods provide a deeper understanding of the aviation industry's intricacies that quantitative approaches may not fully capture.

Understanding Passenger Experiences is pivotal for airlines, as passenger satisfaction directly influences customer loyalty and brand reputation. However, satisfaction is often shaped by subjective factors such as perceptions of comfort, safety, and service quality, which cannot be fully quantified. Qualitative methods, including in-depth interviews and focus groups, allow airlines to gather rich, detailed data on passenger preferences and concerns. Such insights enable airlines to design and implement strategies to enhance the overall travel experience, improving competitiveness in the market (Han et al., 2019).

Enhancing Safety Protocols is another critical application of qualitative research in aviation. Safety forms the foundation of aviation operations, with human factors playing a pivotal role in maintaining high safety standards. Through qualitative research, industry professionals can explore the cognitive and emotional dimensions of critical safety-related activities, such as pilot decision-making, air traffic controller communication, and crew coordination. Techniques like ethnography and narrative analysis help uncover behaviors and patterns that contribute to or mitigate safety risks, offering valuable insights for improving training programs and operational protocols (Dekker, 2017).

The aviation workforce, which includes pilots, cabin crew, and ground staff, operates in high-pressure environments that demand resilience, teamwork, and adaptability. Employee Engagement and Organizational Culture significantly influence morale, job satisfaction, and retention. Qualitative research is instrumental in understanding the factors that affect these aspects. For example, case studies and semi-structured interviews can reveal how leadership styles, workplace policies, and organizational culture impact employee well-being and productivity. Such insights are essential for creating supportive environments that foster collaboration and efficiency (Gillen et al., 2020).

As the aviation industry continues to evolve with technological advancements, Driving Innovation has become a priority for sustaining competitiveness. Emerging technologies such as artificial intelligence (AI), biometrics, and green energy solutions require thoughtful integration into existing systems. Qualitative research provides a platform for understanding stakeholder perceptions of these innovations, ensuring that new technologies are accepted and effectively utilized. By employing participatory research methods, companies can collaborate with customers and employees to co-create solutions, enhancing the likelihood of successful implementation and long-term benefits (Stamper, 2021).

1.3 Objectives of the Chapter

This chapter aims to achieve three core objectives, emphasizing the importance of qualitative research in the aviation industry.

The first objective is to highlight the value of qualitative methods in addressing aviation challenges. By exploring how qualitative research contributes to solving real-world problems, the chapter will demonstrate its role in improving passenger experiences, ensuring safety, enhancing employee engagement, and driving innovation. These applications showcase the practical relevance of qualitative approaches in meeting industry demands.

The second objective is to explore the complexities of the aviation sector through a qualitative lens. The chapter will delve into case studies and existing literature to illustrate how qualitative methods uncover insights that quantitative approaches may overlook. This includes understanding subjective experiences, contextual factors, and human behaviors that shape the aviation industry's operations.

Finally, the chapter seeks to emphasize the practical implications of qualitative research. By offering actionable recommendations, it will highlight how qualitative insights can inform decision-making, enhance operational efficiency, and support sustainable growth. This approach ensures that the findings are not only theoretical but also applicable to real-world aviation challenges.

By achieving these objectives, the chapter underscores the indispensable role of qualitative research in navigating the complexities and opportunities within the aviation industry. Through this exploration, readers will gain a comprehensive understanding of how qualitative methods contribute to the sector's growth, safety, and innovation.

2. UNDERSTANDING QUALITATIVE RESEARCH IN AVIATION

The aviation industry is a dynamic and multifaceted sector that demands a nuanced understanding of its operations, stakeholders, and challenges. This chapter explores the role of qualitative research in addressing the complexities of aviation by emphasizing its exploratory nature, relevance to human experiences, and ability to complement quantitative approaches.

2.1 Definition and Key Features of Qualitative Research

Qualitative research is a methodological approach focused on exploring and understanding phenomena through an in-depth examination of experiences, behaviors, and social interactions within specific contexts. Unlike quantitative research, which emphasizes numerical data and statistical analysis, qualitative research prioritizes exploratory and interpretative methods to generate rich, nuanced insights.

One of the defining characteristics of qualitative research is its focus on subjective experiences and the contextual factors shaping them. This approach acknowledges the complexity and variability of human behavior, offering a lens to examine the emotional, cognitive, and cultural dimensions of experiences. For instance, qualitative methods such as interviews, focus groups, and ethnographic studies enable researchers to capture the lived realities of participants, providing insights into motivations, perceptions, and decision-making processes.

The exploratory nature of qualitative research is particularly valuable in sectors where understanding human interaction, behavior, and context is critical. In aviation, this translates into uncovering the underlying factors that influence passenger satisfaction, safety protocols, and organizational culture, which are often shaped by non-quantifiable elements such as attitudes, emotions, and communication patterns.

2.2 Unique Characteristics of the Aviation Industry

The aviation industry operates in a highly complex environment shaped by technological advancements, operational challenges, and stringent regulatory requirements. These factors create a unique ecosystem where the interplay between human and technical dimensions significantly influences outcomes.

One of the defining features of the aviation sector is its high complexity, arising from the integration of cutting-edge technologies, intricate operational systems, and compliance with international safety and regulatory frameworks. For instance, the development and deployment of new technologies such as autonomous systems or artificial intelligence must align with safety standards while addressing stakeholder concerns. This demands a research approach capable of capturing the nuances of technological integration within real-world contexts.

Another critical characteristic is the interplay between human factors, safety, customer satisfaction, and organizational efficiency. Human factors such as pilot decision-making, crew coordination, and air traffic control communication play a pivotal role in ensuring operational safety. Similarly, customer satisfaction is influenced by both technical aspects, such as flight punctuality, and subjective factors, such as the perceived quality of service.

The aviation workforce also adds to this complexity. Employees operate in high-pressure environments that require teamwork, resilience, and adaptability. Understanding these dynamics necessitates a focus on both individual experiences and collective interactions, which qualitative research is well-suited to explore.

2.3 Why Qualitative Research?

The unique demands of the aviation industry highlight the limitations of solely relying on quantitative methods, which often focus on measurable outcomes without delving into the underlying causes or contextual factors. Qualitative research addresses this gap by providing a deeper understanding of behavioral patterns, decision-making processes, and stakeholder experiences.

Insights into Behavioral Patterns and Decision-Making - Aviation involves critical decision-making at multiple levels, from strategic planning to real-time operations. Pilots, air traffic controllers, and ground staff often face scenarios requiring rapid judgment under pressure. Qualitative research techniques, such as narrative analysis or ethnography, allow researchers to study these decision-making processes in their natural context. By uncovering the cognitive and emotional dimensions of these decisions, qualitative studies contribute to designing better training programs and operational protocols.

Exploration of Experiences of Stakeholders - Stakeholders in the aviation industry include passengers, crew members, management, and regulatory bodies, each with distinct experiences and expectations. Passengers, for example, have subjective perceptions of comfort, safety, and service quality, which directly impact customer satisfaction and loyalty. Qualitative methods such as focus groups or in-depth interviews enable researchers to explore these perceptions, providing actionable insights for airlines to enhance the travel experience (Han et al., 2019).

Similarly, qualitative research sheds light on the experiences of employees, who are central to ensuring safety, efficiency, and customer satisfaction. Studies exploring organizational culture, job satisfaction, and employee engagement help identify factors that contribute to or hinder workforce productivity. For instance, semi-structured interviews with cabin crew may reveal how leadership styles and workplace policies impact morale and performance.

Addressing Gaps Left by Quantitative Studies - While quantitative research excels in measuring and analyzing trends, it often overlooks the contextual richness and individual variability that influence outcomes. In aviation, qualitative research complements quantitative approaches by addressing questions such as:

- Why do certain safety procedures fail despite compliance with regulations?
- How do passengers perceive changes in technology or service delivery?
- What motivates employees to adopt or resist organizational changes?

By addressing these questions, qualitative research provides a holistic understanding of challenges and opportunities, enabling stakeholders to make informed decisions. For instance, participatory research methods involving both employees and customers can guide the successful implementation of innovative technologies, such as biometric boarding systems or sustainable fuel initiatives (Stamper, 2021).

3. KEY AREAS WHERE QUALITATIVE RESEARCH IS ESSENTIAL

Qualitative research plays an indispensable role in uncovering the complexities of the aviation industry, addressing areas where human behavior, cultural dynamics, and subjective experiences are critical. This chapter focuses on the key domains where qualitative approaches contribute to enhancing understanding, improving operational practices, and guiding decision-making. These areas include passenger experience, safety management, organizational culture, innovation, and regulatory frameworks.

3.1 Passenger Experience and Service Quality

Passenger experience is a cornerstone of airline operations, directly impacting customer satisfaction, loyalty, and brand perception. Understanding passenger perceptions, preferences, and grievances is vital for tailoring services to diverse customer needs.

Understanding Passenger Perceptions and Preferences - Qualitative methods, such as in-depth interviews, focus groups, and ethnographic studies, enable airlines to explore passengers' subjective experiences. For instance, passengers' perceptions of comfort, safety, and service quality are influenced by numerous factors, including seating arrangements, inflight entertainment, and interactions with the cabin crew.

Insights gained from such studies can help airlines design customer-centric strategies to enhance satisfaction (Han et al., 2019).

Exploring Cultural and Regional Differences - The global nature of the aviation industry necessitates an understanding of cultural and regional differences in air travel expectations. Passengers from different cultural backgrounds may have varying preferences for meal options, service styles, and inflight communication. Qualitative research enables a granular understanding of these variations, helping airlines cater to diverse demographics effectively. For example, a study using participant observation could reveal specific cultural nuances that inform service delivery, enabling airlines to build stronger relationships with their clientele.

3.2 Human Factors and Safety Management

Safety is the top priority in aviation, and the role of human factors in maintaining safety cannot be overstated. Qualitative research provides a framework for investigating the attitudes, behaviors, and decision-making processes of pilots, crew, and ground staff, shedding light on safety challenges and opportunities for improvement.

Investigating Attitudes Toward Safety Protocols - Through methods such as semi-structured interviews and narrative analysis, qualitative research helps uncover how aviation personnel perceive and implement safety protocols. These studies often reveal gaps between formal procedures and real-world practices, highlighting areas for training or procedural enhancement (Dekker, 2017).

Decision-Making in High-Stress Environments - The aviation industry operates in high-stress conditions, where decisions must be made quickly and accurately. Qualitative research, particularly case studies and ethnographic methods, offers insights into the cognitive and emotional factors influencing decision-making during emergencies. For example, understanding how pilots prioritize tasks during adverse weather conditions can inform the development of simulation-based training programs tailored to real-life challenges.

3.3 Organizational Culture and Leadership

The aviation industry relies heavily on effective leadership and a strong organizational culture to drive performance and ensure safety. Qualitative research illuminates the interplay between leadership styles, workplace culture, and organizational responses to crises.

Role of Leadership Styles in Aviation Companies - Leadership in aviation plays a crucial role in shaping employee morale, job satisfaction, and performance outcomes. Qualitative methods, such as case studies and focus groups, enable a deeper understanding of how different leadership styles, such as transformational or transactional leadership, impact the workforce. For instance, interviews with aviation leaders can reveal how they motivate teams and foster collaboration in high-pressure environments (Gillen et al., 2020).

Organizational Responses to Crises - The aviation industry is highly susceptible to crises, including pandemics like COVID-19 and geopolitical tensions. Qualitative research helps analyze how organizations navigate such disruptions. By exploring employee and management perspectives through interviews and thematic analysis, researchers can identify best practices for crisis management and resilience building. These insights are invaluable for developing strategies to adapt to future challenges.

3.4 Innovation and Sustainability

The aviation industry's push toward innovation and sustainability presents both opportunities and challenges. Qualitative research sheds light on stakeholder perspectives, adoption hurdles, and strategies for integrating new technologies and practices.

Stakeholder Perspectives on Green Aviation Initiatives - Sustainability has become a key focus in aviation, with initiatives such as sustainable aviation fuels and electric aircraft gaining traction. Qualitative methods, including participatory research and focus groups, capture the views of stakeholders, including passengers, employees, and policymakers, on these initiatives. For example, a study might explore passenger willingness to pay for greener travel options, providing airlines with actionable insights into market trends (Stamper, 2021).

Adoption Challenges for New Technologies - The introduction of new technologies, such as autonomous aircraft or biometric systems, often faces resistance due to concerns about safety, reliability, and ethical implications. Qualitative research facilitates an understanding of these concerns by engaging stakeholders through focus groups or deliberative workshops. Insights from such studies can guide the development of communication and training strategies that promote technology acceptance.

3.5 Policy and Regulatory Frameworks

The aviation industry is governed by a complex web of policies and regulations aimed at ensuring safety, efficiency, and environmental compliance. Qualitative research is instrumental in exploring the impact of these frameworks and identifying areas for improvement.

Insights into the Impact of Regulations on Operations - Regulatory compliance often imposes significant operational challenges for airlines, from meeting safety standards to addressing environmental mandates. Qualitative studies, such as policy analysis and stakeholder interviews, help uncover how regulations influence day-today operations. For instance, interviews with airline executives can provide a nuanced understanding of the financial and logistical implications of new emission standards.

Stakeholder Perspectives on Compliance and Reform - Regulations affect a wide range of stakeholders, including passengers, employees, and industry bodies. Qualitative research methods, such as deliberative forums or focus groups, enable the exploration of stakeholder views on regulatory compliance and potential reforms. These insights are critical for policymakers aiming to balance safety, efficiency, and industry growth.

Qualitative research is essential in addressing the multifaceted challenges and opportunities within the aviation industry. By focusing on areas such as passenger experience, human factors, organizational culture, innovation, and regulatory frameworks, qualitative approaches provide invaluable insights that quantitative methods may overlook.

Through in-depth exploration of human behaviors, perceptions, and interactions, qualitative research informs strategies for improving service quality, enhancing safety, fostering innovation, and navigating complex regulations. As the aviation industry continues to evolve, leveraging qualitative research will be vital for ensuring its sustainable growth, operational efficiency, and adaptability to future challenges.

4. METHODOLOGICAL APPROACHES IN AVIATION RESEARCH

The complexity of the aviation industry requires methodological approaches that can capture its multifaceted nature, including technological operations, human interactions, and regulatory frameworks. Qualitative research methodologies, such as interviews, focus groups, ethnography, case studies, and content analysis, offer robust tools for exploring these dimensions. This chapter examines these methods and their application in aviation research, highlighting their strengths and challenges.

4.1 Interviews and Focus Groups

Collecting Rich Data from Stakeholders - Interviews and focus groups are widely used qualitative research methods in aviation studies, offering a means to gather in-depth insights from various stakeholders, including employees, passengers, regulators, and industry leaders. Through semi-structured interviews, researchers can delve into participants' perceptions, attitudes, and experiences. For instance, interviews with pilots can provide nuanced understandings of decision-making under stress, while focus groups with passengers can explore customer expectations regarding service quality. These methods enable the collection of diverse perspectives that contribute to a holistic understanding of aviation challenges and opportunities (Creswell & Poth, 2018).

Challenges in Accessing Aviation Professionals - One significant challenge in employing these methods is gaining access to aviation professionals, who often operate in high-security and time-sensitive environments. Pilots, air traffic controllers, and regulatory personnel work under tight schedules, limiting their availability for research. Additionally, corporate policies may restrict employees from participating in studies without prior approval. Researchers must navigate these barriers through effective stakeholder engagement and clear communication about the study's objectives and confidentiality measures.

4.2 Ethnography and Participant Observation

Observing Operations in Real-Time - Ethnography and participant observation involve immersive research within specific aviation settings, such as airports, control towers, and aircraft. These methods enable researchers to observe workplace dynamics, operational workflows, and interactions between personnel in real-time. For example, ethnographic studies of cabin crews may uncover the subtle ways team members coordinate during emergencies, contributing to a deeper understanding of human factors in aviation safety (Jordan, 2019).

Understanding Workflow Processes - Participant observation is particularly valuable for examining workflow processes and identifying inefficiencies or stress points. By spending extended periods in aviation environments, researchers can gather rich, contextualized data that quantitative methods may overlook. For instance, observing ground staff during aircraft turnarounds can reveal operational bottlenecks and inform strategies for improving turnaround times. However, ethnographic research requires prolonged engagement, making it resource-intensive and challenging in highly regulated aviation settings.

4.3 Case Studies

Examining Airline Responses to Events - Case studies provide a comprehensive means to analyze airline responses to specific events, such as accidents, weather disruptions, or public relations crises. By combining multiple data sources, including interviews, historical records, and organizational reports, case studies offer a detailed narrative that captures the complexity of these situations. For instance, a case study of an airline's response to a major accident might explore the interplay between crisis communication, regulatory compliance, and operational adjustments (Yin, 2018).

Combining Historical Data with Qualitative Methods - The case study approach often integrates historical data with qualitative methods to construct a complete picture of past events. For example, analyzing how airlines adapted their operations during the COVID-19 pandemic can provide valuable lessons for managing future disruptions. Researchers can use interviews with airline executives and staff to complement archival data, ensuring that both organizational and personal perspectives are represented.

4.4 Content and Discourse Analysis

Analyzing Marketing Materials and Policy Documents - Content and discourse analysis involve examining texts, such as marketing campaigns, policy documents, and public statements, to uncover underlying themes, narratives, and biases. In the aviation industry, these methods can be applied to assess how airlines position themselves in competitive markets or how regulatory bodies communicate new policies. For example, analyzing airline advertisements can reveal how companies highlight specific features, such as sustainability or luxury, to attract different customer segments (Krippendorff, 2018).

Assessing Media Narratives Around Aviation Issues - The aviation industry is often the focus of intense media scrutiny, particularly following accidents or controversies. Discourse analysis of media coverage can help researchers understand how aviation issues are framed and perceived by the public. For instance, studies of media narratives surrounding aircraft emissions can reveal societal attitudes toward aviation's environmental impact. These insights can inform industry strategies for managing public relations and engaging with stakeholders on critical issues.

4.5 Advantages and Challenges of Qualitative Approaches

Advantages - The qualitative methods discussed above offer significant advantages for aviation research. They enable a contextualized understanding of complex issues, provide insights into human behaviors and organizational dynamics, and complement quantitative data with rich, detailed narratives. These methods are particularly well-suited for exploring emerging challenges, such as adapting to sustainability demands or integrating advanced technologies.

Challenges - However, qualitative approaches also face challenges in the aviation context. Accessing research participants can be difficult due to time constraints, security concerns, and corporate policies. Additionally, the resource-intensive nature of methods like ethnography requires careful planning and substantial investment in time and effort. Researchers must also navigate ethical considerations, such as maintaining participant confidentiality and obtaining informed consent in highly regulated environments.

The aviation industry's complexity necessitates diverse methodological approaches that can capture its technological, human, and regulatory dimensions. Qualitative methods, including interviews, focus groups, ethnography, case studies, and content analysis, provide invaluable tools for exploring these dimensions. Each method offers unique strengths, from gathering stakeholder insights to analyzing organizational responses to crises.

While these approaches present challenges, such as access barriers and resource demands, their ability to uncover nuanced insights makes them indispensable for aviation research. By leveraging qualitative methodologies, researchers can inform strategies that enhance operational efficiency, improve safety, and address emerging challenges in the dynamic aviation sector.

5. CHALLENGES AND ETHICAL CONSIDERATIONS IN AVIATION RESEARCH

Qualitative research in the aviation industry provides invaluable insights into complex and dynamic aspects of operations, human factors, and organizational culture. However, conducting such research is fraught with challenges and ethical considerations that demand careful navigation. This chapter examines three key dimensions: access and data sensitivity, ethical concerns, and practical limitations, with a focus on their implications for qualitative research in aviation.

5.1 Access and Data Sensitivity

Difficulty in Accessing Proprietary and Sensitive Operational Information - One of the primary challenges in aviation research is gaining access to proprietary and sensitive information. The industry is highly regulated, with strict protocols governing the confidentiality of operational data, employee performance records, and passenger feedback. Airlines, airports, and regulatory bodies are often hesitant to share such data due to concerns about commercial confidentiality, competitive advantage, and potential legal liabilities.

Accessing information on safety protocols or pilot decision-making processes may require navigating multiple layers of approval. Researchers may need to collaborate closely with organizations to build trust and demonstrate the value of their study. However, this process can be time-consuming and does not always guarantee access. Establishing non-disclosure agreements (NDAs) and emphasizing the secure handling of data can mitigate these barriers to some extent (Creswell & Poth, 2018).

Balancing Transparency with Confidentiality - Once access is granted, researchers face the challenge of balancing transparency with the need to protect sensitive information. Transparency is critical for ensuring the credibility and replicability of findings, but disclosing too much detail may violate agreements with participants or organizations. For example, revealing specifics about operational inefficiencies in a particular airline may damage its reputation or lead to regulatory scrutiny.

To address this, researchers must anonymize data and provide aggregated results that do not identify individuals or organizations. Using pseudonyms and redacting identifiable details in publications are common strategies to maintain this balance. Furthermore, researchers must communicate their intentions clearly to stakeholders to align expectations and foster cooperation.

5.2 Ethical Concerns

Protecting Participant Anonymity and Consent in a High-Stakes Industry - The aviation industry is inherently high-stakes, with significant implications for safety, customer satisfaction, and employee well-being. Researchers must ensure that participants are fully informed about the study's purpose, potential risks, and their rights before obtaining informed consent. However, securing genuine consent can be challenging in hierarchical organizations where employees may feel pressured to participate.

Pilots or cabin crew may fear repercussions if they discuss operational weaknesses or workplace grievances. This underscores the importance of voluntary participation and protecting participant anonymity. Researchers should establish clear protocols for de-identifying data and reassure participants that their contributions will not be linked to their identities (Jordan, 2019).

Managing Potential Conflicts of Interest - Qualitative research often involves close collaboration with stakeholders, creating opportunities for conflicts of interest. For instance, an airline funding a study on passenger satisfaction may expect favorable results that align with its branding. Researchers must navigate these dynamics carefully to maintain their integrity and objectivity.

Mitigating conflicts of interest requires transparency in reporting funding sources and adhering to ethical guidelines established by institutions and professional organizations. Researchers should also involve independent reviewers to ensure that their methodologies and findings are robust and unbiased. Clear communication about the study's objectives and limitations can help set realistic expectations for stakeholders.

5.3 Practical Limitations

Resource Constraints in Conducting Fieldwork - Conducting qualitative research in aviation often involves resource-intensive activities, such as fieldwork at airports, interviews with employees, and participant observation during flights. These activities require substantial financial investment, time, and logistical planning. For academic researchers or small research teams, these constraints can limit the scope and depth of their studies.

Conducting an ethnographic study of airport operations may involve long periods of observation, requiring researchers to secure funding for travel, accommodation, and access permissions. Additionally, the unpredictable nature of aviation operations—such as delays or emergencies—can disrupt research plans, further straining resources (Yin, 2018).

Navigating Bureaucratic Barriers and Resistance - The aviation industry's regulatory environment introduces bureaucratic barriers that can impede research progress. Gaining access to sites like airport control towers or maintenance hangars often involves extensive paperwork and approval processes. Researchers may also encounter resistance from industry insiders who perceive academic studies as intrusive or irrelevant to operational goals.

Building rapport with industry stakeholders is essential for overcoming these barriers. Researchers should demonstrate how their work aligns with the organization's objectives, such as improving customer satisfaction or enhancing safety protocols. Establishing long-term partnerships with aviation companies can also facilitate smoother access for future studies.

5.4 Strategies for Addressing Challenges and Ethical Considerations

Building Trust and Collaboration - Establishing trust with aviation organizations and participants is fundamental to overcoming access barriers and ethical concerns. Researchers can build credibility by presenting clear research proposals, adhering to agreed timelines, and maintaining open communication throughout the study. Collaborative approaches, such as involving stakeholders in the research design process, can foster mutual understanding and support.

Adopting Flexible Research Designs - Given the unpredictable nature of aviation operations, researchers should adopt flexible research designs that allow for adjustments in response to changing circumstances. For instance, if fieldwork is disrupted due to operational emergencies, researchers can pivot to other methods, such as virtual interviews or document analysis.

Prioritizing Ethical Oversight - Ethical oversight is crucial for ensuring that research practices align with established guidelines. Institutional Review Boards (IRBs) and ethics committees play a vital role in reviewing study proposals, identifying potential risks, and suggesting safeguards. Researchers should engage with these bodies early in the planning process to address ethical concerns proactively.

Qualitative research in the aviation industry presents unique challenges and ethical considerations that require thoughtful navigation. Accessing proprietary and sensitive information, balancing transparency with confidentiality, and addressing ethical concerns such as anonymity and consent are critical for conducting responsible research. Practical limitations, including resource constraints and bureaucratic barriers, further complicate the research process.

Despite these challenges, qualitative research remains indispensable for understanding the nuanced dynamics of the aviation industry. By employing strategies such as building trust, adopting flexible designs, and prioritizing ethical oversight, researchers can overcome obstacles and contribute valuable insights that enhance safety, efficiency, and stakeholder satisfaction in this high-stakes sector.

6. CASE STUDIES ILLUSTRATING THE IMPORTANCE OF QUALITATIVE RESEARCH

Qualitative research has proven instrumental in navigating the complexities of the aviation industry, providing insights that have informed critical decisions across passenger experiences, crisis management, and organizational adaptation. This chapter highlights three case studies where qualitative methods played a pivotal role in addressing key challenges and driving innovation.

6.1 Passenger-Centric Innovations

Improving In-Flight Services Through Qualitative Feedback - A leading international airline, referred to here as Airline X, faced declining customer satisfaction scores despite maintaining competitive prices and a broad route network. To understand the underlying causes, Airline X employed qualitative research methods, including in-depth interviews and focus groups, to gather insights directly from passengers.

The study revealed that passengers valued personalized services, cultural inclusivity, and better communication during delays over the tangible aspects of travel, such as meals or seating arrangements. For example, passengers traveling on long-haul flights expressed dissatisfaction with the lack of culturally relevant meal options and inadequate in-flight entertainment tailored to diverse demographics.

Armed with these insights, Airline X implemented several changes:

Customized Meal Options: The airline introduced menus catering to dietary preferences based on the cultural backgrounds of its passengers.

Enhanced Entertainment Systems: Regional and multilingual content was added to cater to a wider audience.

Proactive Communication Strategies: Cabin crew received training to improve communication during delays or unexpected disruptions.

The airline also embraced passenger-driven co-creation by involving frequent travelers in pilot programs for testing new services. As a result, Airline X saw a significant increase in customer satisfaction scores within a year, alongside improved loyalty and positive brand perception. This case demonstrates the value of qualitative feedback in tailoring services to meet passenger expectations, providing a competitive edge in a crowded market.

6.2 Crisis Management During the COVID-19 Pandemic

Analyzing Responses to Pandemic-Related Disruptions - The COVID-19 pandemic caused unprecedented disruption in the aviation industry, grounding fleets and creating widespread uncertainty for employees and passengers. To manage this crisis, qualitative research was employed by several airlines to understand the experiences and perceptions of key stakeholders.

One notable case involved a regional airline, referred to as Airline Y, which faced operational and reputational challenges due to sudden lockdowns and evolving health regulations. Airline Y conducted semi-structured interviews and digital surveys with three stakeholder groups:

Employees: Pilots, cabin crew, and ground staff shared their concerns about job security, workplace safety, and mental health during the pandemic.

Passengers: Frequent flyers provided feedback on their willingness to travel and perceptions of safety protocols, such as mask mandates and social distancing.

Industry Partners: Collaborations with airport authorities and health officials were evaluated to align operational strategies with public health guidelines.

Key findings included:

Employee Sentiments: Many employees reported heightened stress due to uncertainty about layoffs and inadequate communication from management.

Passenger Insights: Safety emerged as a top priority, with passengers favoring airlines that implemented stringent cleaning protocols and contactless processes.

Operational Gaps: Discrepancies between airline policies and airport enforcement of health measures caused confusion and dissatisfaction.

Using these insights, Airline Y launched several initiatives, such as frequent town hall meetings to improve internal communication, a comprehensive "Fly Safe" campaign to reassure passengers, and coordinated efforts with airports to standardize health protocols.

The airline's ability to adapt swiftly based on qualitative research findings helped it regain passenger trust and maintain employee morale during a highly volatile period. This case highlights the critical role of qualitative research in managing crises by addressing the nuanced concerns of diverse stakeholders.

6.3 Organizational Adaptation to Sustainability Goals

Insights Into Environmental Strategies - As sustainability becomes a key priority in aviation, airlines are under increasing pressure to adopt green initiatives, such as reducing carbon emissions and investing in sustainable technologies. A European airline, referred to as Airline Z, serves as a case study for how qualitative research informed its strategy to achieve sustainability goals.

Airline Z conducted a series of interviews and focus groups with employees, executives, and external stakeholders, including environmental organizations and regulators, to understand their perspectives on sustainability. Key objectives included:

Identifying Barriers: Exploring resistance or challenges in adopting sustainable practices among employees and management.

Assessing Perceptions: Understanding how stakeholders viewed the airline's existing efforts and where they expected improvements.

Co-Creating Solutions: Collaborating with employees and external partners to design actionable environmental initiatives.

Findings from the qualitative study revealed several critical insights:

Employee Perspectives: Many employees expressed skepticism about the feasibility of transitioning to electric aircraft or adopting biofuels without compromising operational efficiency.

Executive Concerns: Senior leaders identified cost constraints as a significant barrier but acknowledged the reputational benefits of sustainability.

External Expectations: Environmental organizations emphasized the importance of transparency and measurable impact, advocating for the airline to adopt more ambitious carbon-offset programs.

In response, Airline Z developed a comprehensive sustainability roadmap that included:

Employee Training Programs: Workshops and awareness campaigns to educate staff on the importance of sustainability.

Investments in Green Technology: Partnerships with manufacturers to explore hybrid-electric aircraft.

Transparency Initiatives: Publishing annual sustainability reports to track progress and build trust with external stakeholders.

These measures not only improved the airline's environmental performance but also enhanced its reputation as a forward-thinking industry leader. The case underscores how qualitative research can uncover hidden barriers and foster collaboration in achieving complex goals like sustainability.

The three case studies illustrate the indispensable role of qualitative research in addressing critical challenges and opportunities within the aviation industry. Whether improving passenger experiences, navigating crises, or advancing sustainability, qualitative methods provide the depth and context needed to understand diverse perspectives and drive actionable change.

Passenger-centric innovations at Airline X demonstrate how qualitative feedback can refine service offerings to align with customer expectations. The crisis management efforts of Airline Y reveal the importance of stakeholder engagement in navigating turbulent times, while Airline Z's sustainability journey highlights the role of collaborative strategies in achieving long-term goals.

These cases reinforce the value of qualitative research in the aviation sector, where the interplay of human factors, operational complexities, and external pressures demands nuanced, context-rich insights. By leveraging qualitative methods, aviation stakeholders can better address the challenges of today while preparing for the opportunities of tomorrow.

7. IMPLICATIONS AND FUTURE DIRECTIONS

The aviation industry is a complex ecosystem, where qualitative research provides valuable insights to address multifaceted challenges and uncover opportunities for growth. This section explores the practical implications for air transport management and examines the potential for advancing qualitative research in the sector.

7.1 Practical Implications for Air Transport Management

Improving Customer Satisfaction - One of the primary goals of qualitative research in aviation is to enhance customer satisfaction by addressing passenger needs, grievances, and expectations. Through interviews, focus groups, and ethnographic observations, researchers can identify key factors influencing passenger perceptions of airline services.

Understanding the cultural nuances that affect travel preferences enables airlines to tailor their offerings to diverse customer bases. Regional adaptations in meal services, entertainment options, and onboard announcements are examples of customer-centric solutions informed by qualitative research (Jordan, 2019).

Similarly, qualitative feedback from passengers during flight disruptions can guide the development of proactive communication strategies. Effective responses to unforeseen delays or cancellations not only improve customer satisfaction but also bolster brand loyalty.

Enhancing Operational Efficiency - Operational inefficiencies, such as delayed turnarounds or miscommunication between teams, can significantly impact the profitability and reputation of an airline. Qualitative methods like participant observation and case studies allow researchers to analyze workflow processes, identify bottlenecks, and propose actionable solutions (Creswell & Poth, 2018).

Ethnographic studies of ground handling teams have revealed inefficiencies in baggage transfer procedures and crew scheduling. Addressing these issues based on qualitative insights can lead to smoother operations, reduced costs, and improved on-time performance.

Boosting Employee Engagement - Employee satisfaction and engagement are critical for operational success in aviation. Qualitative research can explore employees' perceptions of their work environment, leadership styles, and organizational policies.

Interviews with cabin crew may reveal frustrations with insufficient rest periods or lack of professional development opportunities. Addressing such concerns can improve morale, reduce turnover, and enhance service quality (Stamper, 2021).

Engaging employees in co-creating workplace policies, informed by qualitative research, fosters a sense of ownership and collaboration. This approach aligns with modern management practices that emphasize participatory decision-making and transparency.

Influencing Policy-Making and Regulatory Reforms - Qualitative research also informs policy-making and regulatory decisions by capturing the perspectives of stakeholders affected by aviation policies. For example, interviews with industry professionals can shed light on the challenges of adhering to environmental regulations or adapting to evolving safety protocols.

Qualitative studies exploring public perceptions of noise pollution or airport expansion projects can guide regulatory bodies in balancing community concerns with industry requirements. By integrating stakeholder feedback into the decision-making process, policymakers can design more equitable and effective frameworks.

7.2 Potential for Advancing Qualitative Research in Aviation

Integrating Qualitative Insights with Quantitative Data - One promising direction for aviation research lies in the integration of qualitative insights with quantitative data, enabling a mixed-methods approach. While quantitative data provides measurable indicators such as passenger load factors or delay statistics, qualitative research adds depth by exploring the reasons behind these trends.

Customer satisfaction surveys often rely on numerical ratings, but follow-up interviews can uncover the underlying emotions and expectations driving those scores. Similarly, analyzing accident reports through both statistical modeling and qualitative case studies enhances the understanding of human factors involved in safety incidents (Krosnick & Presser, 2017).

The combination of qualitative and quantitative methods offers a holistic perspective, helping stakeholders address challenges more effectively and make evidence-based decisions.

Leveraging Emerging Technologies - Emerging technologies, particularly artificial intelligence (AI), are transforming the landscape of qualitative research in aviation. AI tools can assist in collecting, organizing, and analyzing large volumes of qualitative data, making the research process more efficient and scalable.

Natural Language Processing (NLP): AI-driven NLP tools can analyze open-ended responses from surveys, social media posts, or interview transcripts to identify recurring themes and sentiments. For instance, an airline could use NLP to assess passenger feedback across multiple channels, gaining real-time insights into customer satisfaction.

Sentiment Analysis: Sentiment analysis algorithms can evaluate public opinions on social media regarding flight delays, environmental initiatives, or customer service, enabling airlines to respond proactively to emerging issues.

Predictive Analytics: By combining qualitative insights with machine learning models, researchers can predict passenger behavior and preferences. For instance, analyzing historical feedback and booking patterns can help airlines design targeted marketing campaigns or optimize route planning.

Despite these advancements, the human element remains crucial in interpreting qualitative data. Researchers must ensure that AI tools complement, rather than replace, their critical thinking and contextual understanding.

Addressing Underexplored Areas - While qualitative research has made significant contributions to aviation, several areas remain underexplored. For instance, the experiences of marginalized groups, such as travelers with disabilities or employees from underrepresented backgrounds, warrant greater attention.

Similarly, the rise of autonomous and electric aircraft presents new challenges and opportunities for qualitative inquiry. Understanding stakeholder perspectives on these emerging technologies can inform their successful adoption and integration into existing aviation ecosystems.

Building Interdisciplinary Collaborations - Advancing qualitative research in aviation also requires collaboration across disciplines. Researchers from psychology, sociology, engineering, and business management can bring diverse perspectives to address complex issues. For example, psychologists can contribute insights into human factors, while engineers can provide technical expertise on operational challenges.

Interdisciplinary collaborations not only enrich the research process but also ensure that findings are relevant and actionable for a wide range of stakeholders.

The implications of qualitative research for air transport management are profound, influencing customer satisfaction, operational efficiency, employee engagement, and policy-making. By addressing these areas, qualitative methods help aviation stakeholders navigate the complexities of a dynamic and competitive industry.

Looking ahead, the integration of qualitative insights with quantitative data and the adoption of emerging technologies like AI present exciting opportunities for advancing research methodologies. However, these advancements must be complemented by efforts to address underexplored topics and foster interdisciplinary collaborations.

As the aviation industry continues to evolve, qualitative research will remain an indispensable tool for understanding the human dimensions of air travel, guiding innovations, and shaping a more sustainable and inclusive future.

8. CONCLUSION

The case studies explored in this chapter underscore the vital role of qualitative research in navigating the dynamic challenges faced by the aviation industry. Qualitative methods—such as interviews, focus groups, and ethnographic studies—provide deep, context-rich insights that quantitative data alone cannot capture. In areas such as passenger experience, crisis management, and organizational adaptation, qualitative research has proven essential in driving innovation, improving customer satisfaction, and fostering resilience.

For instance, the case of Airline X demonstrates how qualitative feedback can refine in-flight services to better meet passenger expectations, leading to enhanced satisfaction and brand loyalty. The response of Airline Y during the COVID-19 pandemic highlighted how qualitative insights into employee concerns and passenger perceptions were pivotal in crisis management, allowing the airline to swiftly adapt and regain trust. Similarly, Airline Z's efforts to implement sustainable practices underscore how qualitative research can reveal both internal and external barriers, while fostering collaboration to achieve long-term environmental goals. These examples reflect a broader trend in the aviation sector where qualitative research provides the nuanced understanding needed to navigate complex, human-centered challenges.

Despite the evident successes, the aviation industry continues to face evolving challenges, including rapid technological advancements, shifting customer expectations, and increasing pressure to adopt sustainable practices. To remain competitive and adaptable, it is imperative that researchers, practitioners, and policymakers prioritize qualitative methods in addressing these issues. Policymakers should integrate qualitative insights into regulatory frameworks, ensuring that policies reflect the realities and concerns of stakeholders. Practitioners in airlines, airports, and related industries should incorporate qualitative feedback into service design and organizational strategies to create more personalized, responsive, and resilient operations. Finally, researchers are encouraged to continue exploring new qualitative approaches, leveraging technologies such as AI and big data analytics to further deepen understanding and inform decision-making.

The future of the aviation industry holds immense potential for meaningful improvements driven by qualitative insights. By integrating these insights with existing data and technological advancements, the sector can create more customer-centric, efficient, and sustainable practices. In this future, qualitative research will not only inform operational decisions but also contribute to shaping industry policies and strategies. Ultimately, qualitative research offers the tools to transform the aviation sector into a more dynamic, inclusive, and future-ready industry, one that is better equipped to meet the challenges of tomorrow.

REFERENCES

Bryman, A. (2016). *Social research methods* (5th ed.). Oxford University Press.

Creswell, J. W., & Poth, C. N. (2018). *Qualitative inquiry and research design: Choosing among five approaches* (4th ed.). Sage Publications.

Dekker, S. (2017). *The field guide to understanding 'human error* (3rd ed.). CRC Press.

Denzin, N. K., & Lincoln, Y. S. (2017). *The Sage handbook of qualitative research* (5th ed.). Sage Publications.

Gillen, D., Morrison, W. G., & Stewart, C. (2020). *Airline economics in the twenty-first century*. Routledge.

Gillen, M., Sinclair, R. R., & Cleary, P. D. (2020). Leadership in high-stress industries: Addressing employee morale and performance. *Safety Science*, 123, 104561. DOI: 10.1016/j.ssci.2020.104561

Han, H., Yu, J., & Kim, W. (2019). Passenger experience and loyalty in the airline industry: A critical review. *Journal of Travel Research*, 58(4), 564–579. DOI: 10.1177/0047287518797156

Han, H., Yu, J., & Kim, W. (2019). Passengers' experiences and satisfaction in air travel: A qualitative analysis. *Journal of Air Transport Management*, 74, 84–93.

Jordan, B. (2019). Ethnographic research in high-stakes environments: Lessons from the aviation industry. *Journal of Organizational Ethnography*, 8(3), 345–360. DOI: 10.1108/JOE-03-2019-0014

Krippendorff, K. (2018). *Content analysis: An introduction to its methodology* (4th ed.). Sage Publications.

Krosnick, J. A., & Presser, S. (2017). Question and questionnaire design. In Marsden, P., & Wright, J. (Eds.), *Handbook of survey research* (2nd ed., pp. 263–314). Emerald Group Publishing.

Patton, M. Q. (2015). *Qualitative research & evaluation methods* (4th ed.). Sage Publications.

Stamper, B. (2021). Co-creating sustainable aviation solutions: A participatory approach. *Journal of Air Transport Management*, 91, 102022. DOI: 10.1016/j.jairtraman.2021.102022

Stamper, D. (2021). Exploring stakeholder perceptions of innovation in aviation: A qualitative perspective. *Journal of Transportation Innovation*, 12(3), 45–62.

Stamper, D. (2021). Sustainability in aviation: Challenges and opportunities. *Aviation Studies Journal*, 12(2), 45–63. DOI: 10.1080/ASJ-2021-02-0067

Yin, R. K. (2018). *Case study research and applications: Design and methods* (6th ed.). Sage Publications.

Chapter 2 Qualitative Research Methods: Harnessing Interviews, Focus Groups, Observations, and Document Analysis

Nusrat Khan
https://orcid.org/0000-0003-0788-7786
GD Goenka University, India

Fehmina Khalique
https://orcid.org/0000-0002-2519-3174
Lloyd Business School, India

Kartikay Saini
GD Goenka University, India

ABSTRACT

This article discusses the different data collection methods in qualitative research. The data collection method is based on the complexity and depth of human experiences that one needs to capture while doing research. This article explores a range of the most popular qualitative data collection methods, including interviews, focus groups, participant observations, and document analysis. Each of the data collection methods comes with its own sets of detailing unique characteristics, strengths, and potential limitations. This article therefore aims to understand those prerequisites of data collection methods and understand the different data collection methods. It also discusses the context of its suitability for different research designs and objectives. This chapter offers insights into how researchers can strategically select and apply these methods to gather rich, meaningful data. It also aims to equip researchers with

DOI: 10.4018/979-8-3693-7403-0.ch002

the knowledge and skills necessary to effectively navigate the process of qualitative data collection, thereby enhancing the quality and impact of their research.

1. INTRODUCTION

Qualitative research involves various stages of performing the research before reaching to the data analysis. The usage of data collection is actually an important element of qualitative research. Data collection is an extremely critical component of qualitative research. Data collection is an important part of the research process (Taherdoost, 2021). But even before proceeding with data collection, certain prior requirements are to be fulfilled. The data collection method provides the foundation upon which insights and understanding of the entire research are built upon. There is a remarkable difference between quantitative and qualitative research. Qualitative research only focuses on exploring phenomena in-depth in contrast to quantitative research that relies on numerical data only. Through this type of research tries to capture the richness and complexity of human experiences, behaviours, and perceptions. This chapter delves into the various data collection methods which are popularly used in qualitative research. These include interviews, focus groups, observations, and document analysis. There are many other methods also involved however the purpose of this article is to only understand these few. Each of these methods offers a unique advantage for capturing the nuances of social and cultural contexts. By understanding these methods, researchers can make informed decisions about which techniques best align with their research questions, ensuring the integrity, depth, and validity of their findings. However, before even proceeding for the data collection there are certain issues to be dealt with like seeking relevant permissions (Creswell & Poth, 2016) from different authorities/organizations/individuals from where the data is to be collected.

Creswell and Poth (2016) suggest that certain issues need to be addressed which are mentioned below. All these are relevant issues as they all involve ethical considerations across different phases of data collection.

- Visiting the location from where the data is to be collected
- Seeking relevant permissions
- Sampling strategy
- Means for recording information
- Issues in data storage
- Issues while data collection in field

The most important issue that arises at first is that the location should be chosen carefully and it is all the more important that one visits the site in advance and builds the rapport so that the data collection becomes easier and is of a good quality data. In addition, the researcher takes into account different ethical issues that will arise in this data collection. Issues pertaining to the respondents' privacy, welfare and justice should be taken into account (Creswell and Poth, 2016). Data anonymity is required to disguise the details pertaining to the respondents. There are also certain requirements like seeking the permissions by the (IRB) Institutional Review Boards (Milne, 2005; Creswell and Poth, 2016; Brooks, 2013, Pech, Cob, & Cejka, 2007; Cartwright, Hickman, Nelson & Knafl, 2013). This board usually reviews the entire process of the research involving the site permissions, selection of samples (Curtis, Gesler, Smith, & Washburn, 2000) and all the data collection strategies that will be implemented. Finding the right kind of sample involves a well curated strategy in identification of the respondents/fields of visit. It is not at all easy to seek the permissions, planning and other fieldwork stages of data collection. Several challenges are involved in all the stages (Baffour, Francis, Chong & Harris, 2024). It is not necessary to always conduct research only from a single site. One can do this in multiple sites to increase the qualitative research rigor (Jenkins, Slemon, Haines-Saah, & Oliffe, 2018).

Ostrander, (1993) lays down several strategies that can be adopted by the researcher for getting access to the field, building up the rapport (Gill, et al. 2008) with the participants and conducting good interviews. Interviewing the participants is also a very important part and that challenging questions should not be avoided with the elite participants. Suri, (2010) suggests the adaptability of 16 purposeful sampling strategies of Patton's (Patton, 2002) to the process of qualitative research. Nyimbili & Nyimbili (2024) also suggest the use of these 16 strategies originally identified by Patton in 1990, to be used in research. Campbell, et al. (2020) also explains the use of purposive sampling into the research design depending on the varying contexts.

Selection of participants can vary based on the type of research undertaken. Creswell and Poth, (2016) describe different types of ways to select the relevant participants for the study. Below table 1 describes the type of research and the sampling strategy to be deployed while collecting the data.

Table 1. Research study and Sampling Method

Type of Study	Sampling Method
Phenomenological Studies	Criterion Sampling
Grounded Theory Study	Theoretical Sampling
Ethnography	Within-Culture Sampling

Source: Cresswell and Poth, 2016

Sample size details can be taken further studied from Miles and Huberman (1994, p. 28). Types of sampling varies depending upon the purpose of qualitative research undertaken. Table 2 describes the type of research and the sample size chosen by the researchers in their respective research, which serves as an example.

Table 2. Type of Research and sample size

Type Of Research	Sample Size
Narrative Research	1 or 2
Phenomenology	1 to 325
Grounded Theory Study	20 to 30
Ethnography	Single Culture-Sharing Groups
Case Study Research	Four or Five Case Studies

Source: Cresswell and Poth, 2016

Data can be collected in different forms and they vary with the types of information gathered through that. The four forms of data collections are shown in table 3, however this list may be further expanded in other forms as well including email messages, videos, web based platforms, social media, etc.(Cresswell and Poth, 2016).

Table 3. Data and Types of information

Forms of Qualitative Data	Information Gathered
Interviews	Face to face, group online or offline.
Observations	Participants or Non-participants both
Documents	Private documents or public documents
Audio-visual Materials	Pictures or artifacts

Source: Cresswell and Poth, 2016, pg. 227

A focus group is a qualitative research method commonly used in different fields like marketing, social science, and various other fields. This method is widely used to gather insights, opinions, beliefs, perceptions, attitude and feedback (Powell & Single,1996) from a group of people about a particular idea (Gill, Stewart, Treasure & Chadwick, 2008)., generate a new idea (Breen, 2006) subject, different issues, product, concept, service or policy (Krueger, 2014); therapy, decision making, education, organizing, or behaviour change (Morgan, 1996). The method involves a structured discussion. This discussion is led by a moderator and typically involves a small group of participants (usually between 6 to 12 people (Breen, 2006; Billson, 1989; Krueger & Casey, 2015). The goal is to explore participants' perceptions, opinions, (Billson, 1989). Beliefs and attitudes towards the topic of interest in a controlled, yet conversational, environment. It was used in the marketing research in the earlier times (Powell & Single, 1996).

In the early 1930s some social scientist found that the interview method was not that accurate (Krueger & Casey, 2015). They wanted that the focus of interview is primarily of the interviewer and not on the respondents. Therefore, they wanted that the interviewee should be of the central role and they should give their opinions and feedback without any inhibitions. The focus group techniques that are present today are primarily the classic work done by Merton, Fiske & Kendall in 1956. It is also important that if an experience of any particular kind is the point of discussion in an FGD then the participants should have been exposed to that kind of experiences (Breen, 2006).

2. DATA COLLECTION METHODS

2.1 Interviews

Interviews are considered to be one of the foremost form of qualitative data collection method in qualitative research. As it allows for in-depth exploration of participants' different lived experiences, perceptions, and beliefs. It is a social interaction taking place in the form of a conversation which creates knowledge (Brinkmann and Kvale, 2018), also where participants share their rich detailed experiences providing information for the research questions (Byrne, 2001), based on this interaction that takes place between interviewee and interviewer (Cresswell and Poth, 2016). The questions that are designed in this interview method are actually the sub questions of the research. The type of questions depends upon the research framework and objectives of the research. This method involves a direct conversation that is taking place between the researcher and the participant. The interviews can take up different forms like structured, semi-structured, or unstructured. This is entirely dependent on the research objectives. The interview questions and sub questions may also arise depending upon the information that is shared by the participant. Therefore, probing questions are also asked during the interview to strengthen the perspective and throw some more light on the information previously shared. While talking about the categories of the interviews they can be into different categories including semi-structured interviews, unstructured interviews, structured interviews and in-depth interviews (DiCicco-Bloom & Crabtree, 2006).

Interviews are considered to be a "craft" that is developed over a period of time (Creswell & Poth, 2018; Brinkmann and Kvale, 2018). There are seven stages of an interview as per Brinkmann and Kvale, (2018, pg. 41). These seven stages include Thematizing, Designing, Interviewing, Transcribing, Analyzing, Verifying, and Reporting. These stages are to help the researcher in the entire journey of interviewing to reporting. While Thematizing one seeks an answer to the *Why* and *What* of the interview which the researcher intends to take. The *How* of the research is the last option which refers to the way in which the analysis of the interviews is to be done. The *why* is the intended purpose of the study undertaken, *what* refers to the clarifications of the research intended to do before proceeding for the interview, including some prior knowledge of the field.

Designing stage involves the complete picture of the entire research taken into account before proceeding for the field visit. The designing stage should also consider the kind of knowledge the researcher intends to create from the interviews. Third stage interviewing should be carefully planned and conducted with appropriate protocols and guide. Transcribing refers to the transcribing the interview from an oral text to a written text format. Data needs to be prepared accordingly. In the Analysing stage it is important to keep the research objective at the forefront and then proceed for an appropriate analysis. Since the analysis is completely dependent upon the knowledge that is required to be created based on the research objectives. In Verifying stage, validity, reliability and generalizability are to be conducted. Reliability means the consistency of the results from the interview. Validity refers to the data received from the interview is it actually investigating what it was actually meant for. Reporting is the final stage where the data received from the interviews is presented keeping in mind the ethical considerations. The final reporting should also result into a product that is readable by the audience intended.

Sampling in the interview-based methods of qualitative research is again divided into four different approaches (Robinson, 2014). These include the first approach of defining the data universe by including the inclusion and exclusion criteria beforehand. Second approach is of deciding upon the sample size, third approach involves selecting an appropriate sampling strategy, and the last approach is the sample sourcing.

As per Creswell and Poth, (2018), the final procedures to be adopted for interviewing include the following.

- Determining the types of questions to be asked in the interviews to the participants. The questions should be open ended where the participants understand the meaning intended for which the question was seeked
- Purposefully and carefully selecting the interviewees so that the appropriate information is gathered for knowledge creation
- Determining the mode of interviews
- Using appropriate tools for recording the interview
- Use of an interview protocol guide (Silverman,2008) throughout the interview
- Refining the interview questions based on the pilot test
- Selecting an appropriate place for the interviews to take place without any disturbances (Gill, et al. 2008), so that the recording can be done
- Obtaining the required consents from the participants to voluntarily participate in the research
- Follow the appropriate interview procedure
- Doing transcribing the interviews appropriately (Silverman, 2008).

While conducting the interviews it is also important that (Alshenqeeti, 2014) the value, trust, meaning and wording of the interviews should be paid attention to. Value refers to the words spoken by the participant. Trust revolves around the concept that each and every detail provided in the interviews are accurate and honest. Meaning refers to what the interviewer is trying to convey through this interview. Wordings refer to the words used by the interviewer in the interviews as questions. The best interviews are considered when the questions asked are small and description by the interviewee is more detailed.

Videoconferencing via Skype (Sullivan, 2012), Zoom meetings, Google Meet or any other means of conducting interviews can also be used while conducting the interviews. Use of interview method of data collection has been quite common amongst the researchers. Various researchers have used interview methods for qualitative research (Girija, et al. 2024 ;) Also there are various skills that are required by the researcher to conduct the interviews in the most effective manner (Gill, et al. 2008). These skills include having a neutral body language, active listening, being silent at different points so that more accurate data can be extracted from the respondents are all necessary (Gill, et al. 2008).

There is a slight variation in the type of interviews. The Structured interviews have a predefined set of questions. The semi-structured interviews use an interview guide providing the flexibility and the unstructured interviews are open-ended. The unstructured interviews allows for natural conversation to flow between the participant and the interviewer. All these types of interviews have their own unique benefits. Each of them can provide such deep insights into complex issues or capturing rich, descriptive data that quantitative methods cannot provide (Creswell & Poth, 2018).

It is important to understand the purpose of each interview type and when and where they are used.

- **Unstructured interviews**

This type of interview method is both flexible and open-ended method of data collection widely used interview method of qualitative research. These type of interview methods do not use any predetermined questions, unlike structured or semi-structured interviews. Such interviews allow the conversation to flow naturally which is based on the participant's responses. For example in a health setting this approach facilitates in-depth exploration of participants' experiences. It also helps in understanding the perceptions and emotions involved while also providing rich and detailed data. When there is complex or sensitive topics, researchers tend to use this unstructured interviews method of data collection to delve deeper into the area of research. Researchers also encourage trust and rapport-building between the interviewer and the participant to collect any kind of sensitive information. If any inexperience person is handling the interviews then however, the lack of a fixed framework can present challenges. Several difficulties can arise which can create difficulty in maintaining focus or comparing data across participants. Despite posing these challenges, unstructured interviews can provide invaluable information for gaining insights into phenomena/experiences of situations that are not well understood, making them a cornerstone of grounded theory and other exploratory research methodologies. This type of method is also popular amongst the ethnographers (DiCicco-Bloom & Crabtree, 2006).This method of data collection via un-structured interviews can be combined with field notes as well to consolidate the research undertaken (DiCicco-Bloom & Crabtree, 2006).

- **Semi-structured interviews**

These types of interviews are solely conducted and used as a sole method of data collection. It is usually conducted at a dedicated time and place outside of the routine events (DiCicco-Bloom & Crabtree, 2006). This type of interview methods are most widely used interview methods in qualitative research and it is conducted with either individuals or in groups. The questions arise during the interview through the interactions that happen. They are in depth and can last from half an hour to several hours, depending upon the need of the interviews. Such interviews when done for the groups may result in focus group interviews where the information is seeked from participants about a particular thing from a set of people who are ready to share their knowledge about an experience or of a situation. Such interviews may result in depth of information.

- **Individual in-depth interviews**

This type of interviews are conducted in health based experience and health care systems (DiCicco-Bloom & Crabtree, 2006).This can involve issues pertaining to pain perception and impact of a medicine or attitude towards pain. Typically, in-depth interviews are semi-structured, where an interview guide may be used with an open-ended questions. This allows for the interview conversation to grow. The data that is collected is often analysed through thematic analysis, providing comprehensive insights that inform research objectives. It is also important to understand that one needs to build a rapport with the interviewee to gain the trust and respect, where the interviewee feels comfortable sharing the experiences and attitudes as they have experienced it where truth is revealed (DiCicco-Bloom & Crabtree, 2006).

The biggest challenge of this method of data collection is at the analysis level (Alshenqeeti, 2014). While the interview has been conducted but the manner through which this has been analysed can impact the entire quality, reliability and validity (Hofisi, et al. 2014) of the interview (Alshenqeeti, 2014). Interviews need not be just reported but they need to be justified in the light of the research question for which the information is seeked for and thereby justifying the information that has been just collected (Neuman, 2007). Interviews need to be systematically curated to seek the exploratory information (Jain, 2021).

2.2 Focus Groups

Focus groups is an interaction that happens within a group of participants' along with the moderator for discussing a particular issue. These issues can range from marketing, health, strategy, issues related to a particular topic or any such matter. The participants participating within this group are from a similar background. These participants can range from 6 to 20. There are two types of FGD (focus group discussions) one where the moderator shares a questions and seeks the opinion form each of the member of the group either verbally or in a written format and second where the participants discuss on an open ended questions raised by the moderator in an in-depth manner (Cox, et al. 1976). As per Morgan, (1996) Focus groups are defined as a method of data collection, data sources are the interactions that happen within the focus group discussions and third the role of a researcher in this data collection method.

Focus groups were primarily used in marketing but later on its usage was included in social sciences as well (McLafferty, 2004). The uses have been identified in different fields including Health care (McLafferty, 2004), Personal Health Data; patient experience; end-of-life communication skills, innovative teaching model, marketing, Customer attitude, etc...(Powell & Single,1996; Wong, 2008; Spencer, Sanders, Whitley, Lund, Kaye & Dixon, 2016; Pauling, et al. 2018; Gillett, et al. 2016; White & Thomson 1995; Goldman & McDonald 1987, Greenbaum 1993 in Morgan, 1996; Cox, Higginbotham & Burton (1976); Hinkes & Christoph-Schulz, 2019; Weyrich, et al. 2008, Khan, et al. 2023)

Focus group comprises of participants, recorder and a moderator. Moderator helps facilitate the group members for an insightful, in-depth and meaningful discussions. Role of a recorder is to ensure recording the entire discussions (Bader & Rossi 2002). There are few most important things to keep in mind while selecting the participants. These include selecting the participants only those who have been in the situation for which the Focus Group Discussion data is intended to be collected, they should have experienced the phenomenon, should not be volunteers, participants should be from a single organization and should be a representative sample of the group (Bader & Rossi 2002).

The budget and timelines are the deciding factors to decide how many Focus Group Discussions are to be conducted (Bader & Rossi 2002). Size of the focus groups participants can range from10 to 12. Table 4 represent the method of selecting the number of members for the Focus Group Discussion from different sizes of the organization.

Table 4. Selection of number of members Focus Group Discussion

Small organization	**10-100 employees**	**50-100% of employees**
Mid-sized organizations	100-500 employees	20-50% of employees
Large organizations	1000+ employees	10-25% of employees

Source: Bader & Rossi, 2002

These lists of participants are carefully selected by the researcher and handed over to the manager as a complete list. Refer pg. 17, of Bader & Rossi, (2002) for the sample invitation letter that is sent out to the participants informing about the various details that the participants needs to know before the conduct of the Focus Group Discussion. A facilitator is the person who conducts the Focus Group Discussion session and also writes the agenda for it. The facilitator can be from internal or external of an organization. Facilitator ensures that all participants of the Focus Group Discussions participate in the discussions and they themselves should be asking open ended questions, probing questions while ensuring that the discussion of the Focus Group Discussion revolves around the theme of discussion only (Bader & Rossi, 2002).

The Focus Group Discussion agenda should involve these five different things like Introduction, Warm-up period, Writing option, Question period and Summary (Bader & Rossi, 2002, pg.23). Introduction refers to introducing to each other, making the participants welcome and relaxed, explaining the purpose of gathering and making the participants comfortable. In the warm up period, the participants gets familiar with the simple questions to more complex questions. Writing option refers to that the participants write their personal experiences on paper of the topic that is being discussed. Question period refers to asking 2-3 questions to the participants in the order of enquiry and importance. Every question requires certain time period and duration. Therefore, the discussions should keep that in mind the time that is allocated to each question in the order of importance. There could possibly be some sensitive or complex topics which might require more than the specified time. Table 5 provides examples of the qualitative methodology of data collection while using focus group methods.

Table 5. Qualitative Methodology of Data Collection-Focus Groups

Title/Author	Methodology	Numbers	Participants
"Determinants of physical activity and sedentary behaviour in university students: a qualitative study using focus group discussions"/ Tom Deliens, Benedicte Deforche, Ilse De Bourdeaudhuij and Peter Clarys	Focus group Discussions	Seven	University Students
"The role of information and communication technologies in a university learning environment"/ Rosanna Breen, Roger Lindsay, Alan Jenkins & Pete Smith	Focus group Discussions	Ten	University Students
"Focus group interviews as a data collecting strategy"/ Isabella McLafferty	Focus groups	Six	Nursing Lecturers
"Focus groups with young people: A participatory approach to research planning"/ Anna Bagnoli, Andrew Clark	Focus groups	Twenty-Two	Young People Aged 13

2.2.1 Application of focus groups

Breen (2006) suggests that a researcher needs to identify few questions ahead of the research to begin to choose the research group as a research methodology. Breen (2006, pp.465) suggests researchers to ask questions to themselves that is the Focus Group Discussion intended to seek more information from the participants, "*to share and compare their experiences with each other? Develop and generate ideas? Explore issues of shared importance?*" If the answer to all or any one of the questions is a yes, then it is important that Focus Group Discussion is conducted. Focus groups allow researchers or companies to collect rich, detailed data that might not emerge from surveys or other quantitative methods. They can be used for a variety of purposes, including the following as mentioned in the table 6.

Table 6. Use of Focus Groups in different fields

Area of research	Author
Health care, Personal Health Data; patient experience; end-of-life communication skills	Powell & Single,1996; Wong, 2008; Spencer, Sanders, Whitley, Lund, Kaye & Dixon, 2016; Pauling, et al. 2018; Gillett, et al. 2016; White & Thomson 1995
Innovative teaching model	Weyrich, et al. 2008
Communication studies	Albrecht et al 1993, Staley 1990 in Morgan, 1996
Education	Brotherson & Goldstein 1992, Flores & Alonzo 1995, Lederman 1990 in Morgan, 1996
Political science	Delli Carpini & Williams 1994, Kullberg 1994 in Morgan, 1996
Public health	Basch 1987 in Morgan, 1996
Marketing, Customer attitude	Goldman& McDonald 1987, Greenbaum 1993 in Morgan, 1996; Cox, Higginbotham & Burton (1976); Hinkes & Christoph-Schulz, 2019;
In the field of Sociology for topics ranging from: aging, criminology, medical sociology, political sociology, social Movements, and the sociology of work; relationships; sexuality; behaviour in university students; Supporting Families	Knodel 1995, Duncan & Morgan 1994; Sasson, 1995; Morgan & Spanish 1985, McKinlay 1993; Gamson 1992; Cable 1992; Bobo et al 1995 in Morgan, 1996; Wilkinson, 1998; Bagnoli & Clark 2010; Överlien, Aronsson & Hydén, 2005; Deliens, et al. 2015; Deliens, et al. 2014; Weiber, et al. 2016
spread of HIV	Irwin et al 1991 and Kline et al 1992, Pollak et al 1990 in Morgan, 1996
In development of a new measuring instrument	Gray-Vickery 1993, McKinley et al. 1997 in McLafferty, 2004

Focus Group Discussions can be used in combination with any other research method to collect the data from the participants. Depending upon the outcome of the study undertaken, Focus Group Discussion may be either used in isolation or in combination with other qualitative or quantitative methods (Morgan, 1996). It is also important to note that a researcher should know when not to use the Focus Group Discussions. When one has to educate, come to a consensus, to test about any kind of knowledge of a particular topic, a sensitive information which is not supposed to be shared in a group, statistical projections, and when the environment is emotionally charged (Krueger & Casey, 2015).

2.2.2 Key elements of focus groups for planning

- **Participant Selection**: Criteria for choosing participants, including demographic factors, relevance to the research topic, and size of the group. They should also be chosen in case there is a phenomenon that one needs to research and these participants have also been through this similar phenomenon (Breen, 2006). The number of participants can range between 6 to 10 (Mishra, 2016; Powell & Single, 1996). A diverse group of individuals selected because they represent the broader population of interest. Their selection is based on characteristics relevant to the research question. The number of people selected to participate in the FGDs should not be known to each other (Krueger & Casey, 2015; Powell & Single, 1996). This anonymity ensures that the opinions pined in the FGD although based on personal experiences/negative experiences of individuals are shared without any inhibitions. It is important to have a variety of responses from these selected members. Therefore, the researchers choose the group members based on their requirements of the information that they are seeking for. They may use stratified FGDs according to the group characteristics (Krueger & Casey, 2015; Powell & Single, 1996). Member participants should not be chosen based on the recommendations of fellow FGD members (Powell & Single, 1996). Therefore participants should come from diverse areas of background (Bagnoli & Clark 2010).but should be homogenous. It is also suggested that more participants should be recruited rather than a small size (Mishra, 2016).
- **Number and duration of FGDs**: Number of sessions differ and depends upon the complexity of the problem addressed in the FGD. These can range between 1upto 10 sessions. Usually there is no requirement of further conduct of FGDs when the data received in FGD gets repeated and data saturation occurs. Each FGD can last up to 90 to 120 minutes (Powell & Single, 1996). It is also important that the participants of the FGD are well informed in advance (about 10-15 days) of the conduct of this discussions.

- **Logistics:** Selection on organizing venues should be based on the accessibility by all the participants of FGD and as per their convenience (Breen, 2006).
- **Moderator**: A skilled individual who guides the discussion, encourages participation, and ensures that the conversation remains on topic (Powell & Single, 1996). Role of a moderator is very important as the essentials skills like listening, being non-judgemental are all crucial for the free flow of information. The moderator characteristics should be similar with the FGD participants in terms of age, gender and language (Powell & Single, 1996). Along with the moderator is also the note taker who records the mannerisms of FGD participants based on their interactions. It is also the responsibility of the moderator to encourage all participants of the FGD to equally participate in this discussion. Moderator should conduct this FGD with the guided questions in the form of a discussion guide. Also to ensure to each participant that confidentiality of information will be maintained.
- **Discussion Guide**: A plan that outlines the topics and questions the moderator will introduce to stimulate discussion. These are about 5 to 6 open ended questions, usually in form of a semi structured interview. In case the matter of discussion is sensitive then the questions should flow from less sensitive questions to more intense sensitive questions (Powell & Single, 1996). The discussion guide serves as an outline of the entire Focus Group Discussion. Sample of how to begin the Focus Group Discussion with the opening remarks and concluding notes is mentioned in the Methods Guides How to run a focus group. (n.d.).
- Setting: Focus groups are typically held in a comfortable, neutral location that is free from distractions (Powell & Single, 1996). Online platforms are also increasingly used for virtual focus groups. The settings help the participants to participate freely in the discussions without any distractions. Timings of conducting this FGD is equally important. Therefore, the locations to conduct the FGD should be chosen wisely (Breen, 2006).
- **Holding the session**: it is important that the participants meet informally before the discussion begins. In this manner the moderator and note taker both can be introduced to the group and the ice breaking happens (Powell & Single, 1996). All the participants should be facing each other in a circle and moderator should play a minimal role. Moderator should also ensure that any materials [flash cards, leaflets] which will be essential for conducting the FGD should also be present in the room (Breen, 2006). Moderator may bring in a note to record the details of the discussion or a recorder. But recording the discussion or bringing in an outsider to take notes might make the participants uneasy (Breen, 2006).

- **Ethical Guidelines:** Focus group methodology presents unique ethical challenges. These challenges are related to ***consent***, confidentiality, anonymity, and risk of harm. There will be certain difficulties in obtaining clear consent due to the unpredictable nature of group discussions. Secondly there will be potential issues with both ***confidentiality and anonymity***. Also there is ***risk of harm*** from discussing sensitive topics in a public setting. Confidentiality and anonymity, although often used interchangeably, have distinct meanings which are crucial to research ethics. Confidentiality concerns how information is managed and disclosed once obtained by researchers. On the contrary anonymity relates to whether participants can be identified from the data that is provided. ***Privacy*** involves the initial contact and access to information. In focus groups, these issues are particularly significant, especially while dealing with sensitive topics. Here participants share private views publicly and the reactions of the group cannot be predicted. Therefore, while reporting data can be reported anonymously. But the researcher declaring data confidential implies avoiding direct quotes. This therefore highlights the importance of maintaining confidentiality when anonymity cannot be fully assured (Sim & Waterfield, 2019). ***Informed Consent*:** participants should be informed that there will be note taking and recording of the sessions through audio or video recorded to facilitate accurate data analysis and that confidentiality will be maintained throughout.
- **Execution**: It is extremely crucial to hold an informal meeting with all the participants before the FGD begins. This ice-breaking exercise is extremely important for both the moderators and the FGD participants. This meeting helps and facilitates to know each other and helps in relaxing the participants for an open discussion later on (Powell & Single, 1996). Once the formal discussion begins, there should be lively interaction amongst each of the participants. Therefore, the participants of FGD should be seated in a circle to have a lively and face to face interaction. It is also important to note that the moderator or the facilitator should not be at all actively participating, in fact the role of moderator should be the least so that the attention is not focussed on the moderator. Also to be noted that the audio or a video instrument which is used for recording the FGD should not be attracting the attention of the FGD members. In fact, it should be placed in a position that it does not hinders the participants' active discussion at any point in time while the FGD is going on.

The moderator/facilitator should also be aware of the potential for self-censorship and conformity influenced by the "group effect." Participants may "jump on the group bandwagon," adjusting their contributions based on their individual needs and their perceived position within the group's social expectations (Powell & Single, 1996).

2.2.3 Designing focus group questions

There are certain assumptions before the questions are being asked to take into account. These include that the environment where this FGD is conducted is conducive for an open environment where the discussion can be held without any inhibitions and participants can openly discuss the issue. Second the respondents understand the question, knows the answer, and is able to articulate the answer well and this answer is well understood by the moderator as well (Krueger & Casey, 2015). Usually when the conversation begins it may be factual and also directed to the moderator, it is only in the later stage that the conversation is amongst the participants and the moderator is least important. This is the time when actually deeper meanings and objectives of the FGDs are being discussed. But all of this takes considerable amount of time before the actual conversation begins (Krueger & Casey, 2015).

- **Sequence and Flow**: There are certain sequence that needs to be followed for conducting the FGD. The entire sequence can be further divided into multiple stages including welcoming the participants, describing an overview of the topic of discussion, setting up of the ground rules and also the assuring the participants that their information provided in the FGD will be strictly confidential (Breen, 2006). Moderator should begin with the general questions and then move on to the specific questions. How to order questions for natural discussion flow. An email can be sent in advance to the participants for prior information and briefing followed by the consent form (Methods Guides How to run a focus group (n.d.).
- **Types of Questions**: The questions that the moderator asks should be open-ended. They should also ask the probing, and follow-up questions. Directions of questions should be well thought of and be clear and precise. There will be series of questions that needs to be asked. Few can be the beginner questions which might not require that much amount of time while the other questions could be detail oriented. Beginner questions are easy and are meant to make everyone in the group to get into the conversation. The questions that follow later might also require probing questions to be asked and may also require more amount of time (Krueger & Casey, 2015). There will be a mixture of varied types of questions. Beginner questions, interim questions, probing

questions, transition questions and ending questions. The ending questions are usually to summarize the points discussed earlier giving affirmation and reflection on the previous answers. More details and samples of the type of questions that can be asked can be seen in Krueger & Casey, 2015, pp. 121.

2.2.4 Focus Group Analysis

The objective of conducting the FGD leads to the depth of the analysis (Krueger & Casey (2015). Krueger & Casey (2015) says that the FGD analysis has critical qualities like it is systematic, verifiable, sequential, and consequential. Systematic way of the analysis follows a systematic procedure to derive at a result which is well documented and understood. The process of analysis is transparent and is ready for any kind of scrutiny. Verifiable means that the data was analysed not by the moderator's perception but by sufficient amount of data that creates confirmation. It is also very important to that a good analysis begins with good planning for the entire session. This means that the groups must be chosen wisely, recruiting of participants should be based on the rich information that these participants can provide, questions should be planned ahead keeping in mind the objectives of the research and the time that might be required for each of the questions, and moderating also plays a pivotal role. Through moderation it is necessary that the note taking is not just about the phrases written but about the deeper perspectives of the participants and the summary at the end should serve as a verification for the entire discussion.

There are two stages of analysis. Stage I includes coding and classifying the raw data based on the discussion guide questions. These codes can be further grouped based on the responses provided by the participants (Powell & Single, 1996). Stage II involves comparison of original data with the grouped data based on conceptualization. This is the most difficult stage as one needs to understand the diverse views of the participants and also understand the contrast behaviour of the variables under study. This is an inductive process involving the conceptualization of the themes from the analysis (Powell & Single, 1996). NVivo, ATLAS.ti, MAXQDA, QDA Miner are few of the software tools that can help in analysis in transcribing recordings for detailed analysis. These softwares can look for patterns, themes, and insights that emerge from the discussion. Data collection and analysis are simultaneous and the analysis begins even after a single a FGD is conducted. The reason one does that is because while doing the analysis if there is a point where more clarification is required the one can always go back and ask the question (Krueger & Casey, 2015). Therefore, it is important that one goes question after question understanding the deeper meaning of the information provided. Analysing along with the data collection serves an important purpose of improvising on any loopholes or lack of information that could not be collected in the previous FGDs. If one waits for all the FGDs to complete then there are likely chances that some vital information may have been missed out.

Focus groups are a powerful tool for gaining deep insights, but they are most effective when combined with other research methods to provide a comprehensive understanding of the topic at hand (Bader & Rossi, 2002). FGDs provide participants a platform where their opinions are being heard. This chapter structure is designed to offer comprehensive guidance on conducting focus groups, addressing both the theoretical underpinnings and practical applications, enriched with real-world insights. FGDs can also be considered to be used in exploratory research. It is necessary while collecting the data that the moderator to introspect and see if the data is getting any contradictory results from the participants. It is recommended that the moderator ask the participants on the views and why do they differ from each other (Morgan, 1996).

2.3 Observations

This method of data collection is a way in which the researchers try to make observations from the events, behaviours and artefacts in a particular social setting. The researchers try to find meaning associated with these things in the setting (Billups, 2021). Ciesielska, Boström, and Öhlander, (2018) discusses three different types of observations. Participant observation refers to the complete immersion into a particular culture for a longer period of time as a visitor or as a member so that the complete information is gathered of that particular culture. In Non Participant observation the researcher is tries to understand the interactions without any categorizations and observations. In indirect observations the researcher is actually relying on observations of others, documentation, recordings etc. Mulhall, 2003 also describes about the unstructured method of observation and the problems associated with access and field notes in it.

Observation may be used as either a sole qualitative method of data collection or in conjunction with other methods. However there are several protocols that has to be followed while even initiating the observations in the field. This includes taking the required permissions from the relevant field of research. This field of research can be a bank, town, village, organization, and others (Ciesielska, Boström and Öhlander, 2018).Observation method may be chosen over other qualitative methods when under the following conditions (Billups, 2021) as shown in Table 7 below. It shows different situations under which this observation method is used in qualitative research.

Table 7. When to choose Observation method

Unexplored topics
Behaviour of people is not known much
Important to understand the setting due to the nature of research
When it is to be observed the difference between what people say and what they actually behave
Group interaction is important for triangulation

There are several field notes that the researcher is required during the data collection while using the observation method. These observation templates may be referred to in Billups, (2021, pg.138, 139,140). There are several challenges that this observation method poses. Billups, (2021) recommends several things to be considered when choosing an organization as an area of research for observation. Management of time and space, objects, social actors, interactions of people, routines, rituals and episodes of people, are all the important aspects to taken into consideration while going in for observations. Table 8 provides examples of the qualitative methodology of data collection while using observation methods.

Table 8. Qualitative Methodology of Data Collection-Observation

Title/Author	Methodology	Numbers/ Participants	Duration
1. "How are decisions made in cancer care? A qualitative study using participant observation of current practice"/Pola Hahlweg, Martin Härter, Yvonne Nestoriuc, Isabelle Scholl	Observation	54 outpatient consultations	2 weeks
"Qualitative observation in a clinical setting: Challenges at end of life"/ Melissa J. Bloomer, Wendy Cross, Ruth Endacott, Margaret O'Connor, and Cheryle Moss,	Observation	Observe Nurses caring for dying patient	

2.4 Document Analysis

Document analysis has been used as a useful tool for the qualitative research. Document can be any text or visual item that is used for the analysis (Bowen, 2009). The document can contain newspaper, books, articles, academic journal articles, institutional reports, photographs, video, and film (Altheide, 2000; Altheide, Coyle, DeVriese & Schneider, 2008; Bowen, 2009; Morgan, 2022). Other documents can be the agendas, attendance registers, minutes of meetings, manuals, background papers, books and brochures, diaries and journals, event programs, letters and memoranda, maps and charts, press releases, program proposals, application forms, and summaries, radio and television program scripts, organisational or institutional reports, survey data, and various public records (Bowen, 2009, pg. 2).

Bowen (2009) also describes these documents which serve as evidences which are used to consolidate the research findings with the help of multiple such document collected. Therefore, Bowen also describes its five different uses of documents. Firstly, documents provide background information and historical insights of the research undertaken. Secondly, the documents can provide further information that can be seeked via interviews or a particular situation that is to be observed more closely. Thirdly, these documents also serve as extra evidence enhancing the knowledge base of the research undertaken. Fourthly, it serves as evidence to check on any changes that might have occurred. These changes can be tracked and can serve to understand the development in the projects. Fifth documents also serve as a way to verify the findings.

There are several advantages of using the document analysis methods (Bowen, 2009). It is efficient as it provides data selection over data collection (Bowen, 2009). Documents are freely accessible over the internet without seeking necessary permissions from the authors. It is also cost effective. They are also not affected by the research process and remain stable over a period of time.

Document analysis is done in different phases. This can include document skimming, then again re reading and then finally the interpretation of the document. Therefore, it is kind of content and thematic analysis (Bowen, 2009). The evidences are cross examined with other methods of data collection to reach to a conclusion of the research undertaken.

Evaluating the documents carefully and finding the right kind of interpretations are also necessary. The purpose of document analysis is to find out if the documents are comprehensive enough to provide sufficient support to the research that it intends to find. To ensure that the document is sufficient and is helping out in triangulation of the research, it is imperative to understand the reason for which it was initially procured (Bowen, 2009). The matter of concern should also be if the author had used primary sources or secondary sources to write this document. Also, if the context is related then other proofs should also be collected for cross evaluating. Therefore, when collecting the document proofs it is necessary to focus on the quality of documents that provide sufficient and complete evidence of the information that is to be found rather than the number of documents. While finding the appropriate documents the researcher should try and balance between sensitivity and objectivity. Table 9 provides examples of the qualitative methodology of data collection while using Document collection methods.

Table 9. Qualitative Methodology of Data Collection-Document Analysis

Title/Author	Methodology
2. "Policy Document Analysis: A Practical Educational Leadership Tool and a Qualitative Research Method"/ Carol Cardno,2018	Document Analysis
"Increasing rigor and reducing bias in qualitative research: A document analysis of parliamentary debates using applied thematic analysis"/ Penny Mackieson, Aron Shlonsky, Marie Connolly	Document Analysis

3. CONCLUSION

There are several key methods involve in data collection in qualitative research. Each of these methods offers a unique insight into human experiences and social phenomena, areas of the research. Interviews provide researchers an opportunity to explore individual perspectives. There are also several ethical considerations and practical challenges encountered during data collection. They also help in providing in depth understanding of participants' thoughts, beliefs, and motivations. They are sometimes used as the only method of data collection and sometimes they are used in conjunction with others. Focus groups, on the other hand, provide a dynamic platform for collective discussion amongst the participants of similar demographics. It encourages an immersive interaction and also reveals group dynamics. Participant observations enable the researcher to immerse themselves in the natural setting of the study, gaining first-hand experience. This provides details of the understanding of the context in which behaviours and interactions occur. Document analysis complements these methods by offering a systematic review of existing materials, providing historical, contextual, or factual data to enrich the findings. Documents are of various variety and provide a comprehensive view of the information that is required by the researcher. Each of these methods has distinct strengths and limitations, making it essential to select the most appropriate technique based on the research question, objectives, and context. Combining these methods can enhance the depth and breadth of the data collected, leading to more robust and comprehensive findings in qualitative research. Triangulation of the study may require a combination of various data collection methods to be used together to come to a conclusion for the research undertaken. Top of Form

REFERENCES:

Alshenqeeti, H. (2014). Interviewing as a data collection method: A critical review. *English linguistics research, 3*(1), 39-45.

Altheide, D., Coyle, M., DeVriese, K., & Schneider, C. (2008). Emergent qualitative document analysis.

Altheide, D. L. (2000). Tracking discourse and qualitative document analysis. *Poetics*, 27(4), 287–299. DOI: 10.1016/S0304-422X(00)00005-X

Armstrong, C. (2021). Key methods used in qualitative document analysis. *OSF Preprints, 1*(9).

Bachiochi, P. D., & Weiner, S. P. (2004). Qualitative data collection and analysis. *Handbook of research methods in industrial and organizational psychology*, 161-183.

Bader, G. E., & Rossi, C. A. (2002). *Focus groups: a step-by-step guide* (3rd ed.). The Bader Group.

Baffour, F. D., Francis, A. P., Chong, M. D., & Harris, N. (2024). Doing qualitative research in a controlled site: Challenges, lessons, and strategies. *Crime and Delinquency*, 70(8), 1921–1942. DOI: 10.1177/00111287221106944

Bagnoli, A., & Clark, A. (2010). Focus groups with young people: A participatory approach to research planning. *Journal of Youth Studies*, 13(1), 101–119. DOI: 10.1080/13676260903173504

Billson, J. M. (1989). Focus groups: A practical guide for applied research. *Chinese Sociological Review*, 7(1), 24.

Billups, F. D. (2019). *Qualitative data collection tools: Design, development, and applications* (Vol. 55). Sage Publications.

Bloomer, M. J., Cross, W., Endacott, R., O'Connor, M., & Moss, C. (2012). Qualitative observation in a clinical setting: Challenges at end of life. *Nursing & Health Sciences*, 14(1), 25–31. DOI: 10.1111/j.1442-2018.2011.00653.x PMID: 22288573

Bowen, G. A. (2009). Document analysis as a qualitative research method. *Qualitative Research Journal*, 9(2), 27–40. DOI: 10.3316/QRJ0902027

Braun, V., & Clarke, V. (2006). Using thematic analysis in psychology. *Qualitative Research in Psychology*, 3(2), 77–101. DOI: 10.1191/1478088706qp063oa

Breen, R., Lindsay, R., Jenkins, A., & Smith, P. (2001). The role of information and communication technologies in a university learning environment. *Studies in Higher Education*, 26(1), 95–114. DOI: 10.1080/03075070123233

Breen, R. L. (2006). A practical guide to focus-group research. *Journal of Geography in Higher Education*, 30(3), 463–475. DOI: 10.1080/03098260600927575

Britten, N. (1995). Qualitative research: Qualitative interviews in medical research. *BMJ (Clinical Research Ed.)*, 311(6999), 251–253. DOI: 10.1136/bmj.311.6999.251 PMID: 7627048

Brooks, S. M. (2013). The ethical treatment of human subjects and the Institutional Review Board process. *Interview research in political science*, 45-66.

Busetto, L., Wick, W., & Gumbinger, C. (2020). How to use and assess qualitative research methods. *Neurological Research and Practice*, 2(1), 14. DOI: 10.1186/s42466-020-00059-z PMID: 33324920

Byrne, M. (2001). Interviewing as a data collection method. *AORN Journal*, 74(2), 233–233. DOI: 10.1016/S0001-2092(06)61533-0 PMID: 11503203

Campbell, S., Greenwood, M., Prior, S., Shearer, T., Walkem, K., Young, S., & Walker, K. (2020). Purposive sampling: Complex or simple? Research case examples. *Journal of Research in Nursing*, 25(8), 652–661. DOI: 10.1177/1744987120927206 PMID: 34394687

Cardno, C. (2018). Policy document analysis: A practical educational leadership tool and a qualitative research method. *Educational Administration: Theory and Practice*, 24(4), 623–640.

Carter, S., & Henderson, L. (2005). Approaches to qualitative data collection in social science. *Handbook of health research methods: Investigation, measurement and analysis, 1*, 215-230.

Cartwright, J. C., Hickman, S. E., Nelson, C. A., & Knafl, K. A. (2013). Investigators' successful strategies for working with institutional review boards. *Research in Nursing & Health*, 36(5), 478–486. DOI: 10.1002/nur.21553 PMID: 23813748

Ciesielska, M., Boström, K. W., & Öhlander, M. (2018). Observation methods. *Qualitative methodologies in organization studies: Volume II: Methods and possibilities*, 33-52.

Cox, K. K., Higginbotham, J. B., & Burton, J. (1976). Applications of focus group interviews in marketing. *Journal of Marketing*, 40(1), 77–80. DOI: 10.1177/002224297604000117

Creswell, J. W., Fetters, M. D., & Ivankova, N. V. (2004). Designing a mixed methods study in primary care. *Annals of Family Medicine*, 2(1), 7–12. DOI: 10.1370/afm.104 PMID: 15053277

Creswell, J. W., & Poth, C. N. (2016). *Qualitative inquiry and research design: Choosing among five approaches*. Sage publications.

Curtis, S., Gesler, W., Smith, G., & Washburn, S. (2000). Approaches to sampling and case selection in qualitative research: Examples in the geography of health. *Social Science & Medicine*, 50(7-8), 1001–1014. DOI: 10.1016/S0277-9536(99)00350-0 PMID: 10714922

De Villiers, C., Farooq, M. B., & Molinari, M. (2022). Qualitative research interviews using online video technology–challenges and opportunities. *Meditari Accountancy Research*, 30(6), 1764–1782. DOI: 10.1108/MEDAR-03-2021-1252

Deliens, T., Clarys, P., De Bourdeaudhuij, I., & Deforche, B. (2014). Determinants of eating behaviour in university students: A qualitative study using focus group discussions. *BMC Public Health*, 14(1), 1–12. DOI: 10.1186/1471-2458-14-53 PMID: 24438555

Deliens, T., Deforche, B., De Bourdeaudhuij, I., & Clarys, P. (2015). Determinants of physical activity and sedentary behaviour in university students: A qualitative study using focus group discussions. *BMC Public Health*, 15(1), 1–9. DOI: 10.1186/s12889-015-1553-4 PMID: 25881120

Denzin, N. K., & Lincoln, Y. S. (Eds.). (2011). *The Sage handbook of qualitative research*. Sage.

DiCicco-Bloom, B., & Crabtree, B. F. (2006). The qualitative research interview. *Medical Education*, 40(4), 314–321. DOI: 10.1111/j.1365-2929.2006.02418.x PMID: 16573666

Easton, K. L., McComish, J. F., & Greenberg, R. (2000). Avoiding common pitfalls in qualitative data collection and transcription. *Qualitative Health Research*, 10(5), 703–707. DOI: 10.1177/104973200129118651 PMID: 11066874

Farooq, M. B., & De Villiers, C. (2017). Telephonic qualitative research interviews: When to consider them and how to do them. *Meditari Accountancy Research*, 25(2), 291–316. DOI: 10.1108/MEDAR-10-2016-0083

Flick, U. (2018). Doing qualitative data collection–charting the routes. *The SAGE handbook of qualitative data collection*, 1-16.

Gibbs, L., Kealy, M., Willis, K., Green, J., Welch, N., & Daly, J. (2007). What have sampling and data collection got to do with good qualitative research? *Australian and New Zealand Journal of Public Health*, 31(6), 540–544. DOI: 10.1111/j.1753-6405.2007.00140.x PMID: 18081574

Gill, P., Stewart, K., Treasure, E., & Chadwick, B. (2008). Methods of data collection in qualitative research: Interviews and focus groups. *British Dental Journal*, 204(6), 291–295. DOI: 10.1038/bdj.2008.192 PMID: 18356873

Gillett, K., O'Neill, B., & Bloomfield, J. G. (2016). Factors influencing the development of end-of-life communication skills: A focus group study of nursing and medical students. *Nurse Education Today*, 36, 395–400. DOI: 10.1016/j.nedt.2015.10.015 PMID: 26530442

Girija, S., Banerji, B., Batra, N., Paruchuru, M., & Yeediballi, T. (2024). Making frugal innovations inclusive: A gendered approach. *Journal of Cleaner Production*, 434, 140182. DOI: 10.1016/j.jclepro.2023.140182

Hahlweg, P., Härter, M., Nestoriuc, Y., & Scholl, I. (2017). How are decisions made in cancer care? A qualitative study using participant observation of current practice. *BMJ Open*, 7(9), e016360. DOI: 10.1136/bmjopen-2017-016360 PMID: 28963286

Hennink, M., Hutter, I., & Bailey, A. (2020). Qualitative research methods. *Sage (Atlanta, Ga.)*.

Hinkes, C., & Christoph-Schulz, I. (2019). Consumer attitudes toward palm oil: Insights from focus group discussions. *Journal of Food Products Marketing*, 25(9), 875–895. DOI: 10.1080/10454446.2019.1693468

Hofisi, C., Hofisi, M., & Mago, S. (2014). Critiquing interviewing as a data collection method. *Mediterranean Journal of Social Sciences*, 5(16), 60–64. DOI: 10.5901/mjss.2014.v5n16p60

Israel, G. D., & Galindo-Gonzalez, S. (2008). Using Focus Group Interviews for Planning or Evaluating Extension Programs: AEC 387/PD036, rev. 5/2008. *EDIS, 2008*(5).

Jain, N. (2021). Survey versus interviews: Comparing data collection tools for exploratory research. *The Qualitative Report*, 26(2), 541–554. DOI: 10.46743/2160-3715/2021.4492

Jenkins, E. K., Slemon, A., Haines-Saah, R. J., & Oliffe, J. (2018). A guide to multisite qualitative analysis. *Qualitative Health Research*, 28(12), 1969–1977. DOI: 10.1177/1049732318786703 PMID: 30024317

Khan, N., Saini, K., Khan, M. A., & Khajuria, G. (2022). Marketing Practices for Fund Raising and Sustainability of NGOs: Focus Group Discussions. *World Journal of Management and Economics, 16*(8).

Khan, T. H., & MacEachen, E. (2022). An alternative method of interviewing: Critical reflections on videoconference interviews for qualitative data collection. *International Journal of Qualitative Methods*, 21, 16094069221090063. DOI: 10.1177/16094069221090063

Kneale, P. (2002). Developing and embedding reflective portfolios in geography. *Journal of Geography in Higher Education*, 26(1), 81–94. DOI: 10.1080/03098260120110386

Krueger, R. A. (2014). *Focus groups: A practical guide for applied research.* Sage publications.

Krueger, R. A., & Casey, M. A. (2015). *Focus groups: a practical guide for applied research* (5th ed.). SAGE.

Krueger, R. A., Casey, M. A., Donner, J., Kirsch, S., & Maack, J. N. (2001). Social analysis: Selected tools and techniques. *World Development*, 36, 4–23.

Kvale, S., & Brinkmann, S. (2018). Doing Interviews. *Sage (Atlanta, Ga.).*

Lo Iacono, V., Symonds, P., & Brown, D. H. (2016). Skype as a tool for qualitative research interviews. *Sociological Research Online*, 21(2), 103–117. DOI: 10.5153/sro.3952

Lopez, V., & Whitehead, D. (2013). Sampling data and data collection in qualitative research. *Nursing & midwifery research: Methods and appraisal for evidence-based practice, 123*, 140.

Mack, N., Woodsong, C., MacQueen, K. M., & Guest, G. (2005). *Qualitative research methods.* Family Health International.

Mackieson, P., Shlonsky, A., & Connolly, M. (2019). Increasing rigor and reducing bias in qualitative research: A document analysis of parliamentary debates using applied thematic analysis. *Qualitative Social Work: Research and Practice*, 18(6), 965–980. DOI: 10.1177/1473325018786996

McLafferty, I. (2004). Focus group interviews as a data collecting strategy. *Journal of Advanced Nursing*, 48(2), 187–194. DOI: 10.1111/j.1365-2648.2004.03186.x PMID: 15369499

Mertens, D. M. (2018). Ethics of qualitative data collection. *The SAGE handbook of qualitative data collection*, 33-48.

Merton, R. K., Fiske, M., & Kendall, P. L. (1956). The focused interview: A manual of problems and procedures. Free Pree. In *Krueger, R. A., & Casey, M. A. (2015). Focus groups: a practical guide for applied research* (5th ed.). SAGE.

Methods Guides How to run a focus group. (n.d.). https://sites.reading.ac.uk/wp-content/uploads/sites/42/2020/06/Focus-Groups-Guide-Final.pdf

Meurer, W. J., Frederiksen, S. M., Majersik, J. J., Zhang, L., Sandretto, A., & Scott, P. A. (2007). Qualitative data collection and analysis methods: The INSTINCT trial. *Academic Emergency Medicine*, 14(11), 1064–1071. DOI: 10.1197/j.aem.2007.05.005 PMID: 17652268

Milne, C. (2005, January). Overseeing research: Ethics and the institutional review board. In *Forum Qualitative Sozialforschung/Forum: Qualitative. Social Research*, 6(1), •••.

Mishra, L. (2016). Focus group discussion in qualitative research. *TechnoLearn: An International Journal of Educational Technology*, 6(1), 1–5. DOI: 10.5958/2249-5223.2016.00001.2

Morgan, D. L. (1996). Focus groups. *Annual Review of Sociology*, 22(1), 129–152. DOI: 10.1146/annurev.soc.22.1.129

Morgan, H. (2022). Conducting a qualitative document analysis. *The Qualitative Report*, 27(1), 64–77.

Moser, A., & Korstjens, I. (2018). Series: Practical guidance to qualitative research. Part 3: Sampling, data collection and analysis. *The European Journal of General Practice*, 24(1), 9–18. DOI: 10.1080/13814788.2017.1375091 PMID: 29199486

Mulhall, A. (2003). In the field: Notes on observation in qualitative research. *Journal of Advanced Nursing*, 41(3), 306–313. DOI: 10.1046/j.1365-2648.2003.02514.x PMID: 12581118

Neuman, W. L. (2007). *Social Research Methods: Qualitative and Quantitative Approaches*. (6 th Ed.) Boston: Pearson.

Nyimbili, F., & Nyimbili, L. (2024). Types of Purposive Sampling Techniques with Their Examples and Application in Qualitative Research Studies. *British Journal of Multidisciplinary and Advanced Studies*, 5(1), 90–99. DOI: 10.37745/bjmas.2022.0419

Onwuegbuzie, A. J., Leech, N. L., & Collins, K. M. (2010). Innovative data collection strategies in qualitative research. *Qualitative Report*, 15(3), 696–726.

Ostrander, S. A. (1993). "Surely you're not in this just to be helpful" Access, Rapport, and Interviews in Three Studies of Elites. *Journal of Contemporary Ethnography*, 22(1), 7–27. DOI: 10.1177/089124193022001002

Överlien, C., Aronsson, K., & Hydén, M. (2005). The focus group interview as an in-depth method? Young women talking about sexuality. *International Journal of Social Research Methodology*, 8(4), 331–344. DOI: 10.1080/1364557042000119607

Owen, G. T. (2014). Qualitative methods in higher education policy analysis: Using interviews and document analysis. *The Qualitative Report*, 19(26), 1–19. DOI: 10.46743/2160-3715/2014.1211

Palinkas, L. A., Horwitz, S. M., Green, C. A., Wisdom, J. P., Duan, N., & Hoagwood, K. (2015). Purposeful sampling for qualitative data collection and analysis in mixed method implementation research. *Administration and Policy in Mental Health*, 42(5), 533–544. DOI: 10.1007/s10488-013-0528-y PMID: 24193818

Patton, M. Q. (1990). *Qualitative evaluation and research methods*. SAGE Publications.

Patton, M. Q. (1999). Enhancing the quality and credibility of qualitative analysis. *Health Services Research*, 34(5 Pt 2), 1189. PMID: 10591279

Patton, M. Q. (2002). *Qualitative research and evaluation methods* (3rd ed.). Sage.

Pauling, J. D., Domsic, R. T., Saketkoo, L. A., Almeida, C., Withey, J., Jay, H., & Hewlett, S. (2018). Multinational qualitative research study exploring the patient experience of Raynaud's phenomenon in systemic sclerosis. *Arthritis Care and Research*, 70(9), 1373–1384. DOI: 10.1002/acr.23475 PMID: 29473715

Pauling, J. D., Domsic, R. T., Saketkoo, L. A., Almeida, C., Withey, J., Jay, H., & Hewlett, S. (2018). Multinational qualitative research study exploring the patient experience of Raynaud's phenomenon in systemic sclerosis. *Arthritis Care and Research*, 70(9), 1373–1384. DOI: 10.1002/acr.23475 PMID: 29473715

Payne, S. (2007). Qualitative methods of data collection and analysis. *Research methods in palliative care*, 139-161.

Pech, C., Cob, N., & Cejka, J. T. (2007). Understanding institutional review boards: Practical guidance to the IRB review process. *Nutrition in Clinical Practice*, 22(6), 618–628. DOI: 10.1177/0115426507022006618 PMID: 18042949

Powell, R. A., & Single, H. M. (1996). Focus groups. *International Journal for Quality in Health Care : Journal of the International Society for Quality in Health Care*, 8(5), 499–504. DOI: 10.1093/intqhc/8.5.499 PMID: 9117204

Robinson, O. C. (2014). Sampling in interview-based qualitative research: A theoretical and practical guide. *Qualitative Research in Psychology*, 11(1), 25–41. DOI: 10.1080/14780887.2013.801543

Silverman, D. (2008). Doing qualitative research: A comprehensive guide. *Sage (Atlanta, Ga.)*.

Sim, J., & Waterfield, J. (2019). Focus group methodology: Some ethical challenges. *Quality & Quantity*, 53(6), 3003–3022. DOI: 10.1007/s11135-019-00914-5

Smith, M., & Bowers-Brown, T. (2010). Different kinds of qualitative data collection methods. *Practical research and evaluation: A start-to-finish guide for practitioners*, 111-125.

Spencer, K., Sanders, C., Whitley, E. A., Lund, D., Kaye, J., & Dixon, W. G. (2016). Patient perspectives on sharing anonymized personal health data using a digital system for dynamic consent and research feedback: A qualitative study. *Journal of Medical Internet Research*, 18(4), e5011. DOI: 10.2196/jmir.5011 PMID: 27083521

Sullivan, J. R. (2012). Skype: An appropriate method of data collection for qualitative interviews? *The Hilltop Review*, 6(1), 10.

Suri, H. (2011). Purposeful sampling in qualitative research synthesis. *Qualitative Research Journal*, 11(2), 63–75. DOI: 10.3316/QRJ1102063

Sutton, J., & Austin, Z. (2015). Qualitative research: Data collection, analysis, and management. *The Canadian Journal of Hospital Pharmacy*, 68(3), 226. DOI: 10.4212/cjhp.v68i3.1456 PMID: 26157184

Taherdoost, H. (2021). Data collection methods and tools for research; a step-by-step guide to choose data collection technique for academic and business research projects. [IJARM]. *International Journal of Academic Research in Management*, 10(1), 10–38.

Then, K. L., Rankin, J. A., & Ali, E. (2014). Focus group research: What is it and how can it be used? *Canadian Journal of Cardiovascular Nursing*, 24(1). PMID: 24660275

Wach, E., & Ward, R. (2013). Learning about qualitative document analysis.

Weiber, I., Eklund, M., & Tengland, P. A. (2016). The characteristics of local support systems, and the roles of professionals, in supporting families where a mother has an intellectual disability. *Journal of Applied Research in Intellectual Disabilities*, 29(3), 197–210. DOI: 10.1111/jar.12169 PMID: 25754531

Weyrich, P., Schrauth, M., Kraus, B., Habermehl, D., Netzhammer, N., Zipfel, S., & Nikendei, C. (2008). Undergraduate technical skills training guided by student tutors–analysis of tutors' attitudes, tutees' acceptance and learning progress in an innovative teaching model. *BMC Medical Education*, 8(1), 1–9. DOI: 10.1186/1472-6920-8-18 PMID: 18400106

White, G. E., & Thomson, A. N. (1995). Anonymized focus groups as a research tool for health professionals. *Qualitative Health Research*, 5(2), 256–261. DOI: 10.1177/104973239500500209

Wilkinson, S. (1998). Focus group methodology: A review. *International Journal of Social Research Methodology*, 1(3), 181–203. DOI: 10.1080/13645579.1998.10846874

Wong, L. P. (2008). Focus group discussion: A tool for health and medical research. *Singapore Medical Journal*, 49(3), 256–260. PMID: 18363011

Wood, L. M., Sebar, B., & Vecchio, N. (2020). Application of rigour and credibility in qualitative document analysis: Lessons learnt from a case study. *The Qualitative Report*, 25(2), 456–470. DOI: 10.46743/2160-3715/2020.4240

Chapter 3
Evaluating Qualitative Data Collection Methods in Aviation Management:
Challenges and Opportunities

Mustafa Uzgör
Muğla Sıtkı Koçman University, Turkey

ABSTRACT

Management in air transport is a discipline that requires the effective and efficient management of complex businesses and processes. Therefore, human interaction is of vital importance at every stage of the management. In this respect, the discipline has a research area closely related to the paradigm of interpreting human subjectivity of qualitative research methods. Especially in an area such as the aviation industry, which has high safety and security standards and a rapidly changing environment, critical business decisions are made quickly, and the importance of qualitative data to guide decision-makers cannot be overstated. However, the qualitative data collection process faces various obstacles such as complexities in data collection and analysis processes, ethical issues, and methodological challenges. Such challenges may appear differently in the aviation context or some new opportunities may emerge. In this study, the effectiveness of qualitative data collection methods in aviation management is discussed, and the challenges faced by these methods are examined.

DOI: 10.4018/979-8-3693-7403-0.ch003

1. INTRODUCTION

As an integral part of the global economy, the air transportation industry continues to evolve despite the crises it encounters. The coordinated efforts of numerous stakeholders provide an irreplaceable service, offering swift transportation and time efficiency. The effective management of critical topics such as safety, security, operational efficiency, and customer satisfaction forms the core task of aviation management. Such topics are highly dependent on the performance of the human. The attitudes, behaviors and experiences of different groups, e.g. passengers, employees, pilots and ground handling personnel affect the performance of the aviation system. Additionally, each airport, airline, and passenger group has its own unique conditions. Analyses or decisions made without considering these conditions may fall short. Moreover, aviation industry has strict regulations and high safety standards. In order to prevent any kind of negative outcomes of potential unsafe acts, an elaborated effort may be required to fully understand the context. Herein, qualitative research methods feature several characteristics that are well aligned with the aviation management studies. In this context, the use of qualitative research methods to understand sectoral issues from a qualitative perspective and develop solutions has been increasingly adopted in aviation management studies in recent years.

Qualitative research can be defined as a type of research in which qualitative data collection tools such as observation, interview and document analysis are used, and social phenomena are examined holistically in their natural environment within the environment they are connected to, where a qualitative design is followed (Yıldırım & Şimşek, 2016). Qualitative research methods provide researchers and decision-makers with deep insights to understand complex and multidimensional problems, offering a suitable framework and guide for addressing research questions that are difficult to explain with numbers. Common qualitative designs can be classified into qualitative case study, ethnography, phenomenology, grounded theory, and narrative inquiry (Creswell, 2013; Merriam & Tisdell, 2016). Research design implies the different strategies of conducting a research in terms of type of research question, sample selection, data collection, data analysis and writing. Since these types of qualitative research possess some features in common, they fall under the umbrella concept of "qualitative." (Maxwell, 2012; Merriam & Tisdell, 2016).

While each discipline may require certain research design to be used, whatever the design or the discipline is, collecting and editing qualitative data is one of the most time-consuming and important stage of a qualitative research. Requiring a data collection tool that is sensitive to underlying meaning when gathering data, a qualitative data consist of direct quotes from interviews about people's experiences, opinions, feelings, and knowledge; detailed descriptions of people's activities, behaviors, and actions recorded through observations; or quotations, excerpts, or entire passages extracted from various types of documents (Patton, 2015). Therefore, the main qualitative data collection methods can be examined in three categories: interviews, observation, and document analysis. In other words, qualitative data collection is about asking questions, watching activities, and reviewing documents (Dey, 1993). Qualitative data has the capability to reveal the meaning behind what is visible, which may not be easily noticed at first sight. As is, it can provide significant benefits in explaining events and understanding their causes, especially in critical work areas where errors cannot be remedied. The aviation sector, which is the focus of this study, is one such field. Thus, in the field of aviation management, qualitative data collection is crucial due to its potential to deepen the understanding of sectoral dynamics and human interactions. Michalski and Bearman (2014)'s study serves as a highly relevant example on this subject. The authors investigated the factors affecting the decision making of pilots who fly in Outback Australia. One dramatic finding they obtained was the identification of pilots flying below the allowed altitude limits in response to the demands and pressures of elderly tourist passengers with impaired vision. Considering that companies are unlikely to report such internal incidents, it is evident how challenging it would be to reach this conclusion based solely on numerical data. The study also revealed other notable findings, such as organizational pressures and social pressures. This example is provided to illustrate how critical qualitative data can be in the aviation industry while acknowledging the power of quantitative data, especially in aviation studies where the quantitative paradigm is dominant.

The aviation industry is a highly regulated, safety and security oriented, and complex sector that requires the management of interrelated operations and interactions with various stakeholder groups. As a result, qualitative data collection plays a critical role in understanding this complex structure and identifying areas for improvement in aviation management. However, in addition to its advantages, qualitative data collection faces certain challenges, such as subjectivity, data reliability, and the need for time and resources. These challenges may take on a different dimension in aviation, where standardization, safety, and security are prioritized. To the best of our knowledge, while the number of papers in aviation management which use qualitative data increases, we found no study that focusing on the advantages and challenges of collecting qualitative data in aviation management. In this context, this book chapter aims to examine the qualitative data collection methods encountered in aviation management, as well as the sector-specific challenges and opportunities that may arise. Evaluations were made by reviewing selected studies from the literature.

2. COLLECTING QUALITATIVE DATA

According to Merriam & Tisdell (2016), the most fundamental types of qualitative data collection are observation, interviews, and document analysis. Despite some derivative methods, these three will be taken into consideration as they are comprehensive. In this respect, there is no unique distinction specific to the aviation sector. In a multidimensional and complex field such as aviation management, these methods will help to better understand the detailed qualitative characteristics of the research unit.

Observation

Observation is a natural activity of daily life, typically carried out unconsciously and unsystematically. However, when systematic, tied to a research question, and controlled for producing reliable results, it becomes a data collection tool (Merriam & Tisdell, 2016). In qualitative research observations, the researcher turns a phenomenon that others pay only superficial attention to into the subject of research and gives it special attention. Observations are generally divided into two categories depending on whether the researcher is directly involved in the event: participant or non-participant observation. Broadly, while a participant observation requires the researcher to be involved in routine daily activities with people in the field; non-participant observation does not need any involvement in the daily activities (LeCompte & Schensul, 2010).

Observations are common in various qualitative research methods, including case studies, ethnographies, and qualitative action research. Ethnographic research, in particular, has found some limited use in aviation studies (Rahman, 2023), and observation methods have been relatively underused in these studies. Nonetheless, observation can also be used to support interviews, another qualitative data collection method (Zafiharimalala et al., 2014). Observations differ from interviews in two key aspects. First, they occur in the natural setting where the phenomenon of interest takes place, rather than in a location specifically designated for interviews. Second, observational data provide a direct, firsthand experience of the phenomenon, unlike the secondhand accounts typically gathered through interviews (Corbin & Strauss, 2007; Merriam & Tisdell, 2016).

Interview

In most types of qualitative research, some or all of the data is collected through interviews. Interviews are more common than other qualitative data collection methods. DeMarrais (2004) defines research interviews as a process in which the researcher and the participant engage in a conversation focused on research-related questions. According to Merriam and Tisdell (2016), interviews are selected when observing a research phenomenon is not possible. Both one-on-one and focus group interviews can be conducted, and these interviews must serve a specific research purpose. Structured interviews have defined boundaries and are non-dynamic, while unstructured interviews take the form of an open conversation. Semi-structured interviews are partially constructed but allow the researcher flexibility to direct the interview (Patton, 2015). The interview questions can be asked to participants either verbally, as in structured interviews, or in written form through open-ended survey forms.

Interviews are suitable for collecting qualitative data related to emotions, thoughts, intentions, and other factors not directly observable through observation. Interviews are necessary in cases where behaviors, feelings, or how people interpret the world around them cannot be observed. Additionally, interviews are required when researching past events that cannot be repeated (Merriam & Tisdell, 2016). For example, while it is impossible to recreate a plane crash at an airport that resulted in hundreds of deaths, the effects of such a crash on airport employees could be the focus of an interview.

Document Reviewing

Documents and artifacts that are part of the research environment are also data sources in qualitative research. In document analysis, the researcher typically does not interfere with or alter the research environment, unlike in interviews or observations (Creswell, 2007). Documents are readily available data sources that can exist in both physical and online formats. The term "document" is often used as a general term encompassing a wide range of written, visual, digital, and physical materials related to the research.

Finding relevant materials is the first step in the document analysis process. Once the documents are found, their authenticity should be assessed. The author, the place, and the date the document was written must be fully identified and verified (McCulloch, 2004). Furthermore, if possible, the conditions under which the document was produced should also be determined. However, when dealing with online documents, the issue of authenticity becomes much more complex. The online version of the phenomenon or person being researched may differ from the actual version. In such cases, the researcher may need to gather data from multiple sources. In field studies, document reviewing can be supported with unstructured interviews and observations (Merriam & Tisdell, 2016).

3. CHALLENGES AND OPPORTUNITIES IN COLLECTING QUALITATIVE DATA FOR THE AIR TRANSPORT MANAGEMENT RESEARCH

Observation

Observation offers a valuable opportunity to access an organization's working environment and spend time there, uncovering the meaning behind individuals' everyday practices. Aviation activities require the collaborative and coordinated efforts of many stakeholders. Through observations, research questions can be answered, particularly about how employees manage stakeholder relationships, the work environment they operate in, and how they resolve crises they encounter. Observation brings the researcher closer to the scene, allowing them to capture real-world problems and observe ongoing struggles, which is particularly important for enriching the research experience in aviation contexts, particularly at the operational level (Liang, 2015).

The air transport sector has a dynamic and complex structure, and specific challenges can arise when collecting qualitative data through observation. These challenges can be related to the unique conditions and processes of the sector given below.

Intense and Dynamic Environment: The air transportation sector operates in a fast-paced and dynamic environment. Especially in major airports, passenger flow and operational activities are very fast, complex, and variable (Puls & Wittmer, 2021; Uzgör, 2024). This can make it difficult for observers to follow real-time changes and understand their effects while collecting data. Due to the intensity, the observer may not be able to focus carefully on a specific process or event, which may affect the quality of the data. Such challenges are more likely to arise, particularly when observing an operational process.

Privacy and Security Constraints: Security measures and privacy policies are very strict in the air transport sector. Observers may face restrictions on access to certain areas, reducing their chances of observing critical processes that need to be examined (Zahari et al., 2023). Obtaining permission to access these areas can cause time delays. For example, gaining permission to observe in security screening zones or operations rooms may be difficult, limiting the data collection process. Furthermore, people under observation may not act according to their routine in anticipation of invasion of privacy.

Complex Interactions Between Various Stakeholders: There are many stakeholders in the air transport sector (airlines, airport operators, security personnel, passengers, concessionaires, etc.). These stakeholders have different expectations and behaviors, which can cause complexity during the observation process (Stephenson et al., 2018).

Researchers may find it challenging to understand the interactions between various stakeholders and manage their impact on the observation process.

Determining Observation Duration: Air transport processes are often long and can evolve quickly. For example, flight delays or operational disruptions can occur suddenly (Farhadi et al., 2014). Therefore, determining the observation duration and accurately recording events can be complex. It may be difficult for observers to comprehensively observe all stages of a particular event or process as all operational events are intercorrelated with each other.

Data Accuracy and Consistency: Ensuring the accuracy and consistency of the observational data collected in the air transport sector can pose challenges. In particular, personal observations and experiences may vary, affecting the consistency of the data (McManners, 2016). Observers must be cautious to ensure that subjective interpretations and personal biases do not affect the accuracy of the data.

Despite these challenges, the observation method also offers several opportunities described below in the air transport sector.

In-depth Understanding: The observation method provides an in-depth understanding of the details of air transport processes. In particular, observable qualitative data about employee behavior and passenger experiences can be collected, which is crucial for understanding problems in the sector and developing solutions (Christou & Kassianidis, 2010).

Real-time Data: The observation method enables real-time data collection, allowing the examination of instantaneous changes and events in a dynamic environment (Patton, 2015). Observing instantaneous events and changes in the air transport sector ensures that the data collected is as current and valid as possible.

Comprehensive and Natural Data Collection: Observation allows data to be collected in a natural environment, helping researchers understand how individuals or processes function in a specific context (Glesne, 2015). In the air transport sector, this method enables the collection of comprehensive data in areas such as flight safety, customer satisfaction, and operational efficiency (Zahari et al., 2023). By observing customer-employee relationships, service disruptions can be identified, and service levels can be improved.

Interview

Air transportation is a service sector characterized by intense human-to-human and human-to-machine interaction. As such, human performance inevitably impacts the performance of the service. When conducting research related to people, the interview method allows the collection of data containing deep insights that are difficult to obtain through secondary data or may not be immediately apparent. For this reason, interview data is often preferred over observation data, especially in aviation studies focusing on safety and security (Michalski & Bearman, 2014). The interview method is a valuable tool for collecting qualitative data to understand various dynamics and perspectives in the sector. Particularly when research questions involve elements that are not immediately apparent, interviewing individuals familiar with the context can help uncover these elements.

The challenges of data collection through interviews in the air transport context are listed below.

Accessibility of Interviewees: Professionals in the air transport sector, particularly senior managers and experts, often have busy work schedules, especially during peak periods. This can create difficulties in scheduling interviews and finding suitable time slots (Elbardan, 2023). Additionally, the likelihood of these individuals responding to interview requests may be low, which could hinder the researcher's data collection process.

Privacy and Security Concerns: Security and privacy are high-priority concerns in the air transport sector due to its critical importance to national interests. It is common for participants to be unwilling to share sensitive information during interviews, or they may be unable to provide certain information due to security policies especially when the interview is recorded (Westrum & Adamski, 2016). These privacy concerns can limit the scope of the data collection process and affect the accuracy of the data. This issue is particularly likely to arise in interviewing aviation professionals in aviation institutions that face strict competition under uncertainties or that are managed by public authorities. It can be more visible when investigating sensitive topics such as safety and security.

Data Consistency: There is a risk that the personal viewpoints and experiences of interviewees may affect the consistency of the data collected. In the air transport sector, profit margins are low, and competition levels are high on certain routes (Wensveen, 2011). Therefore, in group interviews, the inconsistent views of participants working for different competing companies may create dilemmas for the researcher during data analysis.

Complex Terminology and Jargon: The air transport sector is rich in technical terminology and jargon (Kale et al., 2021). During interviews, it may be difficult for the researcher to understand sector-specific terms and establish effective communication with participants using these terms, which may affect the efficiency of the interview process and make it difficult to collect accurate data.

Collecting qualitative data through interviews in the air transport sector offers significant opportunities to gain in-depth insights into the sector's complex dynamics. These are outlined below.

Providing In-depth Insights: The interview method can provide in-depth insights into operational processes, strategies, and experiences in the air transport sector. Detailed interviews with participants enable the acquisition of comprehensive information about the challenges, opportunities, and strategic approaches in the sector (Uzgör & Şengür, 2022). These insights contribute to a better understanding of the complex processes in the sector and add depth to research. Especially in sensitive areas such as safety and security, it is possible to discover contextual insights that might not be revealed through quantitative data alone (Gerede, 2015; Michalski & Bearman, 2014).

Gathering Different Perspectives: The interview method is an effective tool for gathering diverse perspectives from various stakeholders in the sector. Information obtained from different stakeholders, such as airline managers, airport employees, security personnel, and passengers, enables a more comprehensive understanding of different aspects of the sector (Stephenson et al., 2018). This diversity contributes to a more balanced and integrated study.

Uncovering New Problems and Opportunities: New problems and opportunities identified during interviews can contribute to the existing knowledge of the sector. This method allows researchers to discover previously overlooked or unnoticed details on specific topics (Corbin & Strauss, 2007; Creswell, 2013). This can help develop innovative solutions and strategies in the sector. For example, in a study exploring innovative solutions to prevent unsafe incidents caused by foreign object debris (FOD) on airports, interviews with multiple stakeholders could yield contributive and original solutions (Hilburn & Pesmen, 2023; Yamamoto et al., 2016).

Building Relationships and Collaboration: The interview process allows for the development of strong relationships between the researcher and participants. This can both enhance the quality of the data and serve as a foundation for future collaborations and research opportunities (Creswell, 2013). Good relationships with participants contribute to obtaining more reliable and comprehensive data. Additionally, this strengthens university-industry collaborations for future projects and research.

Document Reviewing

Many types of documents are used in aviation. For example, airline annual reports, performance evaluation documents, statistical reports on flight delays and cancellations, environmental management policies, and sustainability reports from airports are frequently used materials in aviation research. Additionally, reports on finance, the environment, safety, and security published by international authorities such as the International Civil Aviation Organization (ICAO) or the European Union Aviation Safety Agency (EASA) also provide important data on the sector.

On the other hand, document reviewing often yields secondary data related to the research subject, revealing that studies in the field are often designed with a quantitative rather than qualitative character. Thus, document reviewing can be a versatile data collection tool. This method allows for in-depth analyses based on past data and existing sources. However, several challenges indicated below may arise when collecting qualitative data through document analysis in the air transport research.

Data Access and Quality: Data access issues can pose a significant barrier. Some data may not be publicly available due to commercial confidentiality and security reasons. For example, performance reports from airports and internal operational documents from airlines may have restricted access (Kurt, 2016). Additionally, the quality and currency of the documents are important considerations. Outdated or inaccurate data can affect the accuracy of the analysis (Miles et al., 2014).

Lack of Standardization in Documents: Documents in the air transport sector may adhere to different standards and formats, making it difficult to compare and analyze data. For example, the performance reporting methods used by different airports and airlines may differ, or it may be challenging to find data from the same year (Uzgör, 2024). This variety can make it difficult to ensure consistency in the data collection process when examining multiple organizations.

Diversity and Integration of Data: Documents in the air transport sector can contain a wide range of information, including operational data, financial reports, safety records, environmental data, etc. Additionally, data may need to be obtained from different sources. This diversity can complicate the integration and analysis of data (Keller, 2016). Integrating data from various sources makes it difficult to analyze the data meaningfully and holistically.

Time constraints: In the field of aviation management, the document analysis method can present additional challenges due to the unique characteristics of the industry. Aviation involves a wide range of operational, regulatory, safety, and environmental documents, many of which are subject to strict confidentiality and security protocols (Lyall-Wilson et al., 2019). Accessing these documents can be time-consuming, as researchers may need to navigate various legal and administrative barriers to obtain sensitive or classified data. This may be the case especially when a researcher focuses on collecting historical data (Yalçınkaya & Yalçınkaya, 2020).

In aviation research, document reviewing can be used alone or in combination with other qualitative data collection methods. It can also be used in mixed-methods studies to support quantitative data, providing opportunities for conducting in-depth analyses, accessing comprehensive data, and performing historical analyses. These opportunities are outlined below.

Ability to Conduct In-depth Analysis: Document analysis provides the opportunity to conduct in-depth analyses in the air transport sector. Historical documents and performance reports, in particular, provide valuable information for understanding past trends and development processes in the sector. Such data allows for the analysis of long-term trends and changes in air transportation (Tian & Caponecchia, 2020).

Access to Comprehensive Data: Document analysis allows access to a wide pool of data in the aviation sector. In particular, publicly available documents offer the opportunity to examine various practices and policies in the sector. Such documents can be used to understand best practices in aviation management and innovations in the sector. Researchers working on topics such as environmental sustainability, carbon footprint management, energy efficiency, and waste management can utilize published reports and documents to conduct in-depth analyses (Musa & Isha, 2021).

Understanding Historical and National Differences: Document analysis provides the opportunity to analyze data covering different historical periods and countries. This offers researchers a valuable tool for comparing different practices and policies in the aviation sector. For example, examining the performance differences between airports and airlines in different countries can help understand how international standards and practices have impacted operations in different times or contexts (Kesore, 2020). Document analysis gives researchers the chance to study past operational practices and analyze how these practices have changed over time.

The challenges and opportunities of qualitative data collection, identified through the literature review in this study, are summarized in Figure 1.

Figure 1. Challenges and opportunities of qualitative data collection in aviation management

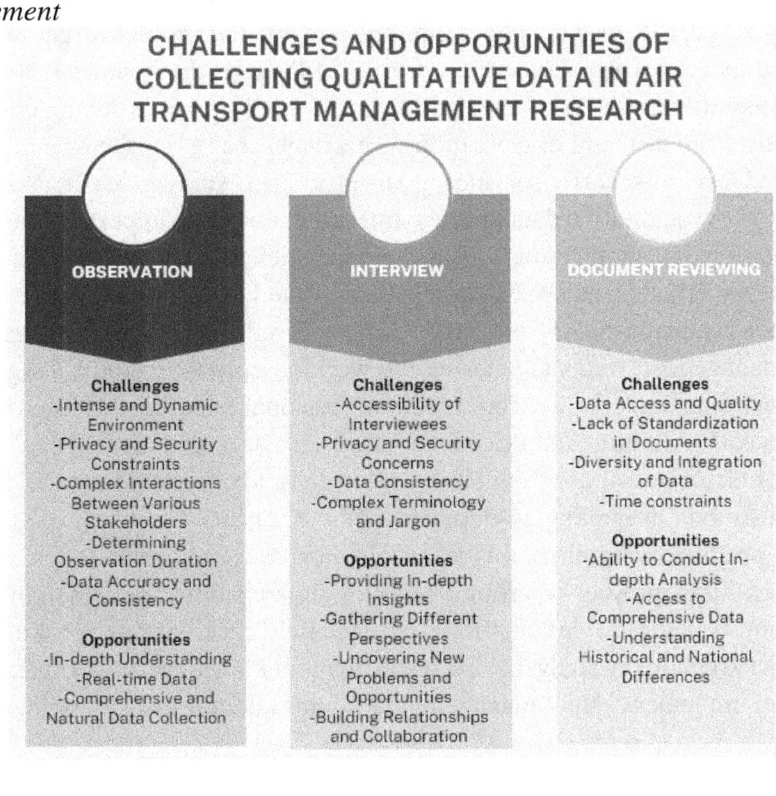

4. CONCLUSION

In this book chapter, the qualitative data collection methods commonly used in air transport management research have been reviewed and evaluated based on the relevant literature. These methods have been assessed in terms of the sector-specific challenges they may pose and the opportunities they present.

It is evident that the use of qualitative data collection methods in air transport management research has increased significantly in recent years. The aviation sector, which operates under high safety and security standards, brings challenges such as access to sensitive data, complex operational environments, and the varying perspectives of numerous stakeholders. However, these challenges are balanced by the potential of qualitative methods—especially interviews, observation, and document analysis—to provide in-depth examinations of human behavior, organizational processes, and stakeholder interactions. These methods enable researchers to explore human and process dynamics that cannot be understood through quantitative data alone. In particular, qualitative data allows access to sensitive and crucial information that cannot be captured through surveys or numerical data, especially in safety and security studies. Michalski and Bearman (2014) presents intriguing findings on factors affecting decision making of pilots, showing the causes of unsafe actions by using interview method. Another interesting research has been carried out by Kayhan et al. (2018). In their paper aiming to reveal the problems that affect the managerial success of airport security services in Türkiye, they primarily found that having different organizations serving in airport security services creates conflict of authority between different organizations. This may act as a risk factor in providing the required level of security services in airports. Their study gives an appropriate example on how qualitative data produce important outputs, supporting the present study. Additionally, another study in which Peksatici and Ergun (2019) employed the interview method reveals that different institutional logics in universities create a gap between academia and the industry. This gap is indicated to pose challenges for aviation management students entering the workforce. In their study, a significant finding regarding students' education and professional lives was obtained through qualitative data. Although interviews are often seen as a sole method in the qualitative research studies, supporting interviews with field observations and document analysis also aids in gaining in-depth insights. Kurt and Gerede (2018)'s study presents a profound example in this regard. In their research, where they conducted interviews with employees of ground handling organizations implementing safety management systems in aviation, they also made field observations by taking notes, and conducted document analysis. As a result, they concluded that most organizations claiming to implement safety management systems adopt it in a symbolic manner.

Although it may be challenging to find a study specifically focusing on the challenges of qualitative data collection methods in aviation, these challenges can generally be found under the limitations sections of studies employing qualitative methods. Accessing the right person by purposeful sampling may be challenging due to the limited number of aviation professionals on certain departments or to their tight schedules especially in busier periods (Kurt & Gerede, 2018). Selecting the right time to collect qualitative data can be suggested in order to avoid time pressures. Another challenge to be mentioned is that since results of qualitative studies based on interview or observation data are not representative of the entire industry, the results can be judged by those who has empirical perspectives or are naive to the qualitative research methods (Creswell, 2007). However, it is apparent that the qualitative research have been gaining grounds among the academia. Supporting to the present study, another challenge is also mentioned by Kistamah and Matsuo (2024). They indicated that the shared perspectives are likely to be affected by the specific beliefs, roles, experiences, and organizational contexts of the interviewees. On the other hand, when ethical concerns, methodological issues, and data management challenges are successfully addressed, qualitative research can offer more comprehensive and context-sensitive solutions for the aviation sector. By integrating qualitative insights with quantitative data, especially in critical areas such as accident investigations, safety reports, or performance evaluations of operational processes, more thorough and in-depth analyses become possible. This approach can help produce more comprehensive and holistic solutions to enhance operational efficiency, improve customer satisfaction, and strengthen safety standards in the sector.

In the use of qualitative data collection methods, it is evident that greater attention must be paid to the reliability and accuracy of qualitative data, given the complexity of the dynamics in the air transport sector, multiple stakeholder interactions, and constantly changing operational conditions. Specifically, challenges related to confidentiality, security, and data accessibility can directly affect researchers' data collection processes. In future, overcoming these challenges by strengthening communication networks, joining education programs or workshops for qualitative studies, disseminating about the qualitative research can enable the development of qualitative research paradigm among aviation management community. Building positive relationships with aviation sector can also be a key in conducting qualitative studies with aviation professionals. Specifically, organizing a qualitative research seminar partnered with aviation sector representatives would be a fruitful experience for both academia and sector. Finally, despite its challenges, the increasing use of qualitative methods in aviation research will pave the way for more comprehensive and detailed studies by facilitating a better understanding of the sector's critical topics and operational dynamics. Future studies can guide qualitative researchers by using new tools to reduce the workloads of data collection, diversifying data collection methods, developing more comprehensive ethical protocols for those who avoid sharing aviation data, and collaborating with the aviation sector.

REFERENCES

Christou, E., & Kassianidis, P. (2010). Adoption of Mobile Commerce in the Air Travel Sector: A Qualitative Survey of Attitudes. *Turizam*, 14(1), 41–52. DOI: 10.5937/Turizam1001041C

Corbin, J., & Strauss, A. (2007). *Basics of Qualitative Research: Techniques and Procedures for Developing Grounded Theory* (3th Editio). *Sage (Atlanta, Ga.).*

Creswell, J. W. (2007). *Qualitative Inquiry & Research Design: Choosing Among Five Approaches* (2nd editio). *Sage (Atlanta, Ga.).*

Creswell, J. W. (2013). *Research Design: Qualitative, Quantitative and Mixed Method Approaches* (4th Editio). *Sage (Atlanta, Ga.).*

DeMarrais, K. (2004). Qualitative interview studies: Learning through experience. In DeMarrais, K., & Lapan, S. D. (Eds.), *Foundations for research* (pp. 51–68). Erlbaum.

Dey, I. (1993). *Qualitative Data Analysis: A User-friendly Guide for Social Scientists* (1st Editio). Routledge.

Elbardan, H. (2023). Interview/Interpretive Approach in Air Travel Research: Research Process, Sampling and Trustworthiness Criteria. In A. Hassan, N. A. A. Rahman, & N. M. Nur (Eds.), *Air Travel Industry* (1st Editio, pp. 87–98). Routledge.

Farhadi, F., Ghoniem, A., & Al-Salem, M. (2014). Runway capacity management - An empirical study with application to Doha International Airport. *Transportation Research Part E, Logistics and Transportation Review*, 68, 53–63. DOI: 10.1016/j.tre.2014.05.004

Gerede, E. (2015). A qualitative study on the exploration of challenges to the implementation of the Safety Management System in aircraft maintenance organizations in Turkey. *Journal of Air Transport Management*, 47, 230–240. DOI: 10.1016/j.jairtraman.2015.06.006

Glesne, C. (2015). *Nitel Araştırmaya Giriş* (A. Ersoy & P. Yalçınoğlu (eds.); 5. Basım). Anı Yayıncılık.

Hilburn, B. G., & Pesmen, B. S. (2023). *Foreign Object Debris Detection System Cost-Benefit Analysis (No. DOT/FAA/TC-22/47).*

Kale, U., Herrera, M., & Nagy, A. (2021). Examining pragmatic failure and other language-related risks in global aviation. *Aircraft Engineering and Aerospace Technology*, 93(8), 1313–1322. DOI: 10.1108/AEAT-03-2021-0081

Kayhan, S., Ergün, N., & Gerede, E. (2018). Research determining issues on the administrative success of security services at civil airports in Turkey. In *Security Journal* (Vol. 31, Issue 2). DOI: 10.1057/s41284-017-0111-4

Keller, R. M. (2016). Ontologies for aviation data management. *AIAA/IEEE Digital Avionics Systems Conference - Proceedings, 2016-Decem*, 1–9. DOI: 10.1109/DASC.2016.7777971

Kesore, L. V. (2020). *Sustainability reporting in the airline industry: a comparative case study analysis of four selected European passenger airlines and their countries of registration on the basis of the airlines' annual reports and sustainability report from 2018*. University of Twente.

Kistamah, H., & Matsuo, M. (2024). Journal of the Air Transport Research Society Above the clouds : Unveiling insights through semi-structured interviews with airline employees from the African, Asian, and North American regions surrounding the COVID-19 pandemic era. *Journal of the Air Transport Research Society*, 3(September), 100047. DOI: 10.1016/j.jatrs.2024.100047

Kurt, Y. (2016). An Overview of Qualitative Research Methods in Aviation Management Research in Turkey. *Proceedings of ADVED 2016 2nd International Conference on Advances in Education and Social Sciences.*

Kurt, Y., & Gerede, E. (2018). An Assessment of Aviation Safety Management System Applications From the New Institutional Theory Perspective. *International Journal of Management Economics and Business*, 14(1), 97–122. DOI: 10.17130/ijmeb.2018137576

LeCompte, M. D., & Schensul, J. J. (2010). *Designing and conducting ethnographic research: An introduction* (2nd ed.). AltaMira Press.

Liang, V. W. (2015). *Engaging with sustainability in everyday transitions: An ethnographic study of an international airport operator*. University of Manchester.

Lyall-Wilson, B., Kim, N., & Hohman, E. (2019). Modeling human factors topics in aviation reports. *Proceedings of the Human Factors and Ergonomics Society Annual Meeting*. DOI: 10.1177/1071181319631095

Maxwell, J. A. (2012). *Qualitative Research Design: An Interactive Approach* (3th Editio). *Sage (Atlanta, Ga.).*

McCulloch, G. (2004). *Documentary research: In education, history and the social sciences* (1st Editio). Routledge.

McManners, P. (2016). The action research case study approach: A methodology for complex challenges such as sustainability in aviation. *Action Research*, 14(2), 201–216. DOI: 10.1177/1476750315597979

Merriam, S. B., & Tisdell, E. J. (2016). *Qualitative Research: A Guide to Design and Implementation* (4th Editio). Jossey-Bass.

Michalski, D. J., & Bearman, C. (2014). Factors affecting the decision making of pilots who fly in outback Australia. *Safety Science*, 68, 288–293. DOI: 10.1016/j.ssci.2014.03.005

Miles, M., Huberman, M., & Saldana, J. (2014). *Qualitative Data Analysis: A Methods Sourcebook* (3rd Editio). *Sage (Atlanta, Ga.)*.

Musa, M., & Isha, A. S. N. (2021). Holistic view of safety culture in aircraft ground handling: Integrating qualitative and quantitative methods with data triangulation. *Journal of Air Transport Management, 92*(May 2020), 102019. DOI: 10.1016/j.jairtraman.2021.102019

Patton, M. Q. (2015). *Qualitative research & evaluation methods: Integrating theory and practice* (4th Editio). Sage publications.

Peksatici, Ö., & Ergun, H. S. (2019). The gap between academy and industry - A qualitative study in Turkish aviation context. *Journal of Air Transport Management*, 79(March), 101687. DOI: 10.1016/j.jairtraman.2019.101687

Puls, R., & Wittmer, A. (2021). Managing airports. In Wittmer, A., Bieger, T., & Müller, R. (Eds.), *Aviation Systems*. Springer. DOI: 10.1007/978-3-030-79549-8_7

Rahman, S. M. A. (2023). Ethnographic Study in Air Travel Research. In A. Hassan, N. A. A. Rahman, & N. M. Nur (Eds.), *Air Travel Industry* (1st Editio, pp. 111–127). Routledge.

Stephenson, C., Lohmann, G., & Spasojevic, B. (2018). Stakeholder engagement in the development of international air services: A case study on Adelaide Airport. *Journal of Air Transport Management*, 71(March), 45–54. DOI: 10.1016/j.jairtraman.2018.06.006

Tian, W., & Caponecchia, C. (2020). Using the Functional Resonance Analysis Method (FRAM) in Aviation Safety: A Systematic Review. *Journal of Advanced Transportation*, 1, 8898903. DOI: 10.1155/2020/8898903

Uzgör, M. (2024). *Havaalanı Çevre Yönetim Performansının Çok Kriterli Karar Verme Yöntemleriyle* Değerlendirilmesi: Türkiye'deki Beş Büyük Havaalanında Bir Uygulama [Doctoral dissertation, Anadolu University].

Uzgör, M., & Şengür, F. (2022). Investigating an underutilized subsidized routes scheme: Underlying reasons and policy recommendations. *Case Studies on Transport Policy*, 10(1), 287–299. DOI: 10.1016/j.cstp.2021.12.010

Wensveen, J. G. (2011). *Air Transportation: A Management Perspective* (7th ed.). Ashgate.

Westrum, R., & Adamski, A. J. (2016). Organizational factors associated with safety and mission success in aviation environments. In Dismukes, R. K. (Ed.), *Human error in aviation* (pp. 475–512). Routledge.

Yalçınkaya, A., & Yalçınkaya, L. A. (2020). The Emergence, Rise, and Fall of Nuri Demirağ as an Entrepreneur: A Historical Analysis within Macro Institutional Theory. *The Journal of American Science*, 18(38), 839–864.

Yamamoto, H., Sattar, R. A., Wang, J., & Mendez, L. F. (2016). *Drone-enabled Foreign Object Debris (FOD) Removal System in ad hoc Situations.*

Yıldırım, A., & Şimşek, H. (2016). *Sosyal Bilimlerde Nitel Araştırma Yöntemleri* (10th ed.). Seçkin Yayıncılık.

Zafiharimalala, H., Robin, D., & Tricot, A. (2014). Why Aircraft Maintenance Technicians Sometimes Do Not Use Their Maintenance Documents: Towards a New Qualitative Perspective. *The International Journal of Aviation Psychology*, 24(3), 190–209. DOI: 10.1080/10508414.2014.918444

Zahari, H. M., Zain, R. M., & Azhar, H. B. (2023). Summary of Qualitative Methods in Air Travel Research. In A. Hassan, N. A. A. Rahman, & N. M. Nur (Eds.), *Air Travel Industry: Theories, Methods and Recent Issues* (1st Editio, pp. 65–74). Routledge.

Chapter 4
Focus Group Research and Applicability of Focus Groups in the Air Transport Studies

Ozlem Colak
https://orcid.org/0000-0003-0813-2561
Gaziantep University, Turkey

ABSTRACT

The focus group approach is one of the key methods in qualitative research and provides a detailed outlook on the research topic and problem. Therefore, this chapter first explores the focus groups and later the application of this methodology in air transport studies. The aim is to understand the nature of the focus groups and explore the areas where this could be applied in air transport research. The focus groups are a significant part of the qualitative research due to providing in-depth and explorative approaches with a dynamic background. This method could also be applied to multiple settings in business studies from management, marketing, organizational culture, and communication to operations and service or product design. The chapter starts with the general outlook of the focus groups and provides key areas to follow in applying this method. Finally, the application of focus groups in air transport and business studies is explored through studies from the literature which could guide future research.

DOI: 10.4018/979-8-3693-7403-0.ch004

1. INTRODUCTION

Qualitative research provides a diverse outlook on the situation compared to quantitative research and requires individual theoretical and practical considerations from question design to writing stage (Qu & Dumay, 2011; Wolcott, 2001). There are also different methods or tools to conduct qualitative research, and some are frequently applied when conducting qualitative research. For example, interviews and observations, content analysis, case studies, or focus groups could be listed as some of these methods (Kvale, 1996; May, 2002; Silverman, 2017). However, this chapter explores focus group research and its applicability in air transport studies. This is due to the gap, as there were not a high number of studies with a methodological application using the focus group design in air transport studies, while the examples are higher and more common in social science or marketing research.

The focus group method has been a complex research tool and has frequently been applied or recommended in social science, human geography, and management studies as observed in the literature (O'Leary, et al., 2023; Martin, et al., 2017). Consequently, this method could be used in a vastly diverse research area such as healthcare (Ranieri, et al., 2019; Kegler, et al., 2023), psychology (Morgan, et al., 2013), brand design (Churchill, 1979), tourism (Hsu, et al., 2011), technology and product development (Seymour, 1987). It has a broad scientific area that can be applied and used as a key research tool. Therefore, the focus group method has proven to be a supportive tool for exploring different research questions in multiple research backgrounds (Fern, 2001, pp. 3-4). However, it is important to highlight that focus groups require individual considerations in design because disciplines have diverse objectives and theoretical backgrounds (Liamputtong, 2011, p. 12).

Air transport is an interdisciplinary field that includes a variety of disciplines; therefore, it is possible to utilize various research theories, methods, or outlooks in its exploration (Doganis, 2010; Graham & Morrell, 2016; Halpern & Graham, 2013). As mentioned, focus groups have also been a tool for diverse research backgrounds in academia and industries. Therefore, this method has a significant potential for air transport research due to its vast applicability. This is especially important considering the increasing level of exploration required in areas such as management, product and service development, sustainability, or human factors in the air transport industry. For example, some researchers focused on exploring marketing topics in the air transport background (Pant, 2022; Lee, et al., 2024). While the focus group approach is not commonly seen as one of the main research tools in air transport studies, it has been applied to explore multiple aspects such as passenger behavior, price sensitivity, or flight crew behavior and safety (Hilditch, et al., 2023; Paraschi, et al., 2024; Yum, et al., 2024; Pauwels, et al., 2024).

It is not possible to deny the benefit of this interactive approach offered by the focus groups (Thelwall & Nevill, 2021; Jung, 2017) and there is an increasing number of studies applying or encouraging this method in air transport research more systematically as observed in the literature (Yum, et al., 2024; Zieba & Johansson, 2022; Han, 2013). This chapter listed some key themes that present applicable areas of focus groups in air transport research: human factors, marketing management and passenger behavior, operation and service design, policy development, and managerial views. Further, these themes and suggestions might guide readers, industry, and researchers in applying the focus group method.

To summarise, this chapter provides practical, theoretical, and ethical guidelines to apply focus groups through examples from other industries and air transport. There are 5 main headings and consequent subheadings. The chapter starts with an introduction to the focus group method, data collection, activity organization, analysis, writing, and ethical approach. Next, the application of focus group research in the air transport industry is explored through the literature. Overall, the aim had been to assess focus group research design followed by the applicability of this methodology in air transport studies while providing guidelines and examples.

2. QUALITATIVE RESEARCH AND FOCUS GROUP STUDIES

The first step of the research is choosing an appropriate methodology to reach the aim and answer research questions. This starts with selecting to use either a qualitative or quantitative approach and the researcher has to choose the most suitable tool to explore a topic (Silverman, 2017; King, et al., 2019). The instruction heading explained that the goal is to explore the qualitative approaches due to the research gap and this chapter focuses on applying one of the key and less implemented qualitative research methods in the air transport industry.

There are multiple qualitative research methodologies, each having individual characteristics and requiring different theories, data collection, and analysis For instance, interviews, observations, case studies, and focus groups could be some of the key qualitative research methods (Qu & Dumay, 2011; Bloor, et al., 2001; Silverman, 2014). Each qualitative research method is based on a particular theoretical background and might require a different approach, data collection, and analysis (Silverman, 2017). Therefore, it is important to understand the focus group method as an individual approach with a detailed exploration.

According to the literature, focus groups started around the 1940s, and researchers continued to apply and improve this model until the current complex methodology was developed. Focus groups became a common model for marketing and psychological research. This new methodological approach diverges from the structured interview method (Bloor, et al., 2001; Silverman, 2014). Structured interviews with theoretical bases provide a fixed approach with one-to-one conversation (Brinkmann & Kvale, 2018) which might provide a less flexible exploration than focus groups. However, focus groups increase interactivity and create a dynamic interview environment with multiple-person dialogues rather than structured interviews. Therefore, it is important to recognize the difference to utilize the opportunity to collect dynamic data in a flexible environment from participants with diverse personalities (Liamputtong, 2011; Brinkmann & Kvale, 2018).

The focus group provides a dynamic system for participants to observe, question, and sometimes reshape the topic and research (Fern, 2001). It is important to explore and analyze situations or products through complex outlooks with multiple views and conversations. Therefore, it is possible to say that this method could provide in-depth data and information on a specific situation (Bloor, et al., 2001). Specifically, it is possible to suggest that this methodological approach could be suitable for air transport studies. This is due to air transport research being frequently shaped around multiple topics around management (Tisdall & Zhang, 2020), finance (Fuerst, et al., 2011), and human factors (Chan & Li, 2022) or areas such as marketing (Hanaysha, 2024), operations (Schultz, et al., 2020), and technology or service development (Bogicevic, et al., 2017). These studies could also focus on various businesses in the industry such as airports (lo Storto, 2018), airlines (Lange, 2019), or other stakeholders. Further, industry and academic studies utilized different qualitative research methods such as interviews (Rhoden, et al., 2008; Sant'Anna & Hilal, 2021) and focus groups while exploring vastly diverse areas such as accidents and human factors or marketing and product development (Hilditch, et al., 2023; Pant, 2022). Overall, the air transport industry could benefit from focus group methodology in the research due to the air transport's interdisciplinary nature.

3. DESIGNING, REPORTING, AND ANALYSING FOCUS GROUPS

In the previous section, it has been mentioned that the decision to choose the methodology is based on applying the most practical method to answer research questions. Consequently, researchers in air transport and other industries applied focus groups to explore people's opinions, and specific behavior, assess actions, and observe a problem's dynamics (Yum, et al., 2024; Boermans, et al., 2024; Coyne, et al., 2024). Most importantly, previous literature shows that the main goal could be assessing the human behavior behind a system (Jittrapirom, et al., 2021), decision-making (Mielinger & Weinrich, 2024), shaping a future service/product (Belton &

Dillon, 2021). Therefore, it might be possible to say that focus groups are applicable in complex settings and aim to provide diverse views in a dynamic background.

A systematic and planned approach is required to design, conduct, and analyze the focus groups. First, the activity step could change depending on the nature and aim of scientific research. The focus groups might involve a discussion setting or interviews and activities to encourage participants to share information depending on the topic (Fern, 2001, p. 3). For instance, marketing or management research requires structured interviews or guidance which the researcher might encourage participants to actively shape the product or system. The goal could be more exploratory and assess the reason behind the behavior in social science-based studies (Liamputtong, 2011). Next, is that qualitative studies need thorough ethics checks and have to apply ethics to each step of the study (Silverman, 2014, pp. 148-149). Another area is that the qualitative data could be scattered and requires a systematic categorization to write the findings (Brinkmann & Kvale, 2018; Krueger, 1998). Lastly, figure 1 presents a systematic outlook to focus group studies and methodological decision steps. The next subheadings explained the pain points in the methodological design stage.

Figure 1. A Systematic Outlook to Focus Group Study (Adapted from the mentioned literature)

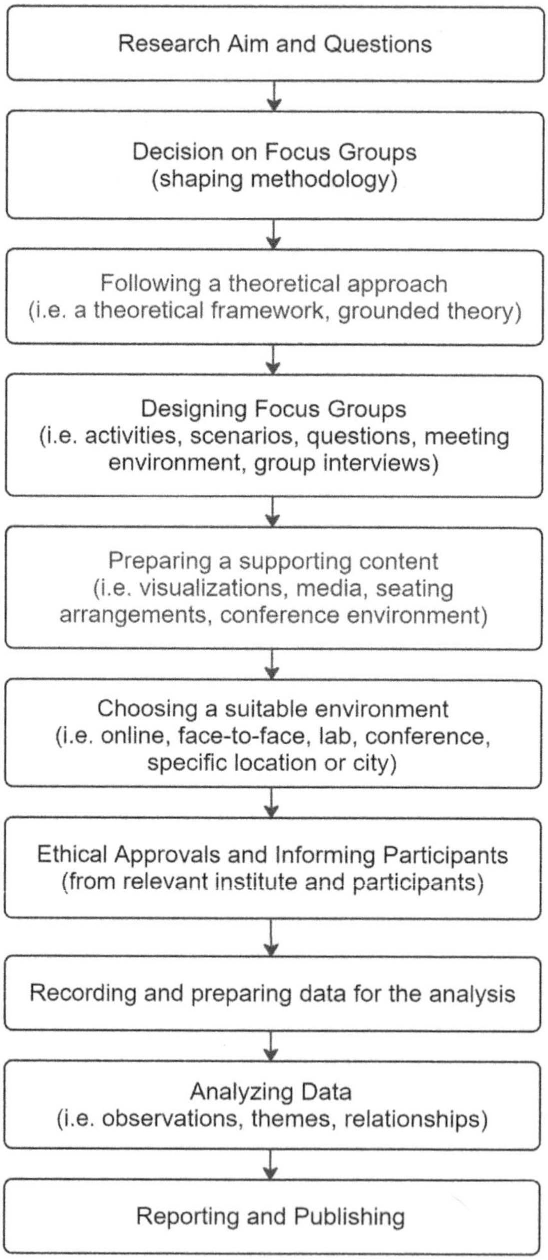

3.1. Ethical Approach

In qualitative research and focus groups, ethics requires significant consideration due to most of the methods involving human participants and personal views (Sim & Waterfield, 2019). The consideration of ethics and research design according to these factors starts before data collection and continues until the end of the research; therefore, each stage requires different ethical considerations regarding the nature of research (Silverman, 2014, p. 161). For example, psychological studies could include participants' personal information, health research could explore participants' health issues (Trimmer, et al., 2024). This is also significant for the air transport industry because information could be private such as staff motivations, safety, or certain behavior analysis (Chan & Li, 2022; Risby, et al., 2022). Therefore, the first step in focus groups is providing an ethically sound research design from start to end (Figure 2).

Figure 2. Ethical Steps in Focus Groups Research Design (Adapted from the mentioned literature)

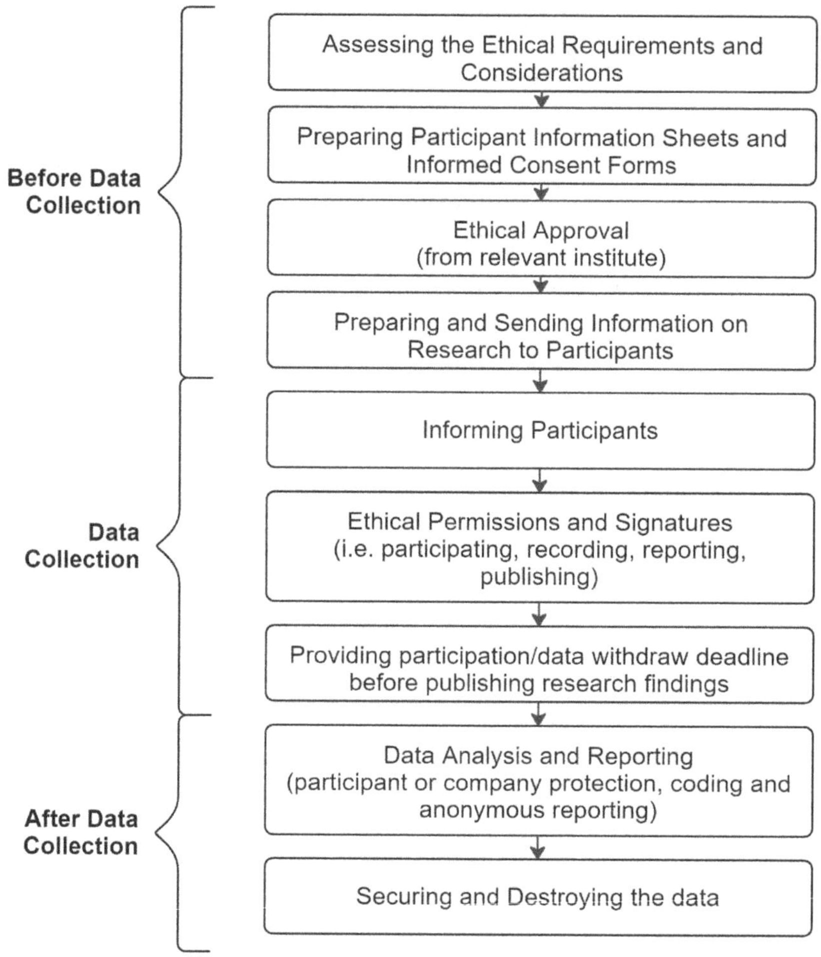

The ethical needs of the planned research have to be considered in planning and activity design. Especially, considering focus groups might involve human interactions and sensitive data depending on the type of study (Ranieri, et al., 2019; Wellings, et al., 2000). Wellings et al. (2019) highlighted that focus group has challenges in providing privacy and no criticism due to the encouragement of discussions in the activity. Therefore, ethical considerations have to be the first action in each stage and the researcher has to obtain ethical approval from the relevant institute (i.e.

university, or organization) and inform participants about the study details before and during the data collection (Silverman, 2014, p. 161).

Next, the analysis and reporting of the results also require confidentiality, and permissions have to be taken from the participants. The coding and anonymity level of the information is parallel to the sensitivity of data, participants, or the research topic (Browne, 2016). In qualitative research studies, participants, identities, locations, or recordings have to be part of the anonymization which could lead to issues in the data analysis or reporting stage (Ryen, 2016; Hilditch, et al., 2023). Also, as seen in previous studies, the researcher collects approval for data to be in the printed media from participants (Ranieri, et al., 2019). Therefore, anonymous reporting is a commonly applied model to protect people's identities and privacy or to control any negativity on parties involved as seen in various qualitative studies (Hennink, 2007; Lincoln, 2009). Overall, it is possible to conclude that ethics are an important part of qualitative research with human participants. Participant permission is crucial, and the researcher could benefit and protect both sides by being clear and detailed, providing research information, focus group actions, having withdrawal options, and being considerate of the participant's consent (Sim & Waterfield, 2019).

3.2. Designing Focus Group Studies

The focus group design stage is flexible; therefore, researchers could create multiple actions and activities for participants to engage with the topic. This is especially important due to the vast options (i.e. activities, moderation) for observation (Fern, 2001, p. 170; Jung, 2017). Firstly, the researcher acknowledges the focus groups as a key tool that is needed to reach study aims. Next, ethical considerations have to be applied due to human nature, participants' safety, and data protection as discussed in the previous heading (see heading 3.1.). For instance, multiple factors have to be considered in the design stage such as participants' reactions to the study, anonymization, question, and activity design. This aspect could be complex and multi-layered (Silverman, 2017, pp. 167-168).

Focus groups have a set of actions in the data collection. Previously studies suggested activity options such as group discussions with different seating arrangements, note-taking or note-sharing activities, visual support through media, and online activities using platforms (Krueger, 1998). It is important to provide a focus group design through activities and discussions. This type of complex case design and moderation is suggested as this could encourage participants to engage and contribute to the discussions and research questions (Moisander & Valtonen, 2006, pp. 72-73). Most importantly, a systematic and planned approach is required to design, conduct, and analyze the focus groups. Each stage will be designed to implement the best activity to examine the topic and participant views which will cover the research topic, reveal the key themes in the study, and answer the research questions (Morgan, 1996; Barbour, 2007; Wolcott, 2001).

3.3. Data Collection and Analysis

The data collection stage requires special consideration for focus groups. First, the literature highlights that the questions or activities have to be designed according to the nature of the research which requires individual considerations for a marketing study, psychological or sociology examination, or medical research (Liamputtong, 2011). Also, depending on the research goals, the focus group study might include interviews and activities to conduct data collection (Krueger & Casey, 2015). However, the researcher could shape activities to create open discussions between participants and provide tasks to encourage knowledge share, having a moderator to observe, support, or lead the discussions and interactions (Moisander & Valtonen, 2006).

Additionally, qualitative studies could be conducted face-to-face or remotely as there are options to conduct meetings via video conferencing through online platforms (Collis & Hussey, 2014). The advancements in technological tools led the data collection channels to evolve and the COVID-19 pandemic has been another factor in the development of the online interviewing approach (Guo, et al., 2024). Furthermore, online interviews and conferencing methods have become a common model for conducting research (Hilditch, et al., 2023; Peasgood, et al., 2023).

Most importantly, this stage includes complex activity structure and moderation in terms of questions, brainstorming, guiding, and ethics as well as the possible reactions or impact of each these could have on the discussions (Moisander & Valtonen, 2006; Catterall & Maclaran, 2006). Overall, the goal is to observe discussions by creating dynamic conversations; therefore, questions, activities, conference settings, and seating arrangements could be designed to collect different views and encourage participant interactions with visual support such as videos, sticky notes, and boards (Krueger & Casey, 2015). For example, Osterwalder & Pigneur (2009) designed the *Business Model Canvas* through an interactive data collection method to shape a common business model framework.

There are also recording options for the data collection. Generally, voice or video recording is beneficial for interviews and focus groups enabling a practical analysis (Barbour, 2007, pp. 76-77). After the data collection, the first step is transcribing the voice record for the data analysis step. For example, the verbatim translation approach requires a word-to-word transcription (Brinkmann & Kvale, 2018, pp. 108-109). Next is the data analysis step which might be supported by varying software models used in the qualitative research such as NVivo, CAQDAS, MAXQDA, or other software (Silverman, 2017; Kuckartz & Rädiker, 2019).

Lastly, the qualitative research analysis is based on interpreting actions, speech, and overall generalization of observations (Fern, 2001, pp. 84-85). In qualitative research books, some of the key analysis methods suggested for qualitative studies are listed as content analysis, grounded theory, narrative analysis, thematic analysis, and discourse analysis (Silverman, 2014; Brinkmann & Kvale, 2018). Most importantly, for a reliable analysis, it is possible to suggest that the researcher is expected to have detailed knowledge of focus group design or moderation steps and follow certain analysis stages.

3.4. Reporting

There are a number of areas to be cautious about when reporting qualitative research findings in interviews and focus groups. First, the coding and anonymity of participants are expected to be protected by creating a coding system. Second, the reporting style is significant for communicating the results and conclusions with readers (Wolcott, 2001; Maxwell, 2012). Multiple factors such as the research aim and questions, data, participant characteristics, or timeline might shape the results. The analysis and reporting could also depend on the theoretical background chosen by the study (Barbour, 2007, p. 149). For example, grounded theory is not based on a specific theory, and the research is expected to create a new theoretical base or an approach as a contribution (Charmaz, 2006, p. 187).

The main data could be interviews, observations, and participants' views, behavior, or interactions in the focus groups (Fern, 2001, pp. 92-93). In previous studies, researchers shape relationships and group themes to analyze focus groups (Varhol, et al., 2024; Jittrapirom, et al., 2021). Further, the thematical analysis method is expected to support the theory while revealing new findings and justifying a final model or factors through this method (Teperi, et al., 2023; King, et al., 2019). On the other hand, the visualization could highlight the similarities, and patterns in the data while focusing on the novelty of these findings (Barbour, 2007, p. 131). In the reporting stage and academic journal articles, researchers list key quotations to present the revealed themes and the key findings (Coyne, et al., 2024; Steen, et al., 2024). Additionally, objectivity and transparency are areas to consider in report writing. Qualitative methodologies have a nature with harder objectivity for researchers compared to quantitative research; therefore, bias is also an area to consider in data analysis (Brinkmann & Kvale, 2018). To sum up, reporting style and thematic analysis could provide a simplified version to communicate complex qualitative data with readers, industry stakeholders, and the scientific community.

4. APPLYING FOCUS GROUPS IN THE AIR TRANSPORT INDUSTRY

In terms of complex methodological decisions, researchers adapted three approaches either applying only focus groups to explore the research topic, mixing qualitative and quantitative methods (Hsu, et al., 2011; Snelson, 2016), or two qualitative research tools (i.e. interviews and focus groups). It is also frequently recommended that the mixed methods approach strengthens a study (Kaplowitz & Hoehn, 2001; Steen, et al., 2024). On the other hand, it could be acknowledged that the application of qualitative research methods such as interviews and focus groups supports an in-depth analysis and interactive knowledge sharing that could solve complex issues in a system, create guidelines, and explore human behavior (Fern, 2001; King, et al., 2019).

As mentioned in the previous sections, the focus group method could be a key tool in multiple research areas. For example, focus groups previously utilized in diverse scientific backgrounds such as medical science, design studies, engineering, education and psychology, human geography, management, marketing, or product development (Langford & McDonagh, 2003; Krueger & Casey, 2015; Vaughn, et al., 1996; Trimmer, et al., 2024; Courage & Baxter, 2005; O'Leary, et al., 2023). Therefore, focus groups could be a research tool for a wide range of scientific topics.

The air transport industry is seen as part of the multi-layered social-technical systems (Salehi, et al., 2021). Therefore, air transport industry-related research could be associated with a wide range of areas from technology and customer relationships (Bogicevic, et al., 2017), financial development of a business (Littlechild, 2018), contract management (Engel, et al., 2018) to security (Rekiel & de Wit, 2013). Additionally, the air transport industry has an unstable and dynamic business environment. The industry dynamics historically have been experiencing significant changes and could be influenced by various types of crises such as health, financial, and security-related events (Kaplanski & Levy, 2010; Adler & Gellman, 2012; Hanson, et al., 2022; Sobieralski, 2020; Hotle & Mumbower, 2021). Overall, as mentioned, the nature of the industry might lead to air transport-related research requiring complex analysis and methodological approaches.

In previous studies, focus groups enabled researchers to understand system-related issues, rationale, or the impact of human behavior (Steen, et al., 2024; Heyes, et al., 2021). Additionally, as seen in the literature, focus groups could provide a unique in-depth analysis and observation opportunity with a flexible design and applicability to diverse topics such as the value observation in organizational safety behavior (Chan & Li, 2022) or airport operation design (Heyes, et al., 2021). Therefore, due to the air transport industry being shaped by multi-layered systems, focus groups could be an applicable methodology for further studies and support in-depth analysis of complex problems as seen in the previous literature.

Firstly, this chapter presented the key steps in the focus group study design. The general methodological application of focus groups for social science, business, and marketing research has been explored. Next, the air transport research that applied the focus groups has been reviewed. In this section, key areas for focus group applications have been grouped under key themes that could be suitable for future air transport-related research. These themes are human factors, marketing management and passenger behavior, operation and service design, policy development, and managerial views. Each of these is also briefly explained in the following parts.

4.1. Human Factors

Safety and security management has been a crucial issue in the air transport industry. Due to the high-risk environment and previous events, there is a continuous need to explore and assess accidents in addition to the potential influence on the business (Gillen & Morrison, 2015; Cox, et al., 2011). For instance, attacks and crises such as 9/11 damaged the air transport industry's finances and led to changes in ongoing operations (Blalock, et al., 2007). Therefore, reporting and investigating the human factor in safety and accident topics have also been a significant and popular area of scientific research (Cusick, et al., 2017; Wiegmann & Shappell, 2003). In the safety and security management literature, human factors might be listed as one of the most highlighted and explored areas in literature. This could be due to human error being seen as the part or sometimes the driver of the accidents, incidents, or issues in the system and operations (Sant'Anna & Hilal, 2021; Teperi, et al., 2023).

The literature in this area has been separated into before and after incidents. Studies explored to reveal a possible weakness or blind area before and asses the cause to prevent repetition (Sant'Anna & Hilal, 2021; Steen, et al., 2024; Kanki, et al., 2019). Further, the nature of the air transport industry necessitated the exploration of the reasons and peoples' actions in the safety area. Most of the studies choose to implement a quantitative approach through a survey to explore the relationship between safety, motivation, and leadership (Chen & Chen, 2014). The quantitative methodology and theoretical background have been a common method to assess the organizational behavior aspect of human factors and safety in aviation (Fogarty & Shaw, 2010; Wang, et al., 2018).

On the other hand, qualitative research methods such as interviews and focus groups have also been utilized to explore the human factors in the transport industry and aviation (Teperi, et al., 2023). These studies suggested that carefully designed research methods, tools, and observations are required to point out the key issues in the system. For instance, focus groups have been frequently used in understanding human psychology, social dynamics, or behavior analysis (Liamputtong, 2011) which psychology and organization have been a key focus in safety studies (Cooper & Phillips, 2004) and aviation-related human factor explorations (Wang, et al., 2018; Koonce, 2002).

It is also possible to say that human factor assessment through focus groups could provide an individual assessment of people and an organization's culture. First, the survey method revealed the significant relationships between environmental factors, perceptions, or behavior (Cooper & Phillips, 2004). Next, applying a purposeful sampling method, and choosing the target participants enable the collection of specific information in qualitative studies (Suri, 2011). As mentioned previously, the aim is to reveal the possible weaknesses before the accidents occur and understand the people, system, organization, and external factors to take possible prevention measures (Kanki, et al., 2019; Steen, et al., 2024). For instance, Coyne, et al. (2024) applied a focus group study to explore the safety aspect of a developing technology. Therefore, as observed in the literature, it might be possible to suggest two types of assessment: either applying a focus group method for an in-depth and thematic analysis or implementing a mixed-method approach (i.e. survey and focus group) to explore the relationships.

4.2. Marketing Management and Passenger Behavior

There are different types of customers depending on the business type in the air transport industry. For instance, airlines, passengers, and retail companies could be listed as the main airport customers (Halpern & Graham, 2013). There could be a number of areas related to market analysis and marketing research that might also implement focus groups or provide improvement through this method. For example, company image and airline product relationship (Koech, et al., 2023), social media (Snelson, 2016) or service quality impact on passenger purchase intentions (Ragab, et al., 2024) could be listed as explored themes in marketing.

The tourism-related research applied focus groups and interviews to examine customer behavior and providers' views which provided insight into brand concept (Ageeva & Foroudi, 2019). However, the air transport-related marketing literature frequently applied complex quantitative models through surveys to analyze consumer views and brand concepts (Rahman & Rahman, 2023; Chen, et al., 2019; Koech, et al., 2023). For example, Hameed et al. (2024) applied a quantitative approach to explore sustainability and brand image for airlines. This study and others (Pabla & Soch, 2023; Xia, et al., 2024) suggested further exploration of the results found on customer relationship and brand aspects through focus groups. Therefore, it is possible to suggest that quantitative market analysis might be further supported with a qualitative approach to explore specific customer behavior. Additionally, some studies assessed certain target customer groups through focus groups to reveal weaknesses in an airline marketing strategy (Westwood, et al., 2000).

Furthermore, in terms of social media management, studies employed qualitative methods (i.e. interviews or focus groups) to obtain users' detailed feedback on the social media platforms (Snelson, 2016). In a highly competitive air transport market, deploying effective and efficient social media campaigns would amplify consumer interest and attention, ultimately leading to increased sales and market share for airline companies (Xia, et al., 2024; Chou, et al., 2023).

Some studies applied focus groups to assess certain customer types aimed to explore less explored areas in airline marketing such as gender factors (Westwood, et al., 2000). Additionally, future investigations through qualitative methodologies have been suggested, such as employing interviews or focus groups, to acquire a comprehensive comprehension of participants' perspectives and attitudes about website characteristics in brand loyalty and word-of-mouth (Hanaysha, 2024). Most importantly, the business environment has been increasingly focusing on understanding consumer behavior (Sheth, et al., 2023) to assess human reaction to developing services (Dichabeng, et al., 2021). Similarly, passenger behavior has been another area to explore for airlines and airports. For example, studies examined areas such as service quality from the airport traveler's side (Fodness & Murray, 2007). To sum up, while quantitative research methods have been commonly applied for data collection in this area for air transport research, it is possible to suggest that the focus groups could support further information and process the dynamic market knowledge.

4.3. Operation and Service Design

Air transport industry-related research could be found in diverse scientific backgrounds such as tourism (Fernández, et al., 2018), social media (Heiets, et al., 2024), architecture and infrastructure (van Oel & van den Berkhof, 2013), public health, or technological development analysis (Khan & Efthymiou, 2021). Most importantly, as a service-focused industry, operations and designs require a collaborative approach and assessment of multiple stakeholders. As could be observed in each of these studies, staff, passengers, regulators, local residents (Hauptvogel, et al., 2024), and other stakeholders could also be a part of the research.

The air transport industry has a complex operational structure and aims to implement a customer-centered service for passengers and efficient designs for operations; therefore, the focus group method is utilized to explore this design aspect. For instance, airport operations (Heyes, et al., 2021), surface access development from the employee perspective (Risby, et al., 2022), technological operation and service developments (Conde, et al., 2022), and new passenger-focused product design (Bogicevic, et al., 2017; Chu, 2001).

Further, focus groups are one of the key methodologies in designing services or products (Henriques & O'Neill, 2023; Tremblay, et al., 2010). In previous studies, it is possible to observe that technological innovations and product design topics frequently used this tool (Seymour, 1987). The service and operational design studies in air transport also started to utilize focus groups as a key assessment tool. For instance, Conde et al. (2022) applied this methodology to explore the development of new data systems in operational efficiency. Another research conducted a focus group study to assess customer expectations in a certain country background when developing a website for target consumers (Chu, 2001). Overall, it is important to acknowledge that air transport could further benefit from the direct or renewed adaptation of the focus group methodology from product and service design literature.

4.4. Policy Development

The air transport industry has a highly regulated environment; therefore, regulatory control and policy implementation are significant parts of the management environment and operational system (Doganis, 2010; Graham, 2023). For instance, key airline or airport operational areas such as security, slot allocation, noise and public relations, legal travel measures, or daily flight activities are controlled and shaped by regulation (Hou, et al., 2021; Niemeier, 2021; Heyes, et al., 2021; Wu, et al., 2018; Oum & Lee, 2002; Blalock, et al., 2007). It is also possible to highlight that policy development has been a significant part of the industry in controlling market problems such as congestion and delays (Fu, et al., 2020) or designing solutions or a regulative environment for new systems (Scheelhaase, et al., 2024).

The latest policy focus has been the sustainability measures, challenges, and policy implementation on designing sustainable systems, finance, and operations (McManners, 2016; Peeters, et al., 2016). Consequently, the policy development aspect in this area might support infrastructure development, encourage and protect investments, provide guidelines, and prepare the industry for upcoming changes (Jaffary & Wiedemann, 2024; Gössling, et al., 2016). There are some studies to explore this area from the passenger side through quantitative analysis (Araghi, et al., 2016) while focus groups have been a common methodology for analyzing the developing sustainability policy system (Kantenbacher, et al., 2018). Additionally, policymaking is an area that requires continuous and long-term communication and discussions with multiple stakeholders and businesses (O'Leary, et al., 2023). In terms of sustainability and policy development, focus groups have been an exploratory tool to assess factors such as public views on the measures taken and people's reactions (Stoll-Kleemann, et al., 2001) and new policy implementations (Ryley, et al., 2010).

To sum up, focus groups have been accepted as a commonly applied tool to collect robust and dynamic observation data on people, markets, and systems (Foroudi, et al., 2017; Thelwall & Nevill, 2021). Therefore, it might be possible to suggest that the application of focus groups in policymaking research provided an in-depth examination of people and regulations in the air transport industry (Ryley, et al., 2010). Additionally, expert views, managerial outlook, and stakeholder assessment could provide a detailed observation of the situation (Henriques & O'Neill, 2023; Gössling, et al., 2016). Further, the aim is to utilize expert and stakeholder views to design new approaches, adaptation processes, or operational changes as seen in the current sustainability and policy research in air transport (Paraschi, et al., 2024; McManners, 2016; Smith, et al., 2017).

4.5. Managerial View

The obtaining managerial view has been a commonly applied model in air transport studies. For example, previous studies preferred to review senior-level management to explore airport revenues (Freathy & O'Connell, 1999), crisis management (Tisdall, et al., 2021), or supplier-side dynamics (Almofeez, et al., 2024). Researchers frequently choose expert views and senior management-level participants when exploring a system, policy, or business model (Donehue & Baker, 2012). There was a mix of expert interviews that included senior-level participants in some studies (Jaffary & Wiedemann, 2024). These studies collected data through interviews to explore the research questions from a managerial view. Thelwall and Nevill (2021) also mentioned finding a lower use of the focus group method with senior manager participants than the interview method in a literature review.

On the other hand, interviews have a two-way observation option while focus groups provide an interactive environment through dynamic activities (Morgan, 1996). Each qualitative research method has unique benefits, and the researcher should decide the best method to reach the aim and suitable fit to reach trustable data and findings (Thelwall & Nevill, 2021; Morgan, 1997). Overall, the management literature presented that the focus groups with senior managers could improve leadership aspects, operational design, crisis, and resilience management.

5. CONCLUSION

Focus group methodology was developed to explore human opinions and behavior in a dynamic background. This model has been commonly applied in social science and used in areas such as human geography, sociology, and criminology to assess human behavior (Liamputtong, 2011). The market or business research has also applied focus groups to analyze customers and asses the company's market position or products. Most importantly, focus groups aim to collect information through in-depth examinations of certain topics (i.e. products or services), group discussions, and observe factors or relationships.

The nature of the research, research aim, and questions have a direct relationship with the methodological decisions and structure. Further, it is important to highlight that the focus groups require a diverse methodological design compared to quantitative research, or other qualitative methods. First, this chapter provided a guideline and highlighted pain points in the general practical application of focus groups, such as ethical considerations, activity and question design, data analysis, and reporting. Each stage needs individual reviewing according to the nature of the scientific area (i.e. marketing, medicine, psychology), research topic and aim, ethics required, type of data or participant type, and data analysis or reporting. Next, the applicability of focus groups in air transport research has been explored. Themes such as complex human factors and organizational dynamics (Steen, et al., 2024), marketing topics, and passenger behavior (Westwood, et al., 2000), or infrastructure design (Risby, et al., 2022) were some of the key areas previously explored by researchers.

On the other hand, after analyzing the literature, it is possible to mention that focus groups might not be a fully utilized method in air transport due to the low number of studies conducted compared to the other frequently used techniques such as surveys or interviews. Also, researchers frequently suggested further exploration of quantitative findings after studies. It is also recommended to implement mixed methods or focus group research to review the specific results further with detailed observations (Risby, et al., 2022; Steen, et al., 2024). Some studies choose to implement a mix of qualitative research methods (Westwood, et al., 2000) while some applied focus groups to create a numerical exploration of a topic (Munoz & Laniado, 2021). Therefore, the use and recognition of focus groups in air transport-related research have been increasing.

The application of focus groups has a significant potential in air transport-related research. In heading 4, after reviewing the air transport-related literature, common focus group applicability areas have been categorized and key themes listed. These were listed as human factors, marketing management and passenger behavior, operation and service design, policy development, and managerial views. Furthermore, the methodological guidelines could be found in other scientific areas according to the research aim. For example, qualitative study application in design studies reveals customer or stakeholder expectations, practical issues, pain points in implementation stages, or key concerns. Focus groups have also focused on creating a dynamic and interactive design environment. Consequently, a practical guideline from design studies might support to improvement of the upcoming technological adaptations for airports and airlines.

Overall, due to the nature of focus groups, this method could be applied to complex issues that require in-depth observations and analysis of human nature or interactions. The focus group activities could provide a dynamic environment for researchers to explore the topics in air transport such as airport operations, airline products, brand image, technological adaptations, and policy making in sustainability. This methodology could be further utilized to explore and reveal the issues and weaknesses in the system or operations, human interactions, organizational problems, and communication. Therefore, it is possible to say that focus groups have a significant potential for application in the air transport industry.

REFERENCES

Adler, N., & Gellman, A. (2012). Strategies for managing risk in a changing aviation environment. *Journal of Air Transport Management*, 21, 24–35. DOI: 10.1016/j.jairtraman.2011.12.014 PMID: 32834688

Ageeva, E., & Foroudi, P. (2019). Tourists' destination image through regional tourism: From supply and demand sides perspectives. *Journal of Business Research. Cilt*, 101, 334–348.

Almofeez, S. ve diğerleri. (2024). Freight forwarder satisfaction, airport customer loyalty, and competitiveness. *Transport Policy. Cilt*, 159, 314–327.

Araghi, Y., Kroesen, M., Molin, E., & Van Wee, B. (2016). Revealing heterogeneity in air travelers' responses to passenger-oriented environmental policies: A discrete-choice latent class model. *International Journal of Sustainable Transportation*, 10(9), 765–772. DOI: 10.1080/15568318.2016.1149645

Barbour, R. (2007). *Doing Focus Groups* (1st ed.). SAGE Publications. DOI: 10.4135/9781849208956

Beirão, G., & Sarsfield Cabral, J. A. (2007). Understanding attitudes towards public transport and private car: A qualitative study. *Transport Policy*, 14(6), 478–489. DOI: 10.1016/j.tranpol.2007.04.009

Belton, O. & Dillon, S., (2021). Futures of autonomous flight: Using a collaborative storytelling game to assess anticipatory assumptions. *Futures,* Cilt 128.

Blalock, G., Kadiyali, V., & Simon, D. H. (2007). The impact of post-9/11 airport security measures on the demand for air travel. *The Journal of Law & Economics*, 50(4), 731–755. DOI: 10.1086/519816

Bloor, M., Frankland, J., Thomas, M., & Robson, K. (2001). *Focus Groups in Social Research*. SAGE Publications Ltd. DOI: 10.4135/9781849209175

Boermans, D. D., Jagoda, A., Lemiski, D. & Wegener, J., (2024). Environmental awareness and sustainable behavior of respondents in Germany, the Netherlands and Poland: A qualitative focus group study. *Journal of Environmental Management,* Cilt 370.

Bogicevic, V., Bujisic, M., Bilgihan, A., Yang, W., & Cobanoglu, C. (2017). The impact of traveler-focused airport technology on traveler satisfaction. *Technological Forecasting and Social Change*, 123, 351–361. DOI: 10.1016/j.techfore.2017.03.038

Brinkmann, S., & Kvale, S. (2018). Doing Interviews. 2nd ed. Los Angeles; London: s.n. DOI: 10.4135/9781529716665

Browne, A. L. (2016). Can people talk together about their practices? Focus groups, humour and the sensitive dynamics of everyday life. *Area*, 48(2), 198–205. DOI: 10.1111/area.12250

Catterall, M., & Maclaran, P. (2006). Focus Groups in Marketing Research. In Belk, R. W. (Ed.), *Handbook of qualitative research methods in marketing* (pp. 255–267). Edward Elgar. DOI: 10.4337/9781847204127.00027

Chan, W. T. K., & Li, W. C. (2022). Investigating professional values among pilots, cabin crew, ground staff, and managers to develop aviation safety management systems. *International Journal of Industrial Ergonomics*, 92, 103370. DOI: 10.1016/j.ergon.2022.103370

Charmaz, K. (2006). *Constructing Grounded Theory: A Practical Guide through Qualitative Analysis* (1st ed.). SAGE Publications.

Chen, C.-F., & Chen, S.-C. (2014). Measuring the effects of Safety Management System practices, morality leadership and self-efficacy on pilots' safety behaviors: Safety motivation as a mediator. *Safety Science*, 62, 376–385. DOI: 10.1016/j.ssci.2013.09.013

Chen, L., Li, Y.-Q., & Liu, C.-H. (2019). How airline service quality determines the quantity of repurchase intention - Mediate and moderate effects of brand quality and perceived value. *Journal of Air Transport Management*, 75, 185–197. DOI: 10.1016/j.jairtraman.2018.11.002

Chou, S., Chen, C.-W., & Wong, M. (2023). When social media meets low-cost airlines: Will customer engagement increase customer loyalty? *Research in Transportation Business & Management*, 47, 100945. Advance online publication. DOI: 10.1016/j.rtbm.2023.100945

Chu, R. (2001). What online Hong Kong travelers look for on airline/travel websites? *International Journal of Hospitality Management*, 20(1), 95–100. DOI: 10.1016/S0278-4319(00)00046-3

Churchill, G. A.Jr. (1979). A Paradigm for Developing Better Measures of Marketing Constructs. *JMR, Journal of Marketing Research*, 16(1), 64–73. DOI: 10.1177/002224377901600110

Collis, J., & Hussey, R. (2014). *Business Research: A practical guide for undersgraduate and postgraduate students* (4th ed.). Palgrave. DOI: 10.1007/978-1-137-03748-0

Conde, J. ve diğerleri, (2022). Applying digital twins for the management of information in turnaround event operations in commercial airports. *Advanced Engineering Informatics,* Cilt 54.

Cooper, M., & Phillips, R. (2004). Exploratory analysis of the safety climate and safety behavior relationship. *Journal of Safety Research*, 35(5), 497–512. DOI: 10.1016/j.jsr.2004.08.004 PMID: 15530924

Courage, C., & Baxter, K. (2005). *Understanding your users: a practical guide to user requirements: methods, tools, and techniques*. Morgan Kaufmann Publishers.

Cox, A., Prager, F., & Rose, A. (2011). Transportation security and the role of resilience: A foundation for operational metrics. *Transport Policy. Cilt*, 18, 307–317.

Coyne, R. ve diğerleri. (2024). Understanding drivers' perspectives on the use of driver monitoring systems during automated driving: Findings from a qualitative focus group study. *Transportation Research Part F: Traffic Psychology and Behaviour. Cilt*, 105, 321–335.

Cusick, S. K., Cortes, A. I., & Rodrigues, C. C. (2017). Commercial Aviation Safety. 6th ed. s.l.:McGraw-Hill Education.

Dichabeng, P., Merat, N., & Markkula, G. (2021). Factors that influence the acceptance of future shared automated vehicles – A focus group study with United Kingdom drivers. *Transportation Research Part F: Traffic Psychology and Behaviour*, 82, 121–140. DOI: 10.1016/j.trf.2021.08.009

Doganis, R. (2010). *Flying Off Course* (4th ed.). Rouledge.

Donehue, P., & Baker, D. (2012). Remote, rural, and regional airports in Australia. *Transport Policy*, 24, 232–239. DOI: 10.1016/j.tranpol.2012.08.007

Engel, E., Fischer, R., & Galetovic, A. (2018). The joy of flying: Efficient airport PPP contracts. *Transportation Research Part B: Methodological*, 114, 131–146. DOI: 10.1016/j.trb.2018.05.001

Eriksson, P., & Kovalainen, A. (2008). *Qualitative methodsin business research.* SAGE Publications Ltd. DOI: 10.4135/9780857028044

Fern, E. F. (2001). *Advanced Focus Group Research.* Sage Publications, Inc. DOI: 10.4135/9781412990028

Fernández, X. L., Coto-Millán, P., & Díaz-Medina, B. (2018). The impact of tourism on airport efficiency: The Spanish case. *Utilities Policy. Cilt*, 55, 52–58.

Fodness, D., & Murray, B. (2007). Passengers' expectations of airport service quality. *Journal of Services Marketing*, 21(7), 492–506. DOI: 10.1108/08876040710824852

Fogarty, G. J., & Shaw, A. (2010). Safety climate and the Theory of Planned Behavior: Towards the prediction of unsafe behavior. *Accident; Analysis and Prevention*, 42(5), 1455–1459. DOI: 10.1016/j.aap.2009.08.008 PMID: 20538101

Foroudi, P., Hafeez, K., & Foroudi, M. (2017). Evaluating the impact of corporate logos towards corporate reputation: A case of Persia and Mexico. *Qualitative Market Research*, 20(2), 158–180. DOI: 10.1108/QMR-05-2015-0043

Freathy, P., & O'Connell, F. (1999). Planning for profit: The commercialization of European airports. *Long Range Planning*, 32(6), 587–597. DOI: 10.1016/S0024-6301(99)00075-8

Fu, X. ve diğerleri, (2020). On-time performance policy in the Chinese aviation market - An innovation or disruption?. *Transport Policy,* Cilt 95.

Fuerst, F., Gross, S., & Klose, U. (2011). The sky is the limit? The determinants and constraints of European airports commercial revenues. *Journal of Air Transport Management*, 17(5), 278–283. DOI: 10.1016/j.jairtraman.2011.03.001

Gillen, D. & Morrison, W. G., (2015). Aviation security: Costing, pricing, finance and performance. *Journal of Air Transport Management,* Cilt 48.

Gössling, S., Cohen, S. A., & Hares, A. (2016). Inside the black box: EU policy officers' perspectives on transport and climate change mitigation. *Journal of Transport Geography. Cilt*, 57, 83–93.

Graham, A. (2013). Understanding the low cost carrier and airport relationship: A critical analysis of the salient issues. *Tourism Management*, 36, 66–76. DOI: 10.1016/j.tourman.2012.11.011

Graham, A. (2013). Understanding the low cost carrier and airport relationship: A critical analysis of the salient issues. *Tourism Management*, 36, 66–76. DOI: 10.1016/j.tourman.2012.11.011

Graham, A. (2023). *Managing Airports: An International Perspective* (6th ed.). Routledge.

Graham, A., & Morrell, P. (2016). *Airport Finance and Investment in the Global Economy*. Routledge. DOI: 10.4324/9781315566498

Guo, D., Ramos, R. L. & Wang, F., (2024). Qualitative online interviews: Voices of applied linguistics researchers. *Research Methods in Applied Linguistics,* 3(3).

Halpern, N., & Graham, A. (2013). *Airport Marketing*. Routledge. DOI: 10.4324/9780203117903

Hameed, I., Chatterjee, R. S., Zainab, B., Tzhe, A. X., Yee, L. S., & Khan, K. (2024). Navigating loyalty and trust in the skies: The mediating role of customer satisfaction and image for sustainable airlines. *Sustainable Futures : An Applied Journal of Technology, Environment and Society*, 8, 100299. DOI: 10.1016/j.sftr.2024.100299

Han, H. (2013). Effects of in-flight ambience and space/function on air travelers' decision to select a low-cost airline. *Tourism Management*, 37, 125–135. DOI: 10.1016/j.tourman.2013.01.008

Hanaysha, J. R. (2024). An examination of the linkages among website characteristics, brand love, and word of mouth in the airline industry. *Transport Economics and Management*, 2, 242–248. DOI: 10.1016/j.team.2024.09.001

Hanson, D., Toru Delibasi, T., Gatti, M., & Cohen, S. (2022). How do changes in economic activity affect air passenger traffic? The use of state-dependent income elasticities to improve aviation forecasts. *Journal of Air Transport Management*, 98, 1–14. DOI: 10.1016/j.jairtraman.2021.102147

Hauptvogel, D. ve diğerleri, (2024). Fairness perspectives of airport residents: A qualitative approach. *Transportation Research Interdisciplinary Perspectives,* Cilt 27.

Heiets, I. et al., (2024). Social media activities of airlines: What makes them successful?. *Journal of the Air Transport Research Society,* Volume 2.

Hennink, M. (2007). *International Focus Group Research: A Handbook for the Health and Social Sciences*. Cambridge University Press. DOI: 10.1017/CBO9780511619458

Henriques, T. A., & O'Neill, H. (2023). Design science research with focus groups – a pragmatic meta-model. *International Journal of Managing Projects in Business*, 16(1), 119–140. DOI: 10.1108/IJMPB-01-2020-0015

Heyes, G., Hooper, P., Raje, F., & Sheppard, J. (2021). The case for a design-led, end-user focused airport noise management process. *Transportation Research Part D, Transport and Environment*, 95, 102847. DOI: 10.1016/j.trd.2021.102847

Hilditch, C. J., Gregory, K. B., Arsintescu, L., Bathurst, N. G., Nesthus, T. E., Baumgartner, H. M., Lamp, A. C. M., Barger, L. K., & Flynn-Evans, E. E. (2023). Perspectives on fatigue in short-haul flight operations from US pilots: A focus group study. *Transport Policy*, 136, 11–20. DOI: 10.1016/j.tranpol.2023.03.004

Hotle, S., & Mumbower, S. (2021). The impact of COVID-19 on domestic U.S. air travel operations and commercial airport service. *Transportation Research Interdisciplinary Perspectives*, 9, 100277. DOI: 10.1016/j.trip.2020.100277

Hou, M., Wang, K. & Yang, H., (2021). Hub airport slot Re-allocation and subsidy policy to speed up air traffic recovery amid COVID-19 pandemic - case on the Chinese airline market. *Journal of Air Transport Management,* Cilt 93.

Hsu, C. H. C., Oh, H., & Assaf, A. G. (2011). A Customer-Based Brand Equity Model for Upscale Hotels. *Journal of Travel Research*, 51(1), 81–93. DOI: 10.1177/0047287510394195

Jaffary, Z. & Wiedemann, M., (2024). Towards net zero aviation: Exploring safe hydrogen refuelling at airports. *Journal of the Air Transport Research Society.*

Jittrapirom, P., Boonsiripant, S. & Phamornmongkhonchai, M., (2021). Aligning stakeholders' mental models on carsharing system using remote focus group method. *Transportation Research Part D: Transport and Environment,* Cilt 101.

Jung, H. (2017). Focus group interaction in evaluation research. *Applied Linguistics Review*, 9(4), 563–587. DOI: 10.1515/applirev-2017-0023

Kanki, B. G., Anca, J., & Chideste, T. R. (2019). Crew Resource Management. 3rd ed. s.l.:Elsevier Academic Press. DOI: 10.1016/B978-0-12-812995-1.00004-X

Kantenbacher, J., Hanna, P., Cohen, S., Miller, G., & Scarles, C. (2018). Public attitudes about climate policy options for aviation. *Environmental Science & Policy*, 81, 46–53. DOI: 10.1016/j.envsci.2017.12.012

Kaplanski, G., & Levy, H. (2010). Sentiment and stock prices: The case of aviation disasters. *Journal of Financial Economics*, 95(2), 174–201. DOI: 10.1016/j.jfineco.2009.10.002

Kaplowitz, M. D., & Hoehn, J. P. (2001). Do focus groups and individual interviews reveal the same information for natural resource valuation? *Ecological Economics*, 36(2), 237–247. DOI: 10.1016/S0921-8009(00)00226-3

Kegler, J. J., Neves, E. T., Lacerda, M. R., & Oliveira, D. C. (2023). Focus group in the development of concepts for a Nursing model: Experience report. *Revista Brasileira de Enfermagem*, 76(4), e20220689. DOI: 10.1590/0034-7167-2022-0689 PMID: 37820149

Khan, N. & Efthymiou, M., (2021). The use of biometric technology at airports: The case of customs and border protection (CBP). *International Journal of Information Management Data Insights,* Volume 1.

King, N., & Horrocks, C. (2010). *Interviews in Qualitative Research.* SAGE Publications.

King, N., Horrocks, C., & Brooks, J. (2019). *Interviews in Qualitative Research* (2nd ed.). SAGE Publications, Inc.

Koech, A. K., Buyle, S., & Macário, R. (2023). Airline brand awareness and perceived quality effect on the attitudes towards frequent-flyer programs and airline brand choice - Moderating effect of frequent-flyer programs. *Journal of Air Transport Management*, 107, 102342. DOI: 10.1016/j.jairtraman.2022.102342

Koonce, J. M. (2002). *Human Factors in the Training of Pilots* (1st ed.). CRC Press.

Krueger, R. A. (1998). *Analyzing & Reporting Focus Group Results.* basım yeri bilinmiyor:SAGE Publications, Inc.

Krueger, R. A. (1998). *Developing Questions for Focus Groups.* SAGE. DOI: 10.4135/9781483328126

Krueger, R. A., & Casey, M. A. (2015). *Focus Groups: A Practical Guide for Applied Research* (5th ed.). SAGE Publications, Inc.

Kuckartz, U., & Rädiker, S. (2019). Analyzing Qualitative Data with MAXQDA. 1 dü. basım yeri bilinmiyor:Springer. DOI: 10.1007/978-3-030-15671-8

Kvale, S. (1996). *Interviews: an introduction to qualitative research interviewing.* SAGE.

Lange, A. (2019). Does cargo matter? The impact of air cargo operations on departure on-time performance for combination carriers. *Transportation Research Part A: Policy and Practice. Cilt*, 119, 214–223.

Langford, J., & McDonagh, D. (2003). *Focus groups: supporting effective product development* (1st ed.). Taylor & Francis. DOI: 10.4324/9780203302743_chapter_12

Lee, J., Yu, J., Radic, A., & Han, H. (2024). Uncovering air traveler purchase behavior: Influence of airline goods product characteristics towards repurchase intention. *Journal of Retailing and Consumer Services*, 79, 103858. DOI: 10.1016/j.jretconser.2024.103858

Liamputtong, P. (2011). *Focus Group Methodology: Principles and Practice* (1st ed.). SAGE Publications Ltd. DOI: 10.4135/9781473957657

Lincoln, Y. (2009). Ethical practices in qualitative research. In Mertens, D., & Ginsberg, P. (Eds.), *The Handbook of Social Research Ethics* (pp. 150–169). Sage Publications. DOI: 10.4135/9781483348971.n10

Littlechild, S. (2018). Economic regulation of privatised airports: Some lessons from UK experience. *Transportation Research Part A, Policy and Practice*, 114, 100–114. DOI: 10.1016/j.tra.2018.01.024

lo Storto, C. (2018). The analysis of the cost-revenue production cycle efficiency of the Italian airports: A NSBM DEA approach. *Journal of Air Transport Management*, 72, 77–85. DOI: 10.1016/j.jairtraman.2018.05.010

Martin, E. A., Winter, S., & Holder, T. (2017). An Exploration of Trainee Practitioners' Experiences When Using Observation. *The Sport Psychologist*, 31(2), 160–172. DOI: 10.1123/tsp.2016-0019

Maxwell, J. (2012). *A realist approach for qualitative research*. SAGE.

May, T. (2002). *Qualitative Research in Action*. SAGE Publications. DOI: 10.4135/9781849209656

McManners, P. (2016). Developing policy integrating sustainability: A case study into aviation. *Environmental Science & Policy. Cilt*, 57, 86–92.

Merton, R. K. & Kendall, P. L., (1946). The Focused Interview. *American Journal of Sociology,* Cilt 51.

Mielinger, E., & Weinrich, R. (2024). Insights into plastic food packaging waste sorting behaviour: A focus group study among consumers in Germany. *Waste Management. Waste Management (New York, N.Y.)*, 178, 362–370. DOI: 10.1016/j.wasman.2024.02.038 PMID: 38430750

Moisander, J., & Valtonen, A. (2006). *Qualitative marketing research: A cultural approach*. SAGE Publications. DOI: 10.4135/9781849209632

Moisander, J., & Valtonen, A. (2006). *Qualitative marketing research: A cultural approach*. SAGE Publications. DOI: 10.4135/9781849209632

Morgan, D. (1996). Focus groups. *Annual Review of Sociology. Cilt*, 22, 129–152.

Morgan, D. L. (1997). *Focus Groups as Qualitative Research* (2nd ed.). SAGE Publication. DOI: 10.4135/9781412984287

Morgan, P. B., Fletcher, D., & Sarkar, M. (2013). Defining and characterizing team resilience in elite sport. *Psychology of Sport and Exercise*, 14(4), 549–559. DOI: 10.1016/j.psychsport.2013.01.004

Munoz, C. & Laniado, H., (2021). Airline choice model for international round-trip flights: The role of travelers' satisfaction and personality traits. *Research in Transportation Economics,* Cilt 90.

Niemeier, H.-M., (2021). Is economics good for aviation policy? Some problems in bridging the gap between theory and policy. *Journal of Air Transport Management,* Cilt 96.

O'Leary, S., Lieberman, S., Gulyas, A., Ogilvie, M., Bates, D., Heath, T., Pelz, C., Williams, S., & Shalet, D. (2023). Management actions to address the climate emergency: Motivations and barriers for SMEs and other societal micro/meso-level groups. *International Journal of Management Education*, 21(3), 100831. DOI: 10.1016/j.ijme.2023.100831

Osterwalder, A., & Pigneur, Y. (2009). *Business Model Generation: A Handbook for Visionaries, Game Changers, and Challengers*. Self Published.

Oum, T., & Lee, Y. (2002). The Northeast Asian air transport network: Is there a possibility of creating open skies in the region? *Journal of Air Transport Management. Cilt*, 8, 325–337.

Pabla, H., & Soch, H. (2023). Up in the air! Airline passenger's brand experience and its impact on brand satisfaction mediated by brand love. *Journal of Air Transport Management*, 107, 102345. DOI: 10.1016/j.jairtraman.2022.102345

Pant, P. (2022). Technology social practices by Millennials and Gen Z at airport departure terminals. *Tourism Management Perspectives*, 43, 100978. DOI: 10.1016/j.tmp.2022.100978

Paraschi, E. P., Poulaki, L., & Papageorgiou, A. (2024). Sustainability challenges in airlines contemporary environmental management. *Journal of Air Transport Management*, 118, 102616. DOI: 10.1016/j.jairtraman.2024.102616

Pauwels, J., Buyle, S. & Dewulf, W., (2024). Regional airports revisited: Unveiling pressing research gaps and proposing a uniform definition. *Journal of the Air Transport Research Society*, Volume 2.

Peasgood, T., Bourke, M., Devlin, N., Rowen, D., Yang, Y., & Dalziel, K. (2023). Randomised comparison of online interviews versus face-to-face interviews to value health states. *Social Science & Medicine*, 323, 115818. DOI: 10.1016/j.socscimed.2023.115818 PMID: 36940582

Peeters, P. ve diğerleri. (2016). Are technology myths stalling aviation climate policy? *Transportation Research Part D: Transport and Environment. Cilt*, 44, 30–42.

Qu, S. Q., & Dumay, J. (2011). The qualitative research interview. *Qualitative Research in Accounting & Management*, 8(3), 238–264. DOI: 10.1108/11766091111162070

Ragab, H., Polo-Peña, A. I., & Mahrous, A. A. (2024). The effect of airline service quality, perceived value, emotional attachment, and brand loyalty on passengers' willingness to pay: The moderating role of airline origin. *Case Studies on Transport Policy*, 18, 101313. DOI: 10.1016/j.cstp.2024.101313

Rahman, F. & Rahman, M. M., (2023). Analyzing service quality of domestic airlines in an emerging country- Bangladesh by structural equation models. *Journal of Air Transport Management,* Cilt 107.

Ranieri, V., Kennedy, E., Walmsley, M., Thorburn, D., & McKay, K. (2019). Rare but heard: Using asynchronous virtual focus groups, interviews and roundtable discussions to create a personalised psychological intervention for primary sclerosing cholangitis: A protocol. *BMJ Open*, 9(10), e031417. DOI: 10.1136/bmjopen-2019-031417 PMID: 31578198

Rekiel, J., & de Wit, J. (2013). The security system at European airports-Tour d'Horizon. *Journal of Transportation Security*, 6(2), 89–102. DOI: 10.1007/s12198-013-0105-3

Rhoden, S., Ralston, R., & Ineson, E. M. (2008). Cabin crew training to control disruptive airline passenger behavior: A cause for tourism concern? *Tourism Management. Tourism Management*, 29(3), 538–547. DOI: 10.1016/j.tourman.2007.06.002

Risby, J., Guest, S. & Warnock-Smith, D., (2022). A critical analysis of Bristol Airport's employee surface access habits: Developing strategic recommendations for reducing private vehicle usage. *Research in Transportation Business & Management,* Cilt 43.

Ryen, A. (2016). Research ethics and qualitative research. In Silverman, D. (Ed.), *Qualitative Research* (pp. 31–46). Sage Publications.

Ryley, T., Davison, L., Bristow, A., & Pridmore, A. (2010). Public engagement on aviation taxes in the United Kingdom. *International Journal of Sustainable Transportation*, 4(2), 112–128. DOI: 10.1080/15568310802471735

Salehi, V., Veitch, B., & Smith, D. (2021). Modeling complex socio-technical systems using the FRAM: A literature review. *Human Factors and Ergonomics in Manufacturing*, 31(1), 118–142. DOI: 10.1002/hfm.20874

Sant'Anna, D. A. L. M., & Hilal, A. V. G. (2021). The impact of human factors on pilots' safety behavior in offshore aviation companies: A brazilian case. *Safety Science*, 140, 105272. DOI: 10.1016/j.ssci.2021.105272

Sant'Anna, D. A. L. M., & Hilal, A. V. G. (2021). The impact of human factors on pilots' safety behavior in offshore aviation companies: A brazilian case. *Safety Science*, 140, 105272. DOI: 10.1016/j.ssci.2021.105272

Scheelhaase, J., Grimme, W. & Maertens, S., (2024). EU trilogue results for the aviation sector – key issues and expected impacts. *Transportation Research Procedia,* Cilt 78.

Schultz, M. ve diğerleri, (2020). Future aircraft turnaround operations considering post-pandemic requirements. *Journal of Air Transport Management,* Cilt 89.

Seymour, D. T. (1987). Focus groups and the development of new products by technologically driven companies: A comment. *Journal of Product Innovation Management*, 4(1), 50–54. DOI: 10.1111/1540-5885.410050

Sheth, J. N., Jain, V., & Ambika, A. (2023). The growing importance of customer-centric support services for improving customer experience. *Journal of Business Research*, 164, 113943. DOI: 10.1016/j.jbusres.2023.113943

Silverman, D. (2014). *Interpreting Qualitative Data*. SAGE Publications Ltd.

Silverman, D. (2017). *Doing qualitative research* (5th ed.). SAGE.

Sim, J., & Waterfield, J. (2019). Focus group methodology: Some ethical challenges. *Quality & Quantity*, 53(6), 3003–3022. DOI: 10.1007/s11135-019-00914-5

Smith, P., Gaffney, M. J., Shi, W., Hoard, S., Armendariz, I. I., & Mueller, D. W. (2017). Drivers and barriers to the adoption and diffusion of Sustainable Jet Fuel (SJF) in the U.S. Pacific Northwest. *Journal of Air Transport Management*, 58, 113–124. DOI: 10.1016/j.jairtraman.2016.10.004

Snelson, C. L. (2016). Qualitative and Mixed Methods Social Media Research: A Review of the Literature. *International Journal of Qualitative Methods*, 15(1), 1609406915624574. DOI: 10.1177/1609406915624574

Sobieralski, J. B. (2020). COVID-19 and airline employment: Insights from historical uncertainty shocks to the industry. *Transportation Research Interdisciplinary Perspectives*, 5, 100123. DOI: 10.1016/j.trip.2020.100123 PMID: 34173453

Steen, R., Norman, J. E., Bergström, J., & Damm, G. F. (2024). Dark knights: Exploring resilience and hidden workarounds in commercial aviation through mixed methods. *Safety Science*, 175, 106498. DOI: 10.1016/j.ssci.2024.106498

Stoll-Kleemann, S., O'Riordan, T., & Jaeger, C. C. (2001). The psychology of denial concerning climate mitigation measures: Evidence from Swiss focus groups. *Global Environmental Change*, 11(2), 107–117. DOI: 10.1016/S0959-3780(00)00061-3

Suri, H. (2011). Purposeful Sampling in Qualitative Research Synthesis. *Qualitative Research Journal*, 11(2), 63–75. DOI: 10.3316/QRJ1102063

Teperi, A. M., Paajanen, T., Asikainen, I., & Lantto, E. (2023). From must to mindset: Outcomes of human factor practices in aviation and railway companies. *Safety Science*, 158, 105968. DOI: 10.1016/j.ssci.2022.105968

Thelwall, M., & Nevill, T. (2021). Is research with qualitative data more prevalent and impactful now? Interviews, case studies, focus groups and ethnographies. *Library & Information Science Research*, 43(2), 101094. DOI: 10.1016/j.lisr.2021.101094

Tisdall, L., & Zhang, Y. (2020). Preparing for 'COVID-27': Lessons in management focus – An Australian general aviation perspective. *Journal of Air Transport Management*, 89, 101922. DOI: 10.1016/j.jairtraman.2020.101922 PMID: 32901185

Tisdall, L., Zhang, Y., & Zhang, A. (2021). COVID-19 impacts on general aviation – Comparative experiences, governmental responses and policy imperatives. *Transport Policy*, 110, 273–280. DOI: 10.1016/j.tranpol.2021.06.009 PMID: 36567696

Tremblay, M. C., Hevner, A. R., & Berndt, D. J. (2010). *The Use of Focus Groups in Design Science Research. %1 içindeA. Hevner & S. Chatterjee, düz. Design Research in Information Systems: Theory and Practice. basım yeri bilinmiyor.* Springer.

Trimmer, R. E., Mandy, W. P., Muntoni, F., & Maresh, K. E. (2024). Understanding anxiety experienced by young males with Duchenne muscular dystrophy: A qualitative focus group study. *Neuromuscular Disorders*, 34, 95–104. DOI: 10.1016/j.nmd.2023.12.002 PMID: 38159461

van Oel, C. J., & van den Berkhof, F. (2013). Consumer preferences in the design of airport passenger areas. *Journal of Environmental Psychology. Cilt*, 36, 280–290.

Varhol, R. J., Man Ying Lee, C., Hindmarsh, S., Boyd, J. H., Robinson, S., & Randall, S. (2024). Consumer attitudes, barriers and facilitators to sharing clinical data for research purposes: Results from a focus group synthesis. *Heliyon*, 10(14), e34431. DOI: 10.1016/j.heliyon.2024.e34431

Vaughn, S., Schumm, J. S., & Sinagub, J. (1996). *Focus group interviews in education and psychology*. SAGE. DOI: 10.4135/9781452243641

Wang, D., Wang, X., & Xia, N. (2018). How safety-related stress affects workers' safety behavior: The moderating role of psychological capital. *Safety Science. Safety Science*, 103, 247–259. DOI: 10.1016/j.ssci.2017.11.020

Welker, A., & Kamberelis, G. (2023). *Imagining and reimagining focus group research in education* (4th ed.). International Encyclopedia of Education. DOI: 10.1016/B978-0-12-818630-5.11023-1

Wellings, K., Branigan, P., & Mitchell, K. (2000). Discomfort, discord and discontinuity as data: Using focus groups to research sensitive topics. *Culture, Health & Sexuality*, 2(3), 255–267. DOI: 10.1080/136910500422241

Westwood, S., Pritchard, A., & Morgan, N. J. (2000). Gender-blind marketing: Businesswomen's perceptions of airline services. *Tourism Management*, 21(4), 353–362. DOI: 10.1016/S0261-5177(99)00069-2

Wiegmann, D. A., & Shappell, S. A. (2003). *A Human Error Approach to Aviation Accident Analysis: The Human Factors Analysis and Classification System* (1st ed.). Routledge.

Wolcott, H. F. (2001). *Writing Up Qualitative Research* (2nd ed.). SAGE Publications.

Wu, C., Jiang, Q., & Yang, H. (2018). Changes in cross-strait aviation policies and their impact on tourism flows since 2009. *Transport Policy*, 63, 61–72. DOI: 10.1016/j.tranpol.2017.11.008

Xia, L., Xu, Y., Zhang, Y., Jiang, H., & Cui, B. (2024). Impact of airline social media marketing on purchase intention:Evidence from China using PLS-SEM. *Transport Economics and Management*, 2, 249–262. DOI: 10.1016/j.team.2024.09.003

Xia, L., Xu, Y., Zhang, Y., Jiang, H., & Cui, B. (2024). Impact of airline social media marketing on purchase intention:Evidence from China using PLS-SEM. *Transport Economics and Management*, 2, 249–262. DOI: 10.1016/j.team.2024.09.003

Yum, K. K., Choi, J. O., & Hyun, S. S. (2024). A study on the effect of job stress factors on mental health and service sabotage: Focusing on flight attendants in foreign airlines. *Journal of Air Transport Management*, 115, 102541. DOI: 10.1016/j.jairtraman.2024.102541

Zieba, M., & Johansson, E. (2022). Sustainability reporting in the airline industry: Current literature and future research avenues. *Transportation Research Part D, Transport and Environment*, 102, 103133. DOI: 10.1016/j.trd.2021.103133

Chapter 5
Case Study as a Research Design in Air Transport Management

Hilal Tuğçe Lapçın
https://orcid.org/0000-0002-4942-6331
Samsun University, Turkey

ABSTRACT

This book chapter aims to explain how the case study method can be chosen as a research design in the field of air transport and to describe the process that should be followed if it is selected. It highlights the importance of developing clear research questions, selecting appropriate cases and participants, and using multiple data collection techniques such as interviews, observations, and document analysis. The chapter also addresses common biases and the challenges of generalizability and time consumption in case studies. To enhance research quality, it recommends strategies for ensuring construct validity, internal validity, external validity, and reliability. Through structured and iterative analysis, the case study method allows for an in-depth examination of complex phenomena in their real-life contexts.

INTRODUCTION

In 2024, the global airline passenger traffic is projected to reach approximately 9.5 to 9.7 billion passengers. Furthermore, it is anticipated that this figure will double by 2042. It is believed that academic research will develop at a similar pace. Understanding the nature of studies within the air transport industry has become increasingly important. Conducting qualitative research in the aviation industry is essential for formulating context-specific research questions and uncovering context-

DOI: 10.4018/979-8-3693-7403-0.ch005

relevant issues. In the air transportation sector, the case study design is frequently preferred for in-depth examination of events and phenomena. A case study is defined as a research design that considers the phenomenon within its real-life context, addresses situations where the boundaries between the phenomenon and its context are not clearly evident, and utilizes multiple sources of evidence or data (Yin, 1981). In case studies, by focusing on "how" and "why" questions, researchers can deeply investigate a phenomenon or event that they cannot control. To put it simply in just a few basic steps; researchers who choose the case study method in the air transportation sector must first develop their research questions and sub-problems and identify their units of analysis. Subsequently, the processes of determining the case to be studied and selecting participants must be carried out. Finally, the case studies should be concluded with the processes of data collection, analysis, and reporting of findings. There are specific biases associated with case studies. It is useful to be aware of these when choosing the case study method. These include the bias of the case study itself, the lack of generalizability, and the time-consuming nature of the method. Bias in case studies often arises when some researchers, through careless and inattentive work, produce findings or conclusions based on unsound evidence and their subjective opinions (Yıldırım & Şimşek, 2013, p. 315). However, this issue is not unique to case studies. "Bias" can also occur in other techniques (such as surveys, experimental studies, historical analysis, etc.). Regarding generalizability, carefully designed multiple case studies can overcome this limitation. Among the types of case studies are the holistic single-case design, the embedded single-case design, the holistic multiple-case design, and the embedded multiple-case design. However, the ability to make context-specific and sample-specific generalizations in other research techniques also means that they, too, do not allow for generalization beyond a certain context. Another criticism is that case studies are time-consuming. Nevertheless, Yin emphasizes that this is subject-specific, and if the subject permits, data collection can be accomplished in a very short time (Yin, 2017; Yin, 1981). Despite these criticisms, the root of the problem lies not in the case study method but in a poorly designed research process. It is particularly important for young researchers and those with less experience in qualitative research to design their research processes carefully.

Case Study Method

Case study research requires the investigation of a phenomenon within its real-life, contemporary context or environment (Yin, 2009). While some researchers argue that case study is more of a choice than a research design (e.g., Stake, 2005), many scholars consider it a research strategy (Merriam, 2013; Yin, 2009). Case studies are original works that deeply and longitudinally examine a person, event, or institution, rather than experimental studies where findings are obtained through complex statistical analyses. A case study can be conducted in a short time frame, such as a few weeks, or in long-term studies lasting one or two years. It aims to make in-depth descriptions with focused questions, observations, and, if available, written sources and documents, while creating patterns and making interpretations based on logical inferences on a thematic basis (Paker, 2015). The characteristics of case study research are outlined below (Creswell, 2018).

- A case study begins with the identification of a specific case. The key point here is that the case should be definable or limited by specific parameters, such as a particular time and place.
- A qualitative case study can be designed to illustrate a unique situation, as well as to better understand a particular issue or problem.
- Providing an in-depth understanding of the case is a distinguishing feature of a good qualitative case study.
- Themes and topics can be organized by the researcher within a chronology, cross-case analysis can be conducted based on similarities and differences among cases, or a theoretical model can be presented.

Types of Case Studies

There are several types of case studies. These designs allow for in-depth analysis in qualitative research and can be applied in various contexts.

Table 1. Types of case studies

Scholar	Types of case study
Stake (2005)	Instric Instrumental Collective
Merriam (1998)	Disciplinary orientation Overall intent Multiple case study
Bassey (1999)	Theory-seeking and theory testing Story-telling and picture drawing Evaluative
Yin (1984)	Holistic Single-Case Design Embedded Single-Case Design Holistic Multiple-Case Design Embedded Multiple-Case Design
Yin (1994)	Exploratory Explanatory Descriptive

In response to criticisms of case studies, it is suggested that Yin's recommendations from 1984 be preferred in the air transport sector. Yin emphasizes the necessity of managing a structured and standardized case study process. This study supports that perspective. In this context, the types of case studies as outlined in Yin's (1984) framework are described below.

- Holistic Single-Case Design (Type 1): This design involves a single unit of analysis. It is used to confirm or refute a theory or study unique cases that fall outside general standards (Yıldırım & Şimşek, 2013).
- Embedded Single-Case Design (Type 2): This design is used when there are multiple sub-units within a single case (Yıldırım & Şimşek, 2013).
- Holistic Multiple-Case Design (Type 3): In this design, multiple cases can each be considered holistic on their own, and each case is analyzed holistically before being compared. Data should be collected on the same dimensions in each case (Yıldırım & Şimşek, 2013).
- Embedded Multiple-Case Design (Type 4): This design includes multiple cases, each with its sub-units. To successfully conduct comparisons, standardized processes should be followed during data collection and analysis (Yıldırım & Şimşek, 2013).

Case Study Process in Air Transport Management

The air transport sector has a complex and dynamic structure. Case studies, as a qualitative research method, facilitate an in-depth examination of the air transport sector in Turkey by employing multiple data collection techniques, allowing for systematic information gathering about how the sector operates and functions. Utilizing qualitative research methods is crucial for understanding the events and phenomena within this sector. The air transport industry is characterized by constantly changing technologies, regulations, and market dynamics. In comprehending this complexity, the use of case studies, which allow for an in-depth examination of events, becomes essential. Qualitative research is a necessary method for understanding the perceptions, motivations, satisfaction, and experiences of managers in the air transport sector. Moreover, case studies in the air transport sector enable in-depth analysis of events and phenomena through the use of multiple data sources. The following are the main steps that can be followed when conducting a case study:

- Developing research questions and sub-problems
- Identifying the unit of analysis
- Selecting the case to be studied
- Choosing the individuals to participate in the research

- Collecting data and relating the collected data to the sub-problems
- Reporting the findings

First, the issues and phenomena to be addressed in the air transportation sector must be identified. The reasons that give rise to these issues should be articulated. Subsequently, the research problem and sub-problems should be formulated. It is crucial to express the context in detail here. It is recommended to elaborate on factors such as the strict regulations in Turkey's air transportation sector, the high level of government control, and the significant environmental uncertainty. The air transportation sector in Turkey is subject to both national and international regulations. In addition to international civil aviation organizations such as IATA and ICAO, national bodies like DHMI and SHGM closely monitor sector regulations. Key areas such as flight schedules, permits, and fare pricing are strictly regulated and controlled. In this context, clarifying industry-specific terms enhances the understanding of the sector, particularly in qualitative research, allowing for a more comprehensive analysis of the situation. In addition to these statements, it is necessary to formulate research questions and sub-questions in line with the objectives. The most appropriate questions for case studies are "how" and "why" questions (Yıldırım & Şimşek, 2013). For example, a research question like ("Author statement") can be a starting point. The case study method is used to overcome the challenges related to the multidimensional nature and the difficulty of measuring many concepts studied in the air transport sector (Ratnasingam, 2005; Seppänen et al., 2007). Case study topics allow for an in-depth examination, enabling a comprehensive analysis of the concepts. Yin (2017) emphasizes the importance of following the same approach for each case within this framework. A standardized form designed for one case should be replicated and used in other cases (Ratnasingam, 2005). The research question can be designed as open-ended to explore the topic. For instance, in a study aimed at examining airport-airline relationships through the lens of resource dependency theory, a sub-research question might be: ("Author statement") After this stage, the unit of analysis should be clarified. The situation expressed as an issue in the air transportation sector must be clearly and explicitly articulated as the unit of analysis. The unit of analysis is one of the fundamental distinguishing characteristics of the case study. What criteria were considered in the selection of cases? On what grounds were these criteria established? Answers to these questions should be provided, and the stages of determining the cases based on these criteria must also be completed. Emphasizing the context is particularly important in conducted studies. For example, in a study conducted within the scope of Turkey, the economic, social, technological, political, legal, and international environments in which the selected cases exist directly influence the context. For example, case preferences may vary based on strategies, geographic locations, passenger numbers, business models, and/

or organizational structures. Participant selection is also of critical importance. Guba and Lincoln (1982) stated that the following conditions; purposeful sampling and thick description. Theoretical/Purposeful Sampling involves selecting a sample at the maximum level to ensure that as much data as possible can be collected and to create the most favorable conditions for theory generation. The goal is to provide a comprehensive set of data for the development of theories. Thick description requires providing sufficient information about the context, with raw data being reorganized according to emerging concepts and themes, without adding any interpretation and staying as true as possible to the nature of the data. Direct quotes are frequently used by researchers for this purpose. Through thick description, readers can mentally visualize the environment from which the data were gathered, making it easier to draw potential conclusions relevant to their own context. This allows for the study's findings and hypotheses to be applied to similar situations in future research. With the completion of these stages, the case study process is finalized.

How to Collect Data in Case Study

Observation, interviews, and document analysis can be used as data collection techniques. The difficulty of these processes is not unique to the research design; similar processes are involved in other research methods. Through qualitative research methods and qualitative data collection techniques such as interviews, observation, and document analysis, phenomena can be presented in a realistic and holistic manner within their natural environments. A pilot study should be conducted during the interview process. The pilot study should also assess the researcher's ability to conduct the interview. It is crucial to identify any issues and obstacles in the interview process as well as to eliminate unnecessary questions in qualitative research. During the interview process, whether conducted online or face-to-face, it is essential to ensure the appropriateness of the environment for the interview and to provide information about the recording process. In the air transportation sector, particularly in interviews conducted with senior management of enterprises, data sharing is often not preferred. This makes it difficult to record the interviews. The voluntary participation of individuals who are willing to share data is critical for the findings of the research. Additionally, the limited availability of senior executives poses challenges for conducting in-depth interviews. Therefore, unnecessary questions should be avoided, and the interview form should include fewer but more relevant questions that can yield more comprehensive answers to the research question. It should also be clarified that some questions cannot be addressed through other data collection methods. All processes should be explained transparently. In the air transportation sector, media reports, flight statistics, and activity reports from institutions and organizations are frequently used as documents. Additionally, documents may include letters, bulletins, notice boards, magazines, books, and official gazettes. Considering these criteria, the questions should be compared with the subject matter, and the research situation should be reviewed to reorganize the questions at this stage. Documents can serve as a sole data source in some studies or be used to support primary data sources. The data collection cycle in qualitative research processes is outlined below (Creswell, 2018);

- Identification of participants involves selecting individuals who meet specific criteria relevant to the research objectives.
- Gaining access and establishing rapport with participants is essential to foster trust and ensure the collection of authentic data.
- Purposive sampling is used to intentionally select individuals or groups that are particularly knowledgeable or experienced with the topic of study.
- Data collection involves gathering information through various methods such as interviews, surveys, or observations.
- Data recording ensures the accuracy and integrity of the information collected, often using tools like audio recordings, field notes, or digital platforms.
- Resolution of field issues refers to addressing any challenges encountered during the research process, such as logistical, ethical, or participant-related concerns.
- Data storage involves securely archiving the collected data in a way that maintains confidentiality and allows for future analysis.

Data Collection Techniques

Qualitative research methods, such as interviews, observations, and document analysis, are employed to present phenomena in their natural settings in a realistic and holistic manner (Creswell, 2018; Yıldırım & Şimşek, 2013). In case study analyses, fieldwork, archival records, oral reports, observations, or any combination thereof may be utilized (Yin, 1981).

Interviews

An interview is an interactive process that occurs between at least two people, during which the researcher seeks to collect data by asking questions to the participants (Glesne, 2015). To determine which questions to focus on within the scope of the research, a broad pool of questions should be created, followed by a preliminary pilot interview. The purpose of this step is to identify potential issues and obstacles in the interview process in qualitative research, as well as to eliminate redundant questions. In many cases, some questions can be addressed through alternative data collection methods. At this stage, utilizing reports and documents from the air transport sector is crucial. Following the review process, the interview questions should be revised in a more concise manner. The goal here is to obtain more data from the participant with fewer questions. If necessary, additional sub-questions should be formulated. The clarity of the interview questions directly influences the effectiveness of the interview process. To ensure validity, feedback should be sought from other experts regarding language use, clarity, and the alignment between the topic and the questions. When interviews are conducted online, participants should be informed about audio-visual recordings, and their consent must be obtained. Participants may also be asked to critically evaluate the interview questions. The research process can be adjusted based on this feedback. A pilot study provides an opportunity to assess the suitability of the research method, determine the relevance of the interview questions to the research objectives, and evaluate the researcher's interviewing skills (Glesne, 2015).

The processes of conducting unstructured or semi-structured open-ended interviews, taking interview notes, and transcribing audio recordings should be carried out transparently (Creswell, 2018).

Document Analysis

Document analysis is included among the data collection techniques of qualitative research methods as a secondary data source (Creswell, 2018). Documents can consist of written materials such as letters, bulletins, notice boards, magazines, books, and official gazettes(Creswell, 2018; Yıldırım & Şimşek, 2013).In some studies, documents may serve as the sole data source, or they may be used to support primary data sources (Yıldırım & Şimşek, 2013). In the aviation industry, aviation media reports, flight statistics, institutional and organizational activity reports, websites, press releases, and social media posts can be examined as documents. In some cases, the credibility of the findings and interpretations derived from interview analyses can be enhanced through the use of such documents. Following a scientific journal and keeping a diary are frequently used in case studies during the research process. In addition to participant diaries, the researcher's diary can also be a significant data source.

Analysis Process

In qualitative research, data analysis focuses on discovering and unveiling hidden information within the collected data (Özdemir, 2010). In the aviation sector, the qualitative research process can be inductive, deductive, or a combination of both approaches. When both methods are employed, researchers, who gather an extensive amount of data during the collection phase (Glesne, 2015), initially adopt an inductive approach to construct codes, categories, and themes from the data. Subsequently, researchers may return to the data to reassess whether additional information is required or if themes need further support. Hence, while the process often begins inductively, it may later follow a deductive approach (Creswell, 2018). The stages of the analysis process (Creswell, 2018) typically include:

- Preparing and organizing the data for analysis,
- Conducting a thorough reading of all data,
- Coding the data,
- Identifying themes and generating descriptions.

Moreover, initial theoretical statement or proposition is formulated. Findings from the initial case are compared with the theoretical statement or proposition. The theoretical statement or proposition is revised. Other details of the case are compared with the revised statement. The revised statement is compared with more cases. This process can be utilized multiple times as needed (Subaşı & Okumuş, 2017). In many case studies, the analysis consists of two main stages: within-case analysis and cross-case analysis. During within-case analysis, each case is treated as a comprehensive entity in its own context. Data are collected, and the researcher strives to learn as much as possible about the contextual variables, which may significantly influence the case. Once analysis for each case is complete, cross-case analysis begins. In inductive qualitative research, the aim of cross-case analysis is to develop abstract concepts across cases. Although specific details may vary between cases, the researcher seeks to formulate a general explanation that applies to all cases (Merriam, 2013).

As Braun and Clarke (2006) suggest, it is essential to familiarize oneself thoroughly with the data throughout the analysis process. Numerous annotations and explanations should be made on the transcript, and critical sections should be highlighted. During the coding phase, an inductive and deductive approach may be simultaneously applied. Once saturation is reached and all data sources have been coded, the process of developing themes from the codes should commence. The goal is to uncover meanings associated with the research questions, and this process should be iterative.

Validity and Reliability

In qualitative research literature, the concepts of credibility, transferability, dependability, and confirmability correspond to the notions of internal validity, external validity, reliability, and objectivity in quantitative research (Merriam,

2013). Lincoln and Guba (1982; 1985) categorized these criteria into four main concepts: credibility, dependability, confirmability, and transferability. Yin (1981) emphasized that to enhance the quality of a research design, four characteristics should be considered: construct validity, internal validity, external validity, and reliability. This study adopts Yin's framework.

- **Construct validity**: One of the primary criticisms of case studies is the lack of construct validity. It is believed that subjective judgments of the researcher may interfere with data collection due to the absence of operational measures in case studies. To mitigate this, multiple data types can be used during the data collection process, a chain of evidence can be established regarding the collected data, and the case study report can be reviewed by the participants from whom the data were collected to obtain their feedback.
- **Internal validity**: Case studies seldom focus on the relationship between a limited number of variables. To enhance internal validity, it is essential to clearly present how conclusions were reached and provide evidence of inferences in a manner that allows other researchers to access it.
- **External validity**: This refers primarily to the generalizability of the research findings. In case studies, "analytic generalization" is used instead of statistical generalization. In analytic generalization, the researcher generalizes findings to a theory rather than a population, as is done in statistical generalization. The findings of a case study can thus contribute to proposing a conceptual model. Selecting cases based on specific criteria, ensuring the validity of findings within the specific cases, and achieving maximum diversity across multiple cases may be preferable.
- **Reliability**: This refers to whether the same or similar results would be obtained if the case study were repeated by another researcher following the same steps. To ensure reliability, errors or biases stemming from the researcher must be minimized. The steps of the research process, the stages of conducting the case study, and the data analysis phases should be presented sequentially. Ethical approval is recommended for the research scope, data collection method, and the semi-structured interview questionnaire to ensure ethical standards are met.

In qualitative research, coding is necessary after data collection (Glesne, 2015) and comparative readings. Coding involves labeling meaningful sections of the collected data, such as words or sentences (Yıldırım & Şimşek, 2013). The process of examining statements that have the potential to answer research questions is called coding (Merriam, 2013). By coding the data, themes and patterns can be identified, comparisons can be made, and theoretical explanations can be developed. There are three types of coding in qualitative research. The first involves coding based on the data collected; the second applies predefined codes to organize the data accordingly, and the third combines both predefined codes and those emerging during the data analysis process (Creswell, 2018).

In the coding process, it is important to involve multiple researchers. The use of analysis software is also recommended as it facilitates the process. As Saldana suggests, the coding process can be divided into first-cycle coding, post-first-cycle tasks, and second-cycle coding. In first-cycle coding, coding should be selected in accordance with the research question. For instance, structural coding, in-vivo coding, and holistic coding are possible options. Structural coding is appropriate for studies with multiple participants that use structured or semi-structured data collection techniques, where hypotheses are tested or where the aim is to create a list or index of main categories or themes through exploratory research (Saldana, 2022). In-vivo coding involves using terms and concepts derived directly from the participants' own words (Saldana, 2022). Holistic coding, on the other hand, seeks to understand the main themes and issues by taking a broader approach to the data, rather than analyzing it line-by-line (Saldana, 2022).

Merriam (2013) provides specific recommendations for the analysis process. According to Merriam (2013), it is important to check whether the grouping lists generated from the first data set also appear in the second data set. Separate lists can be made from the interpretations, terms, and notes developed in the second data set. A comprehensive list of interwoven key concepts derived from each data set can be created. This core list forms a simple outline or classification system reflecting recurring patterns in the study. After identifying these patterns, categories or themes can be formed, which classify the items derived from the data. Categories are conceptual elements that encompass multiple distinct examples and are interconnected. Once a set of preliminary categories has been established from the data, they can be refined and strengthened through further examination of the data, gathering additional relevant information. During this process, the initial categories may undergo some revisions. This refinement and review process continues throughout the writing of the findings (Merriam, 2013).

CONCLUSION

In the air transportation sector, case study design is often preferred for in-depth analysis of events and phenomena. Case studies examine a phenomenon within its real-life context, addressing situations where the boundaries between the phenomenon and its context are unclear, using multiple data sources (Yin, 1981). This method allows researchers to explore "how" and "why" questions that they cannot control.

Although case studies face criticisms such as bias, lack of generalizability, and being time-consuming, these challenges can be mitigated with careful design. Bias is not unique to case studies, and generalizability can be improved with multiple-case studies. The time required for case studies can vary depending on the subject, with data collection sometimes being completed quickly (Yin, 2017; 1981). The success of a case study lies in the careful planning of the research process, particularly for novice researchers. To ensure validity, evaluations should be obtained from experts regarding grammar, clarity, and the relevance of the questions to the topic. In case study analyses, field studies, archival records, oral reports, observations, or any combination of these methods can be utilized. In studies conducted in the air transportation sector, structured case studies should be preferred to minimize bias. Additionally, the multiple case study method can be chosen to enhance generalizability. The case study method aims to overcome the challenge posed by the multidimensional nature of many concepts studied in the air transportation sector and the difficulty of their measurement. This method allows for an in-depth examination of topics. Yin (1984) asserts that to enhance the quality of a research design, four characteristics must be considered: construct validity, internal validity, external validity, and reliability. To achieve construct validity, several steps can be taken: utilizing multiple data types during the data collection process, creating a chain of evidence for the collected data, and having the prepared case study report reviewed by the individuals from whom the data was collected to obtain their feedback. First, it is recommended that the researcher extend the interaction period with the case being studied and increase the number of interviewees to reach a saturation point. Second, the researcher is advised to employ data triangulation. "Data triangulation" is defined as the use of multiple data collection methods during the research process, where the collected data is presented in a manner that supports and confirms one another. Third, sharing the fundamental findings with the participants involved in the research is recommended. Participant validation is particularly important for ensuring the accuracy of the research analysis process. To enhance internal validity in case studies, it is essential to clearly explain how the conclusions were reached and present the evidence for the inferences in a way that others can access. Instead of statistical generalization, case studies can use "analytical generalization." The findings obtained from a case study can allow for the proposal of a certain conceptual model. To ensure reliability, the potential for researcher-induced error or bias must be minimized. The stages of the research process, the stages of conducting the case study, and the stages of data analysis should be presented sequentially. The adherence to ethical principles in the research should also be demonstrated. In qualitative research, data analysis focuses on discovering and unveiling the information hidden within the collected data. The analysis in qualitative research can consist of stages where inductive and deductive processes are applied together. Researchers, who obtain a comprehensive data set during the data collection process, initially adopt inductive methods to construct codes, categories, and themes from the data. In the subsequent process, researchers may return to the data to better support the themes or determine if additional information is needed. Thus, while the analysis process starts inductively, it may follow a deductive approach. During reporting, case studies should be presented in a formal, structured manner to allow for the orderly formation of the study. Efforts should be made to describe the case as accurately and objectively as possible. Even though the specific details of particular cases may vary, the researcher aims to provide a general explanation applicable to each case. The aim should be to uncover meanings related to the research questions within the data set. Therefore, the process should be maintained in an iterative manner.

REFERENCES

Braun, V., & Clarke, V. (2006). Qualitative Research in Psychology Using thematic analysis in psychology Using thematic analysis in psychology. *Qualitative Research in Psychology*, 3(2), 77–101. http://www.tandfonline.com/action/journalInformation?journalCode=uqrp20%5Cnhttp://www.tandfonline.com/action/journalInformation?journalCode=uqrp20. DOI: 10.1191/1478088706qp063oa

Creswell, J. W. (2018). *Qualitative Inquiry and Research Design: Choosing Among Five Approaches* (M. Bütün & S. B. Demir (eds.); 4.). Siyasal Publising.

Glesne, C. (2015). *Introduction to Qualitative Research* (Ersoy, A., & Yalçınoğlu, P., Eds.; 5th ed.). Anı Publishing.

Guba, E. G., & Lincoln, Y. S. (1982). Epistemological and methodological bases of naturalistic inquiry. *Educational Communication and Technology*, 30(4), 233–252. DOI: 10.1007/BF02765185

Lincoln, Y. S., & Guba, E. G. (1985). *Naturalistic Inquiry*. Sage Publishing. https://books.google.com.tr/books?hl=tr&lr=&id=2oA9aWlNeooC&oi=fnd&pg=PA7&dq=Lincoln+ve+Guba+(1985)+&ots=0uquW8Pcxk&sig=wdl6zKzPxh_wwFtsjqMFHDZnAQ0&redir_esc=y#v=onepage&q=Lincoln ve Guba (1985)&f=false

Merriam, S. (2013). *Qualitative Research: A Guide to Design and Implementation* (Turan, S., Ed.). Nobel Publishing.

Özdemir, M. (2010). Qualitative Data Analysis: A Study on the Methodological Problem in Social Sciences. *Eskişehir Osmangazi University Journal of Social Sciences*, 11(1), 323–343.

Paker, T. (2015). *Qualitative Research Methods, Techniques, Analysis, and Approaches* (Seggie, F. N., & Bayyurt, Y., Eds.). Anı Publishing.

Ratnasingam, P. (2005). Trust in inter-organizational exchanges: A case study in business to business electronic commerce. *Decision Support Systems*, 39(3), 525–544. DOI: 10.1016/j.dss.2003.12.005

Saldana, J. (2022). *The Coding Manual for Qualitative Researchers* (A. Tüfekci & S. N. Şad (eds.); 2.). Pegem Publishing.

Seppänen, R., Blomqvist, K., & Sundqvist, S. (2007). Measuring inter-organizational trust-a critical review of the empirical research in 1990-2003. *Industrial Marketing Management*, 36(2), 249–265. DOI: 10.1016/j.indmarman.2005.09.003

Subaşı, M., & Okumuş, K. (2017). Case Study as a Research Method. *Atatürk University Journal of Social Sciences*, 21(2), 419–426.

Yıldırım, A., & Şimşek, H. (2013). *Qualitative Research Methods in Social Sciences* (9th ed.). Seçkin Yayıncılık.

Yin, R. (2017). *Case Study Research and Applications Design and Methods.* Sage Publishing.

Yin, R. K. (1981). The Case Study Crisis: Some Answers. *Administrative Science Quarterly*, 26(1), 58–65. http://search.ebscohost.com/login.aspx?direct=true&db=ufh&AN=5007419&site=ehost-live%5Cnhttp://content.ebscohost.com/ContentServer.asp?T=P&P=AN&K=5007419&S=R&D=ufh&EbscoContent=dGJyMMTo50Seqa44y9f3OLCmr0yeprdSrqq4SLCWxWXS&ContentCustomer=dGJyMOzpr1GwqLRIu. DOI: 10.2307/2392599

Yin, R. K. (2009). *Case study research: Design and method* (Vol. 4). Sage Publishing.

Chapter 6
Trustworthiness in Qualitative Research in the Aviation Industry

Satya Subrahmanyam
https://orcid.org/0000-0003-0441-2742
Holy Spirit University of Kaslik, Lebanon

ABSTRACT

This chapter examines the pivotal role of trustworthiness in qualitative research within the aviation industry, crucial for ensuring the validity and credibility of findings. It outlines strategies to enhance trustworthiness, including triangulation, member checking, detailed documentation, and maintaining objectivity. The chapter also addresses unique challenges in aviation research, such as biases and subjectivity, and emphasizes reflexivity and ongoing reflection. The importance of researcher expertise, transparency, and ethical conduct is highlighted. Ultimately, the chapter underscores trustworthiness as essential for producing credible, valid findings that advance knowledge and safety in aviation research.

1. INTRODUCTION

1.1 Purpose of the Chapter

In qualitative research, trustworthiness is critical to ensure the reliability and validity of findings, particularly in a field as complex and dynamic as the aviation industry. Unlike quantitative methods, qualitative research delves deeply into subjective experiences, contextual nuances, and underlying phenomena. This approach requires rigorous standards to establish the credibility, dependability, confirmability,

DOI: 10.4018/979-8-3693-7403-0.ch006

and transferability of research findings. The aviation industry, characterized by its global reach, regulatory frameworks, and technological advancements, demands a high degree of accuracy and transparency in research. Thus, ensuring trustworthiness is imperative for producing actionable insights that inform policy, operational improvements, and strategic decision-making.

The purpose of this chapter is to explore the concept of trustworthiness in qualitative research within the aviation industry. By examining the principles and frameworks that underpin trustworthiness, this chapter aims to guide researchers in maintaining methodological rigor. Trustworthiness is vital for translating qualitative data into meaningful knowledge that supports advancements in areas such as air traffic management, passenger experience, safety protocols, and environmental sustainability.

1.2 Scope and Relevance

The aviation industry presents unique challenges for qualitative researchers. The highly regulated nature of the sector, its reliance on cross-cultural collaboration, and the integration of advanced technologies create a complex ecosystem for study. Ensuring trustworthiness in qualitative research becomes particularly challenging in contexts involving sensitive issues, such as safety culture, crisis management, and the psychosocial dynamics of airline crews. Furthermore, the industry's inherent global nature necessitates studies that can adapt findings across diverse cultural and operational settings, emphasizing the importance of transferability.

The scope of this chapter extends to addressing how researchers can uphold trustworthiness when studying various aspects of aviation, from pilot behavior and organizational culture to passenger satisfaction and airport operations. In a sector where human factors significantly influence outcomes, ensuring the reliability and validity of qualitative findings is essential. This chapter underscores the relevance of trustworthiness as a foundation for actionable insights and evidence-based practices in aviation research.

1.3 Overview of Key Concepts

Trustworthiness in qualitative research refers to the measures taken to establish confidence in the accuracy, consistency, and applicability of research findings. Lincoln and Guba's (1985) seminal framework identifies four core components of trustworthiness: credibility, dependability, confirmability, and transferability. Each of these components serves as a benchmark for evaluating the quality of qualitative research.

Credibility: This dimension relates to the authenticity of findings and the degree to which they represent the participants' lived experiences. In the aviation industry, credibility is vital for studies on sensitive topics such as error reporting, fatigue management, and crew resource management.

Dependability: Dependability focuses on the consistency and stability of research findings over time. Given the dynamic and evolving nature of the aviation industry, qualitative researchers must adopt rigorous documentation and systematic methodologies to ensure dependability.

Confirmability: Confirmability ensures that research findings are free from researcher bias and are grounded in the data. This is particularly important in aviation studies, where personal biases could compromise the objectivity of findings related to operational decisions or stakeholder perceptions.

Transferability: Transferability assesses the extent to which research findings can be applied to other contexts. The global scope of the aviation industry necessitates findings that are adaptable across cultural and regulatory environments.

Establishing trustworthiness in qualitative research requires researchers to employ specific strategies, including triangulation, member checking, rich data descriptions, and audit trails. These methods help validate findings, minimize bias, and enhance the applicability of research outcomes.

The aviation industry's complexity and global nature underscore the importance of trustworthiness in qualitative research. Researchers must navigate challenges such as cultural diversity, sensitive data, and rapid technological change while maintaining methodological rigor. By ensuring credibility, dependability, confirmability, and transferability, qualitative studies can provide robust and actionable insights that support the industry's continuous improvement.

This chapter sets the stage for a detailed exploration of trustworthiness in qualitative research within the aviation sector. By outlining the purpose, scope, and key concepts, it highlights the critical role of trustworthiness in advancing knowledge and informing practice in this vital field.

2. UNDERSTANDING TRUSTWORTHINESS IN QUALITATIVE RESEARCH

Trustworthiness is a cornerstone of qualitative research, serving to ensure that findings are credible, transferable, dependable, and confirmable. These principles are particularly critical in the aviation industry, where qualitative research informs decision-making in areas such as safety management, customer satisfaction, and organizational behavior. This chapter examines the conceptual framework of trustworthiness in qualitative research and its theoretical underpinnings, emphasizing its relevance to the aviation sector.

2.1 Conceptual Framework

The framework of trustworthiness in qualitative research was first articulated by Lincoln and Guba (1985), who proposed four interrelated criteria: credibility, transferability, dependability, and confirmability. Each criterion addresses a specific aspect of research quality and contributes to the robustness of findings.

Credibility refers to the extent to which research findings accurately reflect participants' experiences and realities. It is analogous to internal validity in quantitative research. In qualitative studies within the aviation sector, credibility is crucial for understanding sensitive phenomena, such as pilot fatigue, passenger perceptions, or organizational safety culture. To ensure credibility, researchers often employ strategies such as:

Prolonged Engagement: Building rapport with participants and immersing oneself in the research setting to gain a deep understanding of the context. For instance, a study on cabin crew stress might involve extended observation periods during flight operations.

Triangulation: Using multiple data sources, methods, or investigators to validate findings. In aviation, this might involve combining interviews with air traffic controllers, surveys of pilots, and analysis of safety records.

Member Checking: Seeking feedback from participants on preliminary findings to ensure interpretations align with their perspectives.

These strategies enhance the authenticity and reliability of qualitative research findings, making them more actionable for stakeholders in the aviation industry.

Transferability addresses the applicability of research findings to other contexts. In the aviation industry, this is particularly relevant due to its global and diverse nature. Transferability does not imply generalizability in the quantitative sense but focuses on providing sufficient contextual information for others to determine if findings are applicable to their settings.

To enhance transferability, researchers must provide:

Thick Descriptions: Detailed accounts of the research context, participants, and processes. For instance, a study on airport customer service might include in-depth descriptions of the airport's size, demographic composition, and operational characteristics.

Purposive Sampling: Selecting participants who can provide rich and varied perspectives on the research topic. For example, in studying airline maintenance staff, researchers might choose individuals with diverse roles and levels of experience.

By ensuring that findings are well-contextualized, researchers enable other aviation professionals to draw meaningful insights from their work.

Dependability pertains to the consistency and stability of research processes over time. It parallels reliability in quantitative research and ensures that the study's design and execution are transparent and replicable. Given the dynamic nature of the aviation industry, dependability is essential for producing robust qualitative research.

Key strategies for ensuring dependability include:

Audit Trails: Maintaining comprehensive records of research decisions, data collection methods, and analysis processes. For example, a study on airline employee morale should document every step from interview protocols to coding frameworks.

Peer Debriefing: Engaging with colleagues or experts to review and critique the research process.

Reflexivity: Acknowledging and documenting the researcher's biases, assumptions, and influence on the study. This is particularly important in aviation research, where the researcher's familiarity with industry norms might shape interpretations.

Dependability ensures that qualitative research in the aviation sector remains transparent and methodologically rigorous.

Confirmability focuses on the objectivity and neutrality of research findings, ensuring they are grounded in data rather than researcher bias. In aviation studies, this criterion is vital for maintaining the credibility of insights into operational and human factors.

To achieve confirmability, researchers can employ:

Data Audits: Having independent reviewers examine data collection and analysis processes to verify that findings align with the data.

Reflexive Journals: Documenting the researcher's reflections, decisions, and thought processes throughout the study. For example, a researcher studying pilot decision-making under stress might record their interpretations and consider how their prior knowledge influences these insights.

Triangulation: Using diverse data sources to ensure findings are not the product of a single perspective.

By emphasizing confirmability, researchers in the aviation sector can produce findings that stakeholders regard as objective and trustworthy.

2.2 Theoretical Underpinnings

The concept of trustworthiness in qualitative research is underpinned by several key theoretical frameworks that emphasize rigor and validity in the research process.

Constructivist Paradigm - The constructivist paradigm forms the foundation for trustworthiness in qualitative research. This paradigm posits that reality is socially constructed and context-dependent, requiring researchers to account for multiple perspectives. Trustworthiness ensures that these perspectives are authentically represented and rigorously analyzed (Lincoln & Guba, 1985). In the aviation industry, constructivism supports the study of complex phenomena such as teamwork in multicultural crews or passenger behavior during flight disruptions.

Hermeneutics and Phenomenology - Hermeneutics and phenomenology offer theoretical insights into the importance of trustworthiness in exploring lived experiences. These approaches emphasize understanding participants' subjective realities through deep engagement and interpretation. For instance, phenomenology informs studies on the emotional experiences of flight attendants during emergencies, requiring credibility and confirmability to authentically represent their narratives (Patton, 2002).

Grounded Theory - Grounded theory underscores the importance of systematic data collection and analysis in building trustworthiness. The constant comparative method and iterative coding processes in grounded theory align with dependability and confirmability. In aviation research, grounded theory might be used to explore how airport security staff develop informal protocols to handle ambiguous situations (Glaser & Strauss, 1967).

Systems Thinking - Systems thinking offers a theoretical lens for addressing transferability in aviation research. By viewing the aviation industry as an interconnected system, researchers can identify patterns and relationships that transcend specific contexts. This approach supports the applicability of findings to diverse settings, such as comparing safety cultures across airlines (Senge, 1990).

Trustworthiness is a vital construct in qualitative research, ensuring the credibility, applicability, consistency, and neutrality of findings. Its relevance is heightened in the aviation industry, where qualitative insights inform critical decisions in areas ranging from safety management to customer experience. By adhering to the principles of credibility, transferability, dependability, and confirmability, researchers can produce robust findings that contribute to the industry's advancement. Theoretical frameworks such as the constructivist paradigm, phenomenology, grounded theory, and systems thinking provide the foundation for understanding and implementing trustworthiness in qualitative methodologies.

By maintaining methodological rigor, aviation researchers can build trust with stakeholders and enhance the utility of qualitative studies in addressing the industry's complex challenges.

3. RELEVANCE OF TRUSTWORTHINESS IN THE AVIATION INDUSTRY

The aviation industry operates in a unique and highly complex environment characterized by safety-critical operations, technological advancements, and a diverse array of stakeholders. In this context, trustworthiness in qualitative research is paramount for generating credible, reliable, and actionable insights. Trustworthy qualitative research enables a deeper understanding of the industry's technical, managerial, and behavioral aspects, contributing to informed decision-making and continuous improvement. This chapter explores the relevance of trustworthiness in the aviation industry by examining its complexities, interdisciplinary nature, and impact on various stakeholders.

3.1 Industry-Specific Dynamics

The aviation industry is a global and multifaceted sector involving numerous interconnected systems and processes. The operational environment is safety-critical, where the smallest oversight can lead to significant consequences. This complexity amplifies the need for trustworthy research to ensure that findings are both credible and applicable.

Complexities of Aviation Operations - Aviation operations are inherently complex due to factors such as technological integration, regulatory requirements, and environmental considerations. For instance, managing air traffic control systems involves sophisticated technologies that require precise coordination among multiple teams across different locations. Similarly, airport operations demand seamless collaboration between airlines, ground handling staff, and regulatory authorities. Qualitative research in these contexts seeks to uncover underlying human and organizational dynamics that quantitative approaches may overlook.

To ensure trustworthiness, qualitative studies in aviation must address these complexities with methodological rigor. For example, exploring safety culture within an airline requires credible and transferable findings to inform training programs and operational strategies across different regions. Dependability is particularly important as researchers analyze evolving practices, such as the integration of artificial intelligence in flight operations.

Stakeholder Diversity - The aviation industry involves a wide range of stakeholders, including airline management, employees, passengers, regulatory bodies, and policymakers. Each stakeholder group brings unique perspectives and expectations that qualitative research must accurately represent. For instance, a study on passenger satisfaction must consider the diversity of cultural backgrounds, travel purposes, and service expectations. Similarly, research on pilot decision-making must account for variables such as training, operational environments, and psychological factors.

To maintain trustworthiness, researchers must use techniques like purposive sampling and triangulation to capture these diverse perspectives comprehensively. Providing thick descriptions of participant experiences and contexts further enhances the transferability of findings, enabling stakeholders to apply insights across various settings within the industry.

Safety-Critical Environments - Safety is the cornerstone of the aviation industry, making the trustworthiness of research particularly critical. Qualitative studies often address sensitive topics, such as error reporting, crew fatigue, and crisis management, where participant honesty and researcher credibility are essential. Findings from such studies influence policies, training programs, and safety protocols, underscoring the need for rigor in ensuring that findings are credible, dependable, and confirmable.

Research on the factors influencing error reporting among pilots requires confirmability to avoid bias and ensure that findings are based solely on participant experiences. Member checking and peer debriefing can be instrumental in achieving this goal, reinforcing the reliability of qualitative insights that inform safety initiatives.

3.2 Interdisciplinary Insights

The aviation industry operates at the intersection of technology, management, and human behavior. This interdisciplinary nature necessitates a comprehensive approach to qualitative research that integrates technical, managerial, and behavioral perspectives.

Technical Aspects - Advancements in technology, such as automation, artificial intelligence, and data analytics, play a critical role in aviation. Qualitative research explores how these technologies influence human factors, decision-making, and organizational processes. For instance, studies on pilot interactions with automated systems can uncover insights into trust, workload management, and situational awareness.

To ensure trustworthiness, researchers must adopt systematic methodologies that capture the complexities of technical environments. Triangulating data from interviews, observations, and system logs can enhance credibility and confirmability, ensuring that findings accurately reflect real-world scenarios.

Managerial Aspects - Effective management is vital for coordinating diverse teams, optimizing operations, and ensuring compliance with regulations. Qualitative research in this domain often examines leadership styles, organizational culture, and communication practices. For example, a study on how airline managers foster a safety-oriented culture requires dependable findings to guide leadership training programs and policy development.

Dependability can be ensured by maintaining detailed audit trails and documenting decision-making processes throughout the research. By doing so, researchers can provide stakeholders with transparent and replicable insights that support managerial improvements in the aviation industry.

Behavioral Aspects - Human behavior significantly influences aviation outcomes, from passenger satisfaction to crew performance. Qualitative studies explore these behavioral dimensions to inform training, customer service, and operational strategies. For instance, research on passenger behavior during flight delays can uncover insights into communication needs and stress management.

To maintain trustworthiness, researchers must prioritize credibility by engaging deeply with participants and validating findings through member checking. Rich data descriptions and reflexive practices further enhance the authenticity of behavioral insights, enabling stakeholders to implement evidence-based interventions.

3.3 Stakeholder Impacts

Trustworthy qualitative research has far-reaching implications for various stakeholders in the aviation industry, including policymakers, airlines, regulatory authorities, and consumers.

Policymakers - Policymakers rely on qualitative insights to design regulations that balance safety, efficiency, and environmental sustainability. For example, research on airport noise pollution might inform policies that address community concerns while supporting operational growth. Trustworthiness ensures that such research is credible and applicable, enabling policymakers to make informed decisions that reflect stakeholder needs.

Airlines - Airlines use qualitative research to improve operations, enhance employee performance, and elevate customer experiences. For instance, findings from studies on cabin crew training can inform the development of programs that promote teamwork and resilience. Dependable and confirmable research findings ensure that airlines can implement effective strategies with confidence.

Regulatory Authorities - Regulatory authorities, such as the International Civil Aviation Organization (ICAO) and Federal Aviation Administration (FAA), depend on trustworthy research to establish safety standards and guidelines. For example, qualitative studies on pilot fatigue can provide valuable insights into rest requirements and scheduling practices. Transferable findings enable regulators to apply research across different jurisdictions and operational contexts.

Consumers - Consumers benefit from qualitative research that informs service enhancements, safety measures, and communication strategies. For instance, studies on passenger experiences during flight disruptions can guide airlines in developing empathetic and effective responses. Credible and confirmable findings ensure that consumer-centric improvements are grounded in authentic experiences and needs.

Trustworthiness is a fundamental attribute of qualitative research in the aviation industry, ensuring that findings are credible, transferable, dependable, and confirmable. The industry's complex dynamics, interdisciplinary nature, and diverse stakeholders amplify the importance of rigor in research methodologies. By addressing the technical, managerial, and behavioral dimensions of aviation, trustworthy qualitative research provides actionable insights that support continuous improvement.

Stakeholders across the aviation sector, from policymakers to consumers, benefit from research that adheres to trustworthiness principles. Through methodological rigor and interdisciplinary integration, qualitative research contributes to safer, more efficient, and customer-centric aviation practices.

4. CHALLENGES IN ESTABLISHING TRUSTWORTHINESS IN AVIATION RESEARCH

The aviation industry presents unique challenges for qualitative research due to its multifaceted and dynamic nature. Ensuring trustworthiness in such studies is critical for deriving insights that are credible, transferable, dependable, and confirmable. However, achieving this rigor requires navigating numerous obstacles, ranging from accessing data and participants to managing contextual and technological complexities. This chapter explores the key challenges in establishing trustworthiness in aviation research and discusses strategies to address them.

4.1 Access to Data and Participants

Navigating Confidentiality and Sensitive Information - One of the primary challenges in aviation research is accessing data and participants while safeguarding sensitive information. Aviation operations involve critical infrastructure, regulatory compliance, and passenger data, which are often classified or protected by stringent confidentiality agreements. Researchers frequently encounter barriers when attempting to collect data on internal processes, such as pilot decision-making during emergencies or error-reporting practices in airlines.

Studies on safety management systems (SMS) require insights into error reports and near-miss incidents, which airlines and regulatory bodies may hesitate to share due to reputational risks or legal constraints (Dekker, 2017). This restricts the depth and breadth of data available, potentially affecting the credibility of research findings.

Ethical Considerations in Participant Recruitment - Recruiting participants from the aviation industry poses additional ethical challenges. Professionals such as pilots, air traffic controllers, and maintenance staff operate in high-pressure environments where participation in research may interfere with their duties. Ensuring voluntary participation while avoiding coercion is essential to uphold ethical standards (Patton, 2002).

Strategies to address these challenges include building trust with organizations through clear communication of research objectives and securing informed consent. Establishing partnerships with industry stakeholders can also facilitate data access while ensuring confidentiality and ethical compliance.

4.2 Contextual Complexity

Variability Across Geographies and Cultures - The global nature of aviation introduces significant variability across geographical regions, cultural norms, and regulatory frameworks. Qualitative research in this context must account for diverse operational practices, languages, and socio-political environments. For example, a study on customer service expectations may yield different findings when conducted in North America versus Asia due to cultural differences in communication styles and service preferences (Hofstede, 2001).

Ensuring transferability under such conditions requires researchers to provide thick descriptions of study contexts and participant demographics. Triangulation of data from multiple locations can also enhance the applicability of findings across diverse settings.

Regulatory Frameworks and Compliance - The aviation industry is heavily regulated, with organizations adhering to standards set by bodies such as the International Civil Aviation Organization (ICAO) and the Federal Aviation Administration (FAA). These regulations vary by region and evolve over time, posing challenges for researchers seeking to generalize findings. For instance, studies on pilot training programs may need to account for differences in certification requirements and procedural norms across jurisdictions.

Researchers must stay informed about relevant regulations and consider their implications when designing studies. Collaborating with regulatory authorities can provide valuable insights and improve the credibility and dependability of research outcomes.

4.3 Researcher Bias

Minimizing Preconceptions and External Influences - Researcher bias is an inherent challenge in qualitative research, arising from preconceptions, personal experiences, or external pressures. In the aviation industry, where complex technical and human factors intersect, bias can influence data collection, analysis, and interpretation. For example, a researcher with prior experience as an airline employee may unconsciously prioritize operational perspectives over customer viewpoints.

To mitigate bias, reflexivity is essential. Researchers must critically examine their assumptions and document their decision-making processes to enhance transparency (Lincoln & Guba, 1985). Techniques such as peer debriefing and member checking can further validate findings by incorporating diverse perspectives and ensuring that results reflect participants' realities rather than the researcher's interpretations.

Balancing Insider and Outsider Perspectives - Aviation researchers often face the challenge of balancing insider knowledge with outsider objectivity. While industry experience can provide valuable context, it may also introduce bias or limit the scope of inquiry. Conversely, researchers without industry expertise may struggle to understand technical jargon or organizational dynamics, affecting the depth of their analysis.

Engaging in collaborative research with industry professionals can help strike this balance. Such partnerships can enhance credibility by combining insider insights with external analytical rigor, ultimately producing more trustworthy findings.

4.4 Dynamic Nature of the Industry

Rapid Technological Advancements - The aviation industry is characterized by continuous technological innovation, from advancements in aircraft design to the integration of artificial intelligence in operations. These rapid changes pose challenges for researchers attempting to capture evolving practices and their implications. For example, a study on pilot interactions with automated systems may quickly become outdated as new technologies and protocols are introduced.

To address this challenge, researchers must adopt flexible and adaptive methodologies. Longitudinal studies can provide insights into trends over time, while iterative data collection and analysis allow for the incorporation of emerging developments. Dependability can be ensured by maintaining detailed audit trails that document how changes in the research context are accounted for in the study.

Unpredictable Events and Disruptions - The aviation industry is highly susceptible to disruptions caused by factors such as natural disasters, geopolitical events, and pandemics. These events can significantly impact research timelines and participant availability. For instance, the COVID-19 pandemic disrupted global aviation operations, making it difficult for researchers to access participants or observe normal operational processes (International Air Transport Association [IATA], 2021).

Researchers must be prepared to adapt their methodologies in response to such disruptions. Remote data collection methods, such as virtual interviews, can help maintain continuity while ensuring participant safety. Providing detailed context about the study period and its challenges further enhances the dependability and transferability of findings.

Establishing trustworthiness in aviation research is a multifaceted endeavor that requires researchers to navigate challenges related to data access, contextual complexity, researcher bias, and the dynamic nature of the industry. By addressing these obstacles with methodological rigor and ethical sensitivity, researchers can produce findings that are credible, transferable, dependable, and confirmable.

The aviation industry's global reach, regulatory intricacies, and technological advancements demand a nuanced approach to qualitative research. Through strategies such as building partnerships, ensuring reflexivity, and adopting adaptive methodologies, researchers can overcome barriers to trustworthiness and contribute valuable insights that inform safety, efficiency, and customer satisfaction in aviation.

5. STRATEGIES FOR ENHANCING TRUSTWORTHINESS IN QUALITATIVE AVIATION RESEARCH

The aviation industry presents a complex, multidisciplinary environment where ensuring trustworthiness in qualitative research is essential for deriving actionable insights. Trustworthiness, characterized by credibility, transferability, dependability, and confirmability, underpins the value of qualitative studies, particularly in an industry that demands precision, reliability, and ethical rigor. This chapter discusses strategies to enhance trustworthiness in qualitative aviation research, focusing on data collection, data analysis, and ethical considerations.

5.1 Data Collection Techniques

Triangulation: Using Multiple Data Sources to Enhance Credibility - Triangulation involves using multiple sources, methods, investigators, or theoretical perspectives to cross-verify data, thereby strengthening the credibility of findings (Denzin, 2017). In aviation research, triangulation can be achieved by integrating data from diverse sources such as interviews with pilots, observational studies in air traffic control towers, and document analysis of safety protocols.

A study examining the effectiveness of Crew Resource Management (CRM) training might combine interviews with crew members, observations of CRM workshops, and analysis of post-training evaluation reports. By comparing these data points, researchers can identify consistent patterns and reduce the likelihood of bias or misinterpretation.

Methodological triangulation, in particular, is valuable in aviation research due to the industry's reliance on both technical data and human experiences. Employing both qualitative and quantitative methods can provide a comprehensive understanding of complex phenomena, such as decision-making under stress.

Member Checking: Engaging Participants to Validate Findings

Member checking, or participant validation, involves sharing preliminary findings with participants to ensure their accuracy and resonance with participants' experiences (Lincoln & Guba, 1985). This technique not only enhances credibility but also fosters trust between researchers and participants.

In the aviation industry, member checking can be used during studies on safety culture. For example, researchers might present summaries of their findings to airline staff and management, allowing them to confirm or refine interpretations. This collaborative approach helps ensure that the research accurately reflects the lived realities of those involved.

Thick Description: Providing Detailed Contextual Information for Transferability - Thick description entails providing rich, detailed accounts of the research context, participants, and findings, enabling others to assess the applicability of the results to different settings (Geertz, 1973). This is particularly important in aviation research, where operational practices can vary significantly across regions and organizations.

A study on passenger satisfaction at an international airport should include detailed descriptions of the airport's facilities, the demographic profile of passengers, and cultural considerations influencing service expectations. Such contextual information allows readers to determine whether the findings can be transferred to other airports with similar characteristics.

5.2 Data Analysis and Interpretation

Audit Trails: Maintaining a Detailed Record of Research Decisions and Processes - An audit trail involves systematically documenting all decisions, procedures, and changes made during the research process. This transparency enhances dependability by allowing others to follow the researcher's thought processes and verify the consistency of the study (Carcary, 2009).

In aviation research, maintaining an audit trail is particularly useful when analyzing complex data, such as incident reports or qualitative feedback from training programs. For example, a researcher studying pilot error might record how data were coded, how themes emerged, and how interpretations were validated. These records enable external auditors or reviewers to assess the dependability and confirmability of the findings.

Reflexivity: Encouraging Researcher Self-Awareness and Neutrality - Reflexivity involves critically examining one's biases, assumptions, and potential influences on the research process (Berger, 2015). In the aviation industry, where technical expertise and organizational hierarchies often intersect, reflexivity is crucial for maintaining objectivity.

Researchers must regularly reflect on how their backgrounds, experiences, and perspectives might shape their interpretations. For instance, an aviation researcher with a technical background might unconsciously prioritize operational efficiency over human factors. Keeping reflexive journals and engaging in peer debriefing can help mitigate these biases, ensuring a more balanced and neutral analysis.

5.3 Ethical Considerations

Protecting Participant Rights and Maintaining Confidentiality - Ethical considerations are paramount in aviation research due to the sensitive nature of the data and the high stakes involved. Participants, often professionals such as pilots, engineers, or air traffic controllers, may be reluctant to share information if they perceive risks to their privacy or careers. Ensuring confidentiality and protecting participant rights are essential for fostering trust and encouraging candid participation.

Key strategies include obtaining informed consent, anonymizing data, and adhering to ethical guidelines set by institutional review boards (IRBs) or aviation regulatory bodies. For example, when conducting interviews with airline employees, researchers should ensure that transcripts are anonymized and stored securely, with access limited to authorized personnel.

Balancing Transparency with Ethical Obligations - While transparency is critical for trustworthiness, researchers must balance it with their ethical obligations to protect sensitive information. In the aviation industry, this often involves navigating trade-offs between providing detailed descriptions for transferability and safeguarding proprietary or confidential data.

A study on maintenance practices might involve analyzing proprietary documents or interviewing staff about organizational policies. Researchers should work closely with stakeholders to determine what information can be disclosed while maintaining confidentiality. Transparency about these limitations in the final report enhances the credibility of the study without compromising ethical standards.

5.4 Integrating Strategies for Comprehensive Trustworthiness

To achieve comprehensive trustworthiness in aviation research, the strategies discussed above must be integrated cohesively. For example, a study examining the impact of automation on pilot performance might use triangulation to combine interviews, simulator observations, and training records. Member checking could validate initial findings, while thick description ensures that the results are contextualized for broader applicability.

During data analysis, maintaining an audit trail and practicing reflexivity would further enhance dependability and confirmability. Ethical considerations, such as anonymizing participant data and obtaining informed consent, should underpin all stages of the research process.

By adopting these strategies, researchers can address the unique challenges of aviation research while ensuring that their findings are credible, applicable, consistent, and objective. This rigor not only strengthens the value of qualitative studies but also contributes to evidence-based decision-making in an industry where safety, efficiency, and stakeholder trust are paramount.

Enhancing trustworthiness in qualitative aviation research requires a multifaceted approach that integrates robust data collection techniques, transparent data analysis processes, and rigorous ethical practices. Strategies such as triangulation, member checking, and thick description ensure that findings are credible and transferable, while audit trails and reflexivity enhance dependability and confirmability. Ethical considerations, particularly regarding participant rights and confidentiality, are essential for maintaining trust and integrity in research.

By implementing these strategies, researchers can overcome the complexities of the aviation industry and produce high-quality qualitative studies that inform policy-making, improve operational practices, and enhance stakeholder satisfaction. The emphasis on trustworthiness not only validates the research process but also reinforces the aviation industry's commitment to excellence and innovation.

6. CASE STUDIES: TRUSTWORTHINESS IN AVIATION RESEARCH PRACTICES

In qualitative research within the aviation industry, ensuring the trustworthiness of findings is paramount to producing valid, reliable, and relevant results. Trustworthiness encompasses four primary dimensions: credibility, transferability, dependability, and confirmability. These dimensions provide a framework for maintaining the rigor and ethical standards essential for qualitative research, particularly in complex sectors such as aviation. This section will explore three case studies to illustrate how trustworthiness is achieved in aviation research. The case studies discussed include: a qualitative study on pilot fatigue and operational safety, an investigation of customer experiences in airport services, and an examination of ground handling staff perspectives on operational efficiency. Through these case studies, practical insights and methodological reflections will also be highlighted to guide future research in the aviation domain.

6.1 A Qualitative Study on Pilot Fatigue and Operational Safety

Pilot fatigue remains a critical safety concern within the aviation industry, as it can significantly impact both operational safety and decision-making during flight operations. A qualitative study conducted by Clark et al. (2020) investigated the perspectives of commercial pilots regarding fatigue management systems and their influence on safety. This study employed semi-structured interviews with a diverse group of pilots and supplemented the interview data with document analysis of existing fatigue-related policies within airlines.

In this case, the credibility of the findings was ensured through the use of triangulation, where multiple data sources (interviews and policy documents) were combined to offer a comprehensive understanding of the issue. Additionally, member checking was employed to engage participants in validating the accuracy and relevance of the findings. The dependability of the research was established by maintaining detailed audit trails, documenting all decisions made during the data collection and analysis process. The study ensured that the findings could be transferred to other aviation contexts by providing a thick description of the participants' experiences and the operational settings in which they worked.

One of the key lessons learned from this study is the importance of ensuring that participants feel comfortable sharing sensitive information. By emphasizing confidentiality and creating an open, non-judgmental environment during the interviews, the researchers were able to collect valuable insights into the pressures pilots face when dealing with fatigue. Furthermore, reflexivity was a critical element throughout the study, as researchers remained conscious of their own biases and how these might influence their interpretation of the data.

6.2 Investigating Customer Experiences in Airport Services Through Focus Groups

Understanding customer satisfaction is essential for enhancing airport services and improving passenger experiences. A qualitative study conducted by Johnson and Miller (2019) explored customer experiences in airports, focusing on aspects such as ease of check-in, security procedures, and overall satisfaction with airport facilities. The study used focus groups to gather in-depth insights from passengers, providing an opportunity to explore attitudes, perceptions, and emotions regarding airport services.

To ensure credibility, the researchers employed triangulation by combining the focus group data with surveys to validate the findings. This approach allowed for a richer, more nuanced understanding of passenger experiences. Member checking was also incorporated, allowing focus group participants to review and confirm the accuracy of the findings. The transferability of the findings was ensured by providing detailed contextual information about the airports involved in the study, the demographic profile of participants, and the services evaluated.

The dependability of the study was maintained by ensuring consistency in the application of focus group protocols, with clear guidelines on the interview process and data collection procedures. Additionally, the researchers kept audit trails of all interactions during the focus groups and how the data were coded and analyzed.

One important lesson learned from this study was the need for flexibility during the focus group discussions. Airport passengers may have varying levels of comfort discussing sensitive issues such as security procedures, so it was crucial to create a safe space for participants to express their views. The use of reflexivity was particularly important in this case to avoid any researcher bias, especially when dealing with passengers' potentially emotional responses to certain aspects of airport service. Additionally, the study highlighted the value of member checking, as feedback from participants ensured that the researchers' interpretations accurately reflected passenger sentiments.

6.3 Ground Handling Staff Perspectives on Operational Efficiency

Ground handling staff play a crucial role in ensuring the smooth operation of flights, but their perspectives on operational efficiency are often underexplored in academic research. A study by Davis et al. (2021) sought to understand the challenges and opportunities related to operational efficiency from the perspective of ground handling staff. The study used semi-structured interviews with staff members working in various roles within ground handling operations, such as ramp agents, baggage handlers, and customer service representatives.

In this research, credibility was ensured through the use of multiple data sources—interviews with staff were complemented by an analysis of operational performance data from the airlines and ground handling companies involved. Triangulation was used to cross-check the interview findings with quantitative performance metrics, providing a more robust understanding of operational inefficiencies. To ensure dependability, the researchers maintained comprehensive audit trails documenting the stages of data collection, coding, and analysis.

In terms of transferability, the findings were applicable to different aviation contexts due to the inclusion of staff from various airports and airlines. The thick description of the operational environments and staff roles allowed the research to have broader applicability. In addition, the study was conducted in a way that minimized researcher bias by using reflexivity during the data analysis process, acknowledging the researchers' own professional backgrounds in aviation and the potential for that to influence their interpretation of the data.

A key lesson learned from this study was the importance of establishing trust with participants. Ground handling staff, like many frontline employees, may be wary of researchers, particularly if the research concerns performance-related issues. The researchers used a participatory approach to data collection, ensuring that participants felt they were contributing to a broader discussion on improving operational efficiency rather than being scrutinized.

6.4 Lessons Learned and Methodological Reflections

Through the three case studies, several key lessons can be drawn regarding the importance of ensuring trustworthiness in aviation research. First, the use of triangulation is essential for enhancing the credibility of findings, particularly when dealing with complex and multi-faceted issues such as pilot fatigue, customer satisfaction, and operational efficiency. By integrating multiple data sources—interviews, surveys, and document analysis—researchers can cross-check findings and increase the validity of their conclusions.

Second, the practice of member checking should be incorporated as part of the data validation process. Engaging participants in reviewing and confirming the findings ensures that the researchers' interpretations align with the participants' experiences and reduces the likelihood of misinterpretation or bias.

Third, thick description is critical for enhancing the transferability of research findings. Providing detailed contextual information about participants, settings, and processes allows researchers to apply findings to other similar contexts within the aviation industry.

Finally, reflexivity and the maintenance of audit trails are fundamental for addressing researcher bias and ensuring dependability. By maintaining self-awareness and documenting every stage of the research process, researchers can ensure transparency and consistency in their approach.

These case studies underscore the importance of maintaining methodological rigor and ensuring trustworthiness in aviation research. By adopting strategies such as triangulation, member checking, and reflexivity, researchers can produce credible, transferable, and dependable findings that provide valuable insights into critical issues within the aviation industry.

7. EVALUATING TRUSTWORTHINESS IN AVIATION RESEARCH

In aviation research, ensuring trustworthiness is not only a methodological necessity but also a means of reinforcing the reliability and applicability of findings in a highly sensitive and safety-critical industry. This chapter focuses on evaluating trustworthiness through criteria-based assessment, feedback mechanisms, and metrics and benchmarks designed to address the unique demands of aviation research.

7.1 Criteria-Based Assessment

Trustworthiness, as established by Lincoln and Guba (1985), is a cornerstone of qualitative research and is evaluated based on four dimensions: credibility, transferability, dependability, and confirmability. These dimensions serve as a comprehensive framework for assessing the rigor and validity of qualitative studies in aviation.

Credibility: To evaluate credibility, researchers employ strategies like triangulation, member checking, and prolonged engagement. Triangulation involves cross-verifying data using multiple sources, such as interviews, focus groups, and document analysis, to ensure findings accurately reflect participants' realities. For example, in a study on aviation safety culture, credibility may be assessed by comparing insights from pilots, cabin crew, and ground staff to identify consistent themes. Member checking, where participants validate the findings, ensures that interpretations align with their experiences, reinforcing credibility.

Transferability: Assessing transferability involves examining the richness and thick descriptions provided in the study. Detailed contextual information, such as operational settings, participant demographics, and cultural nuances, allows readers to determine whether findings are applicable to other aviation contexts. For instance, a study on passenger behavior at a European airport can achieve transferability by providing extensive details on the airport's infrastructure, staffing, and cultural environment, enabling other researchers to adapt the findings to airports in similar settings.

Dependability: Dependability focuses on the consistency and reliability of the research process. Evaluating this dimension requires audit trails, documenting every step of the research, from data collection to analysis. Peer reviews and reflexive journaling can further confirm dependability by ensuring the research approach remains systematic and transparent. For instance, in a study on pilot training effectiveness, maintaining a record of interview protocols and coding decisions ensures the process is replicable.

Confirmability: Confirmability evaluates the objectivity and neutrality of the research. This dimension is often assessed through reflexivity and external audits. Reflexivity involves researchers reflecting on their biases and ensuring that their interpretations are grounded in the data rather than personal assumptions. External audits by independent experts can provide an additional layer of objectivity. In studies involving sensitive topics, such as aviation security, confirmability can be strengthened by cross-checking findings with official records or regulatory guidelines.

7.2 Feedback Mechanisms

Incorporating feedback mechanisms enhances the evaluation of trustworthiness by fostering collaboration and transparency. These mechanisms include peer reviews, participant evaluations, and expert consultations, each of which contributes uniquely to ensuring the rigor of aviation research.

Peer Reviews: Peer reviews are a critical tool for evaluating the robustness of qualitative studies. By inviting subject matter experts to review the research design, data collection, and analysis, researchers can identify gaps or biases in their approach. In aviation research, peers from diverse fields such as engineering, management, and human factors can provide interdisciplinary perspectives, enriching the study's validity. For example, a peer review of a study on air traffic controller stress may reveal overlooked variables, such as cultural or regulatory differences, that could impact the findings.

Participant Evaluations: Engaging participants in evaluating findings ensures the credibility and authenticity of the research. Member checking, where participants review and validate data interpretations, serves as a direct feedback mechanism. Additionally, participant evaluations can extend to assessing the relevance and applicability of the study's conclusions. In a study on passenger satisfaction with airport services, participants could be invited to comment on whether the findings accurately capture their experiences, providing valuable insights for refining the analysis.

Expert Consultations: In complex and dynamic fields like aviation, consulting with industry experts can validate the study's findings and methodologies. Experts can assess whether the research aligns with industry standards and addresses key challenges effectively. For instance, a study on unmanned aerial vehicle (UAV) operations could benefit from consultations with regulatory authorities and aviation safety experts to ensure its findings are both practical and compliant with safety norms.

7.3 Metrics and Benchmarks

Quantifying qualitative rigor in aviation research presents unique challenges, given the interpretive nature of qualitative methodologies. However, researchers have developed metrics and benchmarks to objectively evaluate trustworthiness, ensuring that findings are both credible and actionable.

Coding Reliability: One key metric involves assessing the consistency of data coding. Inter-coder reliability, where multiple researchers independently code the same data and compare results, ensures dependability in thematic analysis. In a study on airline employee engagement, high inter-coder reliability would indicate that the themes identified, such as motivation or job satisfaction, are robust and replicable.

Saturation Benchmarks: Data saturation, a critical concept in qualitative research, serves as a benchmark for ensuring sufficient data collection. Saturation is achieved when additional data no longer generate new themes or insights. In aviation research, a study on customer perceptions of flight delays may set a saturation benchmark based on the number of interviews or focus groups required to cover all relevant perspectives.

Reflexivity Indicators: Reflexivity is a vital component of confirmability and can be evaluated through indicators such as the frequency and depth of reflexive journal entries. These entries should document researchers' reflections on their biases and decisions throughout the study. For example, in a study on pilot fatigue, reflexivity indicators might include journal notes on how researchers addressed their assumptions about workload and rest patterns.

Ethical Compliance Scores: Ethical considerations are integral to trustworthiness, particularly in aviation research involving sensitive or confidential data. Compliance with ethical guidelines, such as obtaining informed consent and maintaining participant anonymity, can be quantified through an ethical compliance score. Researchers can use checklists to ensure all ethical requirements are met, reinforcing the credibility and integrity of their work.

Validation Metrics: Validation metrics, such as the percentage of participants who agree with the study's interpretations during member checking, provide a direct measure of credibility. Similarly, peer review scores based on predefined criteria, such as methodological rigor and relevance, can quantify the quality of the research process.

Evaluating trustworthiness in aviation research requires a multi-faceted approach that incorporates criteria-based assessment, feedback mechanisms, and metrics. The four dimensions of trustworthiness—credibility, transferability, dependability, and confirmability—offer a robust framework for assessing the rigor and reliability of qualitative studies. Feedback mechanisms, including peer reviews and participant evaluations, ensure that findings are both authentic and applicable.

Metrics and benchmarks, such as coding reliability, saturation benchmarks, and ethical compliance scores, provide objective tools for evaluating the quality of research processes and outcomes. By adopting these strategies, aviation researchers can navigate the complexities of their field, producing trustworthy findings that inform policy, improve operations, and enhance safety across the industry.

Adhering to rigorous evaluation practices not only strengthens the credibility of qualitative studies but also underscores the critical role of trustworthiness in addressing the challenges and opportunities within the aviation sector.

8. FUTURE DIRECTIONS AND INNOVATIONS

As the aviation industry evolves, so do the methods and tools available for conducting qualitative research. The future of qualitative research in aviation will likely be shaped by technological integration, cross-disciplinary approaches, and global perspectives. These directions will not only enhance the trustworthiness of qualitative studies but also expand their scope, applicability, and relevance in addressing complex challenges within the industry.

8.1 Technological Integration

The integration of artificial intelligence (AI) and big data analytics into qualitative research methodologies offers transformative possibilities for the aviation industry. These technologies enable researchers to enhance credibility, dependability, and confirmability by automating processes, uncovering patterns, and providing new dimensions to data interpretation.

AI-Driven Data Analysis: AI tools can assist in coding and analyzing qualitative data more efficiently and accurately. For instance, natural language processing (NLP) algorithms can identify recurring themes in interview transcripts or focus group discussions, minimizing researcher bias. In studies on pilot fatigue, AI can analyze textual data alongside physiological metrics (e.g., heart rate or sleep patterns) to provide a more holistic understanding of the issue.

Big Data in Contextual Research: The aviation sector generates massive amounts of data, from passenger feedback to operational logs. Integrating qualitative insights with big data analytics allows researchers to validate findings and enhance transferability. For example, researchers investigating passenger experiences might use qualitative interviews to understand emotional responses while leveraging big data from social media sentiment analysis or customer satisfaction surveys for corroboration.

Virtual Reality (VR) and Simulations: Emerging technologies like VR can revolutionize qualitative data collection by creating immersive scenarios for participants. For example, researchers exploring ground handling efficiency could simulate different operational conditions in a virtual environment, allowing participants to interact and provide insights in real-time. Such approaches not only improve data richness but also enhance participant engagement and authenticity.

Digital Tools for Member Checking: Online platforms can facilitate ongoing engagement with participants, enabling member checking at multiple stages of the research process. These platforms can host interactive dashboards where participants review and provide feedback on preliminary findings, increasing their involvement and reinforcing the study's credibility.

8.2 Cross-Disciplinary Approaches

The complexity of aviation challenges necessitates collaboration across disciplines, bringing together expertise from engineering, psychology, business, and more. Such cross-disciplinary approaches enrich qualitative research by providing diverse perspectives and innovative solutions.

Engineering and Human Factors: Collaboration with engineers and human factors specialists can enhance research on topics such as cockpit design, maintenance protocols, or air traffic management. For instance, qualitative studies on pilot workload could integrate technical insights on system ergonomics, offering a comprehensive understanding of how technology and human behavior interact.

Psychology and Behavioral Sciences: Insights from psychology are crucial for understanding the human dimensions of aviation, such as decision-making, stress, and teamwork. Qualitative research on cabin crew dynamics, for example, can benefit from psychological theories on group behavior and conflict resolution. Collaborating with psychologists ensures that data interpretation accounts for underlying cognitive and emotional factors.

Business and Management Perspectives: The aviation industry operates within a competitive and highly regulated business environment. Collaborating with business scholars can provide qualitative research with frameworks to analyze market trends, customer satisfaction, or operational efficiency. For example, qualitative studies on airport management might integrate concepts from strategic management to explore how leadership styles influence service delivery.

Interdisciplinary Research Design: Cross-disciplinary teams can design studies that address multifaceted research questions. For example, a study on the adoption of sustainable aviation fuels might combine qualitative interviews with stakeholders (pilots, engineers, and policymakers) and quantitative analysis of environmental impact, providing a holistic understanding of both technological and social dimensions.

8.3 Global Perspectives

Aviation is inherently global, making international perspectives essential for enhancing the trustworthiness and applicability of qualitative research. Addressing cultural, regulatory, and operational diversity strengthens the relevance of findings across different contexts.

Cross-Cultural Considerations: Understanding cultural nuances is critical in aviation research, particularly when exploring topics like safety culture, passenger behavior, or employee engagement. Researchers must adopt culturally sensitive methodologies, such as conducting interviews in participants' native languages or tailoring focus group discussions to align with local norms. These practices enhance the transferability of findings to diverse aviation contexts.

Comparative Studies: Comparative qualitative research can uncover best practices and lessons learned across regions. For instance, a study comparing safety protocols at airports in North America, Asia, and Europe might reveal how cultural attitudes toward risk influence operational strategies. Such comparisons not only enhance the credibility of findings but also provide actionable insights for global aviation stakeholders.

International Collaboration: Partnering with researchers and institutions across countries can improve access to data and ensure diverse perspectives. For example, a global study on the impact of COVID-19 on airline operations could benefit from collaborations with researchers in regions disproportionately affected by the pandemic, ensuring that findings are representative and comprehensive.

Addressing Regulatory Diversity: The aviation industry is governed by a complex web of regulations that vary significantly by country. Researchers must account for these variations to ensure the dependability of their studies. For instance, qualitative research on drone operations might consider differences in airspace regulations between the United States and Europe, tailoring methodologies accordingly.

8.4 Emerging Innovations

The future of qualitative research in aviation will also be shaped by emerging innovations, such as participatory research methods, real-time analytics, and ethical AI applications.

Participatory Research: Involving stakeholders, such as pilots, ground staff, or passengers, as co-researchers can enhance the relevance and applicability of findings. Participatory methods empower stakeholders to contribute to research design, data collection, and interpretation, ensuring that studies address real-world challenges.

Real-Time Analytics: Advances in technology enable researchers to collect and analyze data in real-time, providing immediate insights. For example, real-time analysis of passenger feedback during a flight can inform qualitative studies on service improvement.

Ethical AI Applications: As AI becomes integral to qualitative research, ensuring ethical use is paramount. Researchers must prioritize transparency, explainability, and fairness in AI-driven methodologies, particularly when analyzing sensitive aviation data.

The future of qualitative research in aviation lies in embracing technological advancements, fostering cross-disciplinary collaborations, and incorporating global perspectives. By leveraging AI, big data, and innovative tools like VR, researchers can enhance the depth, accuracy, and credibility of their studies. Cross-disciplinary approaches bring diverse expertise, enriching qualitative insights and enabling researchers to address the multifaceted challenges of aviation.

Global perspectives ensure that findings are applicable across cultural and regulatory contexts, reflecting the aviation industry's interconnected nature. By adopting these future directions, qualitative researchers can produce trustworthy studies that inform policy, improve operations, and contribute to the industry's growth and resilience in an ever-changing landscape.

9. CONCLUSION

Trustworthiness is a cornerstone of qualitative research, particularly in the aviation industry, where safety, efficiency, and customer satisfaction are paramount. As discussed throughout this chapter, credibility, dependability, transferability, and confirmability are essential dimensions that ensure the rigor of qualitative studies. These elements not only validate the findings but also enhance their applicability and relevance across diverse aviation contexts, from operational safety to customer experience. The integration of technological innovations, cross-disciplinary approaches, and global perspectives further strengthens the trustworthiness of qualitative research, enabling a more comprehensive understanding of complex aviation challenges.

One of the key takeaways from this discussion is the importance of methodological rigor in qualitative research within aviation. The use of strategies such as triangulation, member checking, and thick description ensures that research findings are grounded in multiple perspectives and data sources. Additionally, reflexivity and audit trails contribute to maintaining objectivity and transparency, addressing potential biases and enhancing the overall credibility of research. By adopting these strategies, researchers can minimize errors and inaccuracies, leading to findings that truly reflect the complexities of the aviation industry.

Ethical considerations are also paramount in aviation research, given the sensitive nature of the industry, from safety protocols to passenger data. Ensuring that participants' rights are protected and that data is handled responsibly is crucial for maintaining public trust and compliance with industry regulations. As research methods evolve with technological advancements, it is essential for researchers to prioritize ethical AI use, confidentiality, and informed consent to ensure that their findings are not only valid but also ethically sound.

The aviation industry stands to gain significantly from trustworthy qualitative research. Policy makers, airlines, and regulatory bodies rely on robust research to make informed decisions about safety, operational efficiency, and customer satisfaction. For instance, understanding the human factors that contribute to pilot fatigue or customer dissatisfaction can lead to the implementation of better policies, the development of safer procedures, and improved service delivery. Furthermore, trustworthy research that is grounded in real-world experiences can influence the design of future technologies, the training of aviation staff, and the way airlines interact with passengers, thereby strengthening the industry's foundation.

As the aviation sector continues to grow and face new challenges, researchers must adopt rigorous, ethical, and innovative methodologies to ensure that their work remains credible and relevant. The future of qualitative research in aviation lies in the continued development of cross-disciplinary collaborations, the integration of cutting-edge technologies, and the emphasis on global research practices. By embracing these strategies, researchers can make valuable contributions to the industry, helping to improve safety, efficiency, and customer experiences worldwide.

In conclusion, qualitative research in aviation, when conducted with integrity and attention to trustworthiness, plays a vital role in shaping the future of the industry. Researchers must remain committed to enhancing the credibility, dependability, transferability, and confirmability of their studies. By doing so, they not only advance knowledge but also contribute to the continuous improvement of aviation policies, practices, and customer experiences.

REFERENCES

Berger, R. (2015). Now I see it, now I don't: Researcher's position and reflexivity in qualitative research. *Qualitative Research*, 15(2), 219–234. DOI: 10.1177/1468794112468475

Braun, V., & Clarke, V. (2019). Reflecting on reflexive thematic analysis. *Qualitative Research in Sport, Exercise and Health*, 11(4), 589–597. DOI: 10.1080/2159676X.2019.1628806

Carcary, M. (2009). The research audit trial: Enhancing trustworthiness in qualitative inquiry. *Electronic Journal of Business Research Methods*, 7(1), 11–24.

Charmaz, K. (2014). *Constructing grounded theory* (2nd ed.). SAGE Publications.

Clark, M., Taylor, J., & Moore, R. (2020). Pilot fatigue and safety: A qualitative study of fatigue management systems in aviation. *Journal of Aviation Safety*, 45(2), 123–134. DOI: 10.1016/j.javiationsafety.2020.03.004

Davis, A., Lee, J., & King, T. (2021). Ground handling perspectives on operational efficiency: A qualitative study. *Aviation Operations Journal*, 38(1), 56–71. DOI: 10.1109/aoj.2021.08.002

Dekker, S. (2017). *The field guide to understanding human error* (3rd ed.). CRC Press.

Denzin, N. K. (2017). *The research act: A theoretical introduction to sociological methods* (4th ed.). Routledge. DOI: 10.4324/9781315134543

Geertz, C. (1973). *The interpretation of cultures: Selected essays*. Basic Books.

Glaser, B. G., & Strauss, A. L. (1967). *The discovery of grounded theory: Strategies for qualitative research*. Aldine.

Hofstede, G. (2001). *Culture's consequences: Comparing values, behaviors, institutions, and organizations across nations* (2nd ed.). Sage.

International Air Transport Association. (2021). The impact of COVID-19 on aviation. Retrieved from IATA website

Johnson, L., & Miller, S. (2019). Exploring customer experiences in airport services: Insights from focus groups. *Journal of Airport Management*, 27(4), 112–124. DOI: 10.1016/j.jairports.2019.07.002

Lincoln, Y. S., & Guba, E. G. (1985). *Naturalistic inquiry*. SAGE Publications. DOI: 10.1016/0147-1767(85)90062-8

Morrow, S. L. (2005). Quality and trustworthiness in qualitative research in counseling psychology. *Journal of Counseling Psychology*, 52(2), 250–260. DOI: 10.1037/0022-0167.52.2.250

Patton, M. Q. (2002). *Qualitative research and evaluation methods* (3rd ed.). SAGE Publications.

Senge, P. M. (1990). *The fifth discipline: The art and practice of the learning organization*. Doubleday.

Shenton, A. K. (2004). Strategies for ensuring trustworthiness in qualitative research projects. *Education for Information*, 22(2), 63–75. DOI: 10.3233/EFI-2004-22201

Yin, R. K. (2011). *Qualitative research from start to finish*. Guilford Press.

Yin, R. K. (2018). *Case study research and applications: Design and methods* (6th ed.). SAGE Publications.

Chapter 7
Qualitative Research Methods in Sustainability Aviation Research

Gülaçtı Şen
Balıkesir Üniversitesi, Turkey

ABSTRACT

While the aviation industry continues to grow rapidly, a significant increase in air traffic, passenger and cargo transportation is expected in the future. While the future studies and goals of the sector stakeholders are planned to support this growth, there is an important issue that the sector should focus on: Sustainability. The sustainability process in aviation is considerably difficult; however, it can be achieved. Therefore, sustainability in aviation is addressed in the first part of the study. In this study, where a case study from qualitative research methods is applied, the importance of qualitative research methods in sustainability research in aviation is examined. The importance of qualitative research methods in the aviation industry, where the focus is on human factors, draws attention. Which of the preferred qualitative research methods is applied in sustainability studies in aviation and the reasons for preferring these methods are examined.

1. INTRODUCTION

Historically, the challenges of integrating economic development, social inclusion and environmental sustainability have been local or regional. However, in the 21st century, these problems have become undeniably global. It is therefore envisaged that through global cooperation, the interconnected crises of poverty, economic instability, social inequality and environmental degradation can be overcome (Jer-

DOI: 10.4018/979-8-3693-7403-0.ch007

emić & Sachs, 2014). In this context, the United Nations Sustainable Development Goals adopted in 2015 provide a more systematic way for the SDG responsibility to work for the whole world (Bexell & Jönsson (2017). All goals emphasizing science, technology and innovation point to the year 2030 (Colglazier, 2015). It is not possible to comment on whether the goals set forth within the framework of social, environmental and economic factors will be achieved in 2030. In terms of resource use, if water, energy and sustainable consumption are similar to today on a global scale, it is thought that resource consumption will most likely increase significantly. It is expected that this situation will lead to international conflicts and more serious effects on ecosystems (Wennersten & Qie, 2018).

The negotiations that have been conducted in line with global cooperation from the past to the present are important in the point that today's sustainability has reached. However, the concept was widely accepted in 1987 by the World Commission on Environment and Development, chaired by the Norwegian Prime Minister Gro Harlem Brundtland and consisting of participants from twenty different countries, with the report titled "Our Common Future" (Brundtland, 1987). The Brundtland Report touches upon the differences between the beginning and the end of the twentieth century; it is stated that human activities, whose effects have been limited on a local scale for centuries, are currently affecting all ecosystems on a global scale (Bozlağan, 2005). According to the World Council on Environment and Development, the definition of sustainability is "A development that is sustainable is when it meets the development needs of the present generation without endangering the ability of future generations to meet their own development needs" (Frangapoulos, 2006). In a broader context, it means meeting our current needs without compromising the possibility of meeting the needs of future generations, that is, using resources responsibly to achieve the always difficult balance between economic growth, environmental care and social well-being (García & Sanz, 2018). While sustainable development is the focal point of development-related discussions, it is considered as an internationally recognized and widely adopted growth model. This model requires individuals and businesses to take into account society, economy, environment and future, in short sustainability activities, when making decisions.

It is not possible to make a single and most appropriate definition of sustainability. Sustainability can be expressed as the society should not use more resources than can be reproduced (Aras & Crowther, 2009). It is important to focus on how and by whom sustainability problems are defined, which solutions are sought to combat them and what kind of effects they have in practice (Pfister, Schweighofer & Reichel, 2016). In this sense, the sustainability phenomenon, which should be examined in different disciplines and different industries, is not a subject that can be solved at once. Sustainability is considered as a global issue for the future of all living things. As a subject that requires interdisciplinary research, its challenges concern the whole world. Social sciences have an important role in framing such social challenges (Diedrich et al., 2011). Therefore, almost all studies conducted on sustainability concern all societies and offer a solution proposal. Once it is accepted that all members of society are responsible for sustainable development, businesses are also expected to take responsibility. The impact of businesses on the natural environment and society can contribute to sustainable development (Baumgartner & Rauter, 2017).

The role of businesses in sustainable development was emphasized at the summits held in Rio in 1992 and Johannesburg, South Africa in 2002. The Johannesburg meeting determined that the desired success could not be achieved in international initiatives regarding the implementation of sustainability principles, and it also brought environmental issues to the forefront of international political discussions and, more importantly, the business world and businesses were accepted as important stakeholders in sustainable development at this summit (Idowu and Louche, 2011, 79). On the other hand, the literature questions whether sustainability can be implemented at the institutional level. There are three main claims underlying this discussion. The first is that sustainability is an ecological and social concept and cannot overlap with organizational boundaries. The second is that sustainability expresses a situation, a state of being and that there cannot be a single sustainability position. It is argued that since there is no single point to be reached, it cannot be said that the point of sustainability has been reached or not. The final argument is that there is no predictable sustainability as a result of the interaction between organizations, individuals, societies and states (Çalışkan, 2012). Sustainability must be maintained in a complex system consisting of interrelated elements of business, society and ecology; it is important to understand these elements, their interactions and the behavior of different elements (Mousavi & Bossink, 2017). Civil aviation is one of the fastest growing global industries and it is not possible to be considered sustainable in the long term in terms of sustainability. Because aviation is based on limited natural resources, it is difficult to achieve sustainable development. Serious work is needed to ensure sustainability in aviation (Ağaçsapan & Çabuk, 2019).

The aviation industry, which is considered a global tool for social and economic well-being, stimulates business activities by providing faster transportation, creating employment, supporting other business sectors, facilitating tourism, and stimulating trade. Therefore, it helps improve the quality of life of society and contributes to the economic growth of countries (Singh et al., 2023). A significant increase in air traffic, passenger and cargo transportation is expected in the future for the sector. While the future studies and targets of the sector stakeholders are planned to support this growth, there is one issue that the sector is at the center of criticism; the sustainable development process. Sustainability requires providing solutions that can consistently support the environmental, economic and social outcomes needed across current and future generations (Ansell, 2023).

It consists of sustainability-related regulations of aviation sector managers (Green Deal, Carbon Accreditation Program and Taxation System), sustainability practices of airline companies and aircraft manufacturers (Solid waste conversion, A320neo, B737max), green aircraft designs (B787, A350) and alternative fuel options (Bio-diesel, hydrogen) (Barutçu and Çolakoğlu, 2024). Considering the problems such as high energy requirement of the sector and related high carbon emissions, noise pollution, environmental pollution related to solid waste and groundwater pollution due to operational activities, the urgency of regulations related to sustainability in aviation emerges. In addition, environmental pollution is increasing as a result of increasing aviation activities (Alpman and Göğüş, 2017). The aviation industry, having a dynamic character in the global development process (Elhmoud & Kutty, 2020), has to find solutions to environmental problems and therefore also in economic and social terms.

Due to the sensitivity of its activities to the environment and human life, sustainability studies in aviation attract attention. In this context, stakeholders associated with the industry (governments, regulatory bodies such as ICAO, IATA, etc., aircraft and parts manufacturers, airlines, airport operators, etc.) publish sustainability reports and determine sustainability targets.Sustainability studies in aviation have increased recently and aviation companies have started to publish sustainability reports. Sustainability reports are the tools that enable companies to communicate the economic, environmental and social aspects of their activities to the public (Ertan, 2018). Sustainability reports clearly reveal the contributions of the aviation sector to sustainability. In addition, the data from the published reports can be productive for qualitative research.

Sustainability reports, which reveal the social responsibility, economic activities, environmental protection, ethical culture and management performance of businesses, are corporate reports (Almashhadani & Almashhadani, 2023). Sustainability reporting plays an important role in closing the gap between organizations and their communities by providing comprehensive information (Abu Khalaf, 2024). Every business that produces products/services for society should prepare its reports meticulously. It is possible for businesses to explain how they affect social, economic and environmental performance and to explain their future goals in this direction to society with sustainability reports. Like the aviation industry, aviation businesses that concern the whole society, the activities and goals of the businesses, sustainability reports should be stated transparently.

In this study, the concept of sustainability is addressed from the perspective of the aviation industry and qualitative studies conducted in aviation sustainability studies are examined. The usefulness of qualitative research methods in addressing social, economic and political problems has been proven in the literature (Barclay et al., 2017). In particular, qualitative methods need to be applied to discover a new research area in social sciences. In aviation sustainability studies, qualitative research methods are used especially in examining the sustainability reports of companies such as airlines, airports, ground handling services, etc. It can be stated that the importance of qualitative research methods in this field will increase in the future in terms of examining sustainability reports in depth and examining the achievement of goals of different aviation companies. Based on this, in this study, sustainability in aviation is first explained, then qualitative studies conducted on sustainability in aviation are conveyed and the importance of qualitative research methods in terms of sustainability in aviation is revealed.

2. SUSTAINABILITY IN AVIATION

Sustainability is one of the most widely used words of the last twenty years. In the twenty-year period, alliances have been established between governments and different international organizations, projects have been developed, organizations have been formed and serious costs have been incurred (Scones, 2010). Therefore, it is possible to say that sustainability will become a concept that is discussed for many different sectors (Demir & Sezgin, 2014:714). One of the most important of these sectors is the aviation sector.

The significant increase in annual passenger traffic in recent years has deepened concerns about the aviation sector, meaning more fuel consumption, higher greenhouse gas emissions and more energy consumption. It leads to climate change through various activities such as aircraft operations, especially those resulting from the burning of fossil fuels in aircraft engines, on the ground or in the upper troposphere (Lee et al., 2021). The environmental impact of aviation occurs because aircraft engines emit noise and particulate gases that contribute to climate change and global dimming (Mrazova, 2014). In addition to noise pollution, resource and waste production, and toxicity effects from chemicals released are other important environmental impacts arising from the aviation industry (Rupcic et al., 2023). These problems cause human health, ecosystems and depletion of natural resources. Sustainability is a multidimensional and complex concept, and according to everyday language it has been expanded and reduced to widespread expansion dimensions (Stiebe, 2023). Although the first thing that comes to mind when it comes to sustainability is environmental protection, the concept of sustainability is actually a holistic approach that includes ecological, economic and social dimensions. It can be said that sustainability is a phenomenon that affects not only ecological but also social, economic, environmental and cultural areas (Werbach, A., 2011). In this respect, Litman (2011), who considered sustainability as 'livability', examined the sustainability effects that directly affect society and revealed the sustainability goals. Sustainability goals are listed as follows;

- Economic goals; Economic efficiency, local economic development, resource efficiency, reasonable cost, operational efficiency
- Social goals; Equality / Justice, human safety, security and health, community development, cultural heritage and protection
- Environmental goals; Preventing and reducing climate change, preventing water, air and noise pollution, protecting non-renewable resources, protecting the atmosphere (Openspace), protecting biodiversity

Sustainability is a development model in which all activities in industries will be directed towards the environment, socially correct and economically feasible (Man & Bogeanu-Popa, 2020). In other words, it is important to address economic, social and environmental sustainability together in order for the asset to be sustainable.

In order to achieve success in economic, social and environmental goals in sustainability, good governance and planning are required. For this, efficient pricing is required within the framework of integrated, comprehensive and all-inclusive planning. Litman suggests that sustainability and livability generally share the same goals, but with slightly different perspectives and priorities. For example; both strive to reduce pollution. However, while sustainability generally focuses on emissions for climate change, focus is on local air and noise pollution for "livability" (Litman, 2011).

As seen in Figure 1, sustainability should be considered in three dimensions as a three-circle model and in order to achieve success, it is necessary to give equal importance to the three dimensions. For example, while sustainability aims for justice in the social dimension, it does so in the specific area of human-nature relations, taking into account the long-term and inherently uncertain future. Therefore, it is necessary to be aware that in addition to sustainability, there are other social goals, including justice goals in other specific areas of human life, such as labor or education (Baumgärtner & Quaas, 2010).

Figure 1. The 3 Circle Model of Sustainability

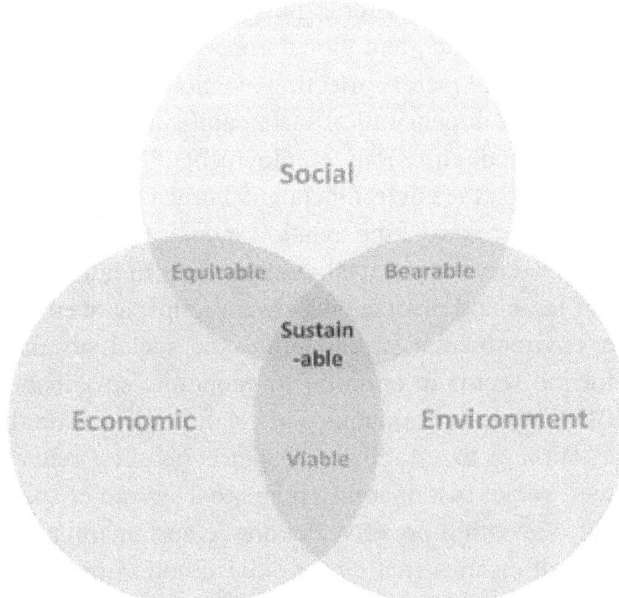

Kaynak: 'Sustainability and Sustainable Development - What is Sustainability and What is Sustainable Development?' https://circularecology.com/sustainability-and-sustainable-development.html

For the development of the aviation sector, it is necessary to take into account the three-circle model, because the issue of sustainable development is becoming increasingly important for the sector in social, ecological and economic dimensions (Machnik-Słomka, 2017). Sustainability management requires businesses to integrate sustainability principles into their processes. The most widely accepted definition of sustainability that has emerged over time is the consideration of economic viability, social responsibility and environmental responsibility together (PWC, 2011). Three important dimensions of sustainability are: 1) environmental sustainability, which consists of natural resource-system dependency; 2) economic sustainability, which refers to economic capacity, economic improvement and financial manageability; c) social compliance, which refers to social righteousness, security, individual health and the superiority of life. Sustainability requires continuous economic development without damaging the environment and causing social inequality. In this context, it is thought that the three dimensions of sustainability can help avoid serious global crises and that more accurate decisions can be made for the future with interdisciplinary approaches (Sikdar, 2004). Sustainability is considered as trying to achieve balance between social, economic and environmental imperatives.

Sustainability in aviation is generally limited to 'reducing environmental impact as much as possible' (ICAO, 2011). Aircraft noise, local air pollution, water use and pollution, waste generation, etc., and global concerns such as climate change are considered as environmental factors affecting aviation.

If we consider the three dimensions of sustainability in terms of the aviation sector; fossil fuel use, hazardous emissions, employment of large and diverse employee groups, negative impact on the environment and human health, and security issues can be expressed. In other words, the aviation industry is at the center of human life and security issues due to its provision of services to international passengers and employment of large and diverse employee groups, as well as environmental issues. Therefore, environmental sustainability and social sustainability are also very important for the sector in addition to economic sustainability (Karaman, Kılıç, & Uyar, 2018). While sustainability in aviation focuses on the preservation of natural capital, aiming to establish a balance between renewable and non-renewable resources, social sustainability refers to a system of social organization that has the role of alleviating poverty. Economic sustainability is related to the production system that ensures that current consumption needs are met without compromising future consumption (Basiago, 1999). Since aviation is also a part of social and cultural life, its sustainability depends on its relationship with society. The relationship between meeting the increasing passenger demand in aviation and protecting the environment must be observed in a balanced manner (Dursun & Aksoy, 2017). Sarrah, Ajmal, and Mertzanis (2021) reveal in their research that four important stakeholders (passengers, Civil Aviation Authority, airlines, and airports) play an important role in defining social responsibility indicators in the aviation sector. They emphasize that the contribution of each stakeholder is a must in order to achieve the determined sustainability goals as a general perception of achieving social sustainability in businesses (Sarrah, Ajmal, & Mertzanis, 2021). Sustainability success in the aviation sector can be achieved by developing systems that can respect the environment, promote economic value, and increase the quality of social life (Alameeri et al., 2017).

Sustainable transportation in terms of environmental, economic and social sustainability can also be called green transportation. Green transportation is a transportation system that does not use natural resources or cause the depletion of natural resources. For example, it is transportation that uses renewable resources or regenerated energy instead of using fossil fuels (Santa vd., 2020). Noise pollution is also considered as one of the important environmental factors in the aviation sector. A sustainable transportation system in aviation takes into account the development needs of society such as safety, human health and ecosystem in terms of social sustainability. In terms of economic sustainability; it is a transportation vehicle that strengthens the economy and is carried out in a fair and efficient manner. In other words, it refers to keeping emissions and waste production within acceptable limits for the planet in sector practices (Rawson & Hooper, 2012).

The aviation sector is at the center of human life and safety issues as it provides services to international passengers and employs a wide and diverse group of employees, in addition to environmental issues. Therefore, in addition to economic sustainability, environmental sustainability and social sustainability are very important for the industry. Sustainability is also addressed in corporate governance studies in businesses. It is seen that it is effective in the design of process and product design, value chains, "business ecosystems" and ultimately markets in businesses (Elkington, 2006). In this context, it can be stated that the publication of sustainability reports in aviation businesses will be effective in the sustainability success of businesses. The survival of businesses in the aviation sector, which plays an important role in economic, social and environmental terms, depends on 'sustainability' activities (Kavacık et al., 2012).

Sustainability reporting means informing the sector stakeholders about the economic, environmental and social improvements of the company, effectiveness and efficiency activities integrated with the strategic elements of the company with qualitative and quantitative information (Koç & Durmaz, 2015). Sustainability reporting is seen as a direct tool for organizations to communicate their corporate profiles, values and performances and is therefore mostly known as a marketing tool (Karagiannis et al., 2019). In addition, it can help reduce the sustainability problems of the aviation sector. In this direction, what aviation authorities and companies should do can be listed as follows (Karaman, Kılıç & Uyar, 2018);

- Boards of directors should encourage companies to integrate sustainability issues into company operations by committing the necessary financial and human resources to participate in sustainability efforts.
- Boards of directors should establish a separate sustainability committee or department that can focus on sustainability issues and reporting practices.
- Regulatory bodies can encourage aviation companies to act in a socially and environmentally responsible manner by proposing legal requirements and providing guidance.

A number of studies have examined independent reports published by businesses in the aviation industry to determine sustainability reporting practices and disclosures. Grosbois (2016) noted that the international nature of the tourism industry, including aviation, has particularly significant impacts on developing and less developed economies, making them vulnerable due to weaker legal and regulatory regimes. Mak and Chan (2006) investigated the environmental reporting practices of companies in the Asia Pacific airline industry. They documented that airlines in more developed countries are more environmentally conscious and devote a large amount of resources to environmental policies. Skouloudis et al. (2012) conducted a qualitative study analyzing the sustainability reports of 13 airports in North America and Europe. They found that sustainability reporting is not yet a widespread practice among international airports and that there is significant variation in reporting practices. Another issue regarding sustainability reporting in aviation is how reporting is evaluated and conducted. The diversity of sustainability performance indicators and measurement differences are the most important reasons for this (Zieba & Johansson, 2022). Measuring sustainability is possible by defining indicators and determining basic subsystems to provide accurate and basic information about the validity of the entire system and each of its components (Bossel, 1999).

In terms of sustainability in aviation, it is possible to state that the importance of implementing sustainable projects at airports has reached a very large dimension (Demirci, Doğan, & Eroğlu, 2023). In particular, the implementation of green airport projects has been observed to increase the performance of businesses in this direction. The green airport system aims to promote environmental best practices to minimize the impact of aviation on the environment and to certify airport members who have outstanding success in environmental projects (Demirkaya & Kuşhan, 2023). In addition to environmental factors, air transportation also has significant economic and social contributions. In order to increase sustainability performance at airports, it is recommended to increase outputs determined as total revenue, number of passengers and number of employees in addition to reducing inputs determined as emissions, water consumption, energy consumption and waste volume (Bulak et al., 2021). Santa et al. (2020) have revealed that there are ten indicators for a sustainable airport in their research to determine the characteristics of a sustainable airport. These indicators, which they also state support the green airport model, are as follows; (1) Noise Reduction (2) Emission Reduction and Air Quality (3) Energy Management (4) Water Management (5) Waste Management (6) Biodiversity protection and land use (7) Cost and economy (8) Quality of the indoor environment (9) Transportation and vehicle control (10) Social and Cultural Dimensions Indicator. Sustainable projects can be applied at every stage of airport operations and sustainability studies can add value to businesses in economic, operational, environmental and social areas.

As the aviation sector grows, the need for sustainability measures will increase in order for aviation to remain a sustainable mode of transportation. Efforts to integrate sustainability into aviation will accelerate (Gemmer, 2019). It is believed that in order to achieve sustainability goals in aviation, new aviation technologies must be developed and air transportation operations must be continuously optimized (Karam et al., 2022). The implementation of these processes is very important for aviation manufacturers and air transportation operators. However, sustainability in aviation is a global problem that concerns the entire world. Therefore, it requires the cooperation of countries, governments, international and national civil aviation enterprises.

3. METHOD

In this study, case study was preferred among qualitative research methods. A case study, which is described as a type of phenomenon occurring in a limited context (Miles & Huberman, 1994), can be expressed as creating a detailed research within the contexts of the phenomena, usually with data collected in a certain period of time, and providing an analysis of the context and processes that illuminate the theoretical issues studied (Hartley, 2004). In this study, where the importance of qualitative research methods in sustainability studies in aviation is conveyed, qualitative research methods are used. Qualitative research methods are required to scan the existing literature, to understand the sustainability phenomenon in aviation in depth, to follow the sustainability processes in the sector, and to achieve the determined goals and success.

As one of the important sectors contributing to global sustainability goals, the aviation industry is addressed within the scope of this study. Using the purposeful sampling model, the aviation sector is examined in depth in terms of three dimensions of sustainability. The aim of the purposeful sampling model is to collect in-depth information about the situation that constitutes the subject of the research (Maxwell, 1996). Therefore, sustainability in aviation is addressed in the first part of the study; the concept of sustainability in aviation is conveyed in terms of the three dimensions of sustainability as social, economic and environmental. Finally, the studies conducted in the literature on the aviation industry in the last decade in the context of economic, social and environmental sustainability are listed. In the studies conducted on sustainability studies in aviation, the subject of the study and the method used in the study are conveyed. Which of the preferred qualitative research methods is applied in sustainability studies in aviation (case study, multi-criteria decision-making technique, etc.) and the reasons for preferring these methods are examined. From this point of view, the importance of qualitative research methods in sustainability studies in aviation is revealed.

4. QUALITATIVE RESEARCH IN SUSTAINABLE AVIATION

Although sustainability studies in aviation have gained momentum in recent years, it is a relatively under-researched area. In general, selective methodologies and techniques have been used to ensure sustainable progress in the sector and to evaluate problems (Singh et al., 2023). When the literature is examined, it is seen that the most commonly used method in the studies is the qualitative research method.

Qualitative research is conducted on a small number of samples that allow for rich and deepening data situations (Trotter, 2012). This method is useful for determining and analyzing the relative perspectives of people with different values and interests on the subject (Voyer et al., 2015). With the more comprehensive, connected information that social science can produce, it is possible to learn more about the possible outcomes of policies before they are implemented (Levontin et al., 2011). In other words, qualitative methods can be used to determine what the research topic means to certain communities (Mertens and Hesse-Biber, 2013). The aviation sector is one of the sectors that is associated with numerous cultural, organizational and technical relationships that are considered to be human-made. It is logical that qualitative research methods are needed for aviation research where people are at the forefront. It has become a necessity to consider qualitative research as a valid component of aviation research methodology, which is effective in adding context and depth to the results of some studies (Ferroff et al., 2012). Unlike quantitative researchers, qualitative researchers do not have to deal with statistical formulas and generalizations, but rather with uncertain situations, complex connections and a constantly deepening data set (Britten, 1995). For this reason, it is more appropriate to use qualitative research methods in sustainability studies in aviation.

Korul (2001) conducted what can be considered as the first airport sustainability research in Turkey by using document analysis, correspondence, interviews and surveys. He aimed to examine the environmental impacts of operations carried out at airports open to international traffic (Adana Şakir Paşa, Ankara Esenboğa, Antalya, Bodrum, Dalaman, İstanbul Atatürk, İzmir Adnan Menderes, Trabzon, Milas, Isparta Süleyman Demirel, Nevşehir Kapadokya, İstanbul Sabiha Gökçen) and Ercan Airport in TRNC and investigated what kind of precautions were taken. Torum and Küçük Yılmaz (2009) used interview and survey methods in their study, which aimed to determine the importance given to sustainability by airports in Turkey and to determine social-environmental-economic sustainability practices. Oto (2011) used document analysis and face-to-face interviews with industry experts in his research, which investigated sustainability practices accepted by international aviation authorities, determined sustainable airport planning, construction and operation rules, and evaluated all of these in the example of Esenboğa Airport.

It can be stated that in qualitative research methods, it is possible to evaluate complex situations from a relative perspective, which helps to understand the sustainability issue, which is important for the aviation industry, more clearly. For example, it can allow airport operators or airline companies to better evaluate the difficulties encountered in the sector (environmental and operational disruptions) or external factors affecting the sector (political issues, international legislation or pandemic, terrorist acts, etc.) in the sustainability goals they set. The use of qualitative research methods and the synthesis of different perspectives are also important for the participation of employees in sustainability processes. From this point of view, the importance of using qualitative research methods in sustainability studies in aviation can be expressed as the guidance, monitoring, and determination of goals and objectives by international and national aviation organizations of sustainability practices in the aviation sector. In addition, it can be said that the examination of the sustainability activity reports published by aviation companies is more possible with qualitative research methods. In order to reveal the importance of qualitative research methods in the aviation sector, sustainability studies in aviation that have been included in the literature in the last ten years are examined. As shown in Table 1, the majority of sustainability studies in aviation use qualitative research methods.

Table 1. Sustainability Studies in Aviation and Research Methods Used

Article	Research Method	Resources
Social sustainability of biojet fuel for net zero aviation.	Qualitative Research Method – Document review	Sharno, M. A., & Hiloidhari, M. (2024).
Qualitative Impact Analysis of Sustainable Aviation Fuels on Aviation Sustainability	Qualitative Research Method - Content analysis	Kanmaz, U. (2024).
Defining Aviation "Skills" to Ensure Effective, Safe, and Efficient Evaluations: A Qualitative Study.	Qualitative Research Method – Case Study	Albelo Ph D, J. L., Cuevas, H. M., Aguiar, M., Piccone, C., Petitt, K., & Villagomez, R. (2022).
Who should hold the baton of aviation sustainability?	Qualitative Research Method – Multi-criteria decision making method	Singh, J., Rana, S., Abdul Hamid, A. B., & Gupta, P. (2023).
Sustainability assessment in aviation industry: A mini-review on the tools, models and methods of assessment.	Qualitative Research Method – Document review	Elhmoud, E. R., & Kutty, A. A. (2020).
Airport sustainability awareness: a theoretical	Qualitative Research Method – Document review	Eid, A., Salah, M., Barakat, M., & Obrecht, M. (2022).

continued on following page

Table 1. Continued

Article	Research Method	Resources
A cognitive mapping approach to analyse stakeholders' perspectives on sustainable aviation fuels.	Qualitative Research Method – Document review	Ahmad, S., & Xu, B. (2021).
Sustainability reporting in airlines industry: evidence from Turkey.	Qualitative Research Method – Document review	Selimoglu, S. K., & Caliskan, A. O. (2018).
Sustainability reporting, materiality, and accountability assessment in the airport industry.	Qualitative Research Method and Quantitative Research Method	Karagiannis, I., Vouros, P., Skouloudis, A., & Evangelinos, K. (2019).
A case for social constructionism in aviation safety and human performance research.	Quantitative Research Method and Qualitative Research Method - Observation based methodology	Ferroff, C. V., Mavin, T. J., Bates, P. R., & Murray, P. S. (2012).
Sustainability reporting in the airline industry: Current literature and future research avenues.	Qualitative Research Method – Systematic Literature Review	Zieba, M., & Johansson, E. (2022).
What impacts sustainability reporting in the global aviation industry? An institutional perspective.	Qualitative Research Method – Systematic Literature Review	Kılıç, M., Uyar, A., & Karaman, A. S. (2019).
A Green Airport model: Proposition based on social and environmental management systems.	Qualitative Research Method – Systematic Literature Review	Santa, S. L. B., Ribeiro, J. M. P., Mazon, G., Schneider, J., Barcelos, R. L., & de Andrade, J. B. S. O. (2020).
Strategies towards a more sustainable aviation: A systematic review.	Qualitative Research Method – Systematic Literature Review	Afonso, F., Sohst, M., Diogo, C. M., Rodrigues, S. S., Ferreira, A., Ribeiro, I., Marques, R., Rego, F., Sohouli, A., Portugal, J., Policarpo, H., Soares, B., Ferreira, B., Fernandes, E.C., Lau, F. & Suleman, A. (2023).
Expanding horizons: a review of sustainability evaluation methodologies in the airport sector and beyond.	Qualitative Research Method – Systematic Literature Review	Jia, X., Macário, R., & Buyle, S. (2023).
Sustainability Analysis in the Aviation Sector; Turkiye.	Qualitative Research Method – Systematic Literature Review	Barutçu, S. B., & Çolakoğlu, Ü. (2024).
Identification of sustainability indicators in the civil aviation sector in Dubai: a stakeholders' perspective.	Qualitative Research Method - Multiple focus groups	Al Sarrah, M., Ajmal, M. M., & Mertzanis, C. (2021).
Airport corporate sustainability: An analysis of indicators reported in the sustainability practices.	Qualitative Research Method – Document review	Koç, S., & Durmaz, V. (2015).

continued on following page

Table 1. Continued

Article	Research Method	Resources
A research on Turkish Airlines sustainability reports.	Qualitative Research Method – Case Study	Karakuş, G. (2023).
Social embeddedness for sustainability in the aviation sector.	Qualitative Research Method – Case Study	Payán-Sánchez, B., Plaza-Úbeda, J. A., Pérez-Valls, M., & Carmona-Moreno, E. (2018).
Examining environmental sustainability practices on energy and climate change at airports.	Qualitative Research Method – Case Study	Rodoplu, H. & Erdoğan, T. (2022).
Firms' capabilities for sustainable innovation: The case of biofuel for aviation.	Qualitative Research Method – Case Study	Mousavi, S., & Bossink, B. A. (2017).
Sustainable development in aviation logistics: Successful drivers and business strategies.	Qualitative Research Method – Case Study	Wu, P. J., & Yang, C. K. (2021).
Green aviation industry sustainable development towards an integrated support system.	Quantitative Research Method - Bibliometric Analysis	Qiu, R., Hou, S., Chen, X., & Meng, Z. (2021).
Bibliometric Analysis on the Place of Sustainability in Aviation in International Literature.	Quantitative Research Method - Bibliometric Analysis	Topçu, G. (2023).
Sustainability reporting in the aviation industry: worldwide evidence.	Quantitative Research Method	Karaman, A. S., Kilic, M., & Uyar, A. (2018).
Sustainability of supply or the planet: a review of potential drop-in alternative aviation fuels.	Qualitative Research Method – Case Study	Rye, L., Blakey, S., & Wilson, C. W. (2010).
Assessment of the impact of material selection on aviation sustainability, from a circular economy perspective.	Qualitative Research Method - Multi-criteria decision analysis	Markatos, D. N., & Pantelakis, S. G. (2022).
What factors predict a consumer's support of environmental sustainability in aviation? A multi-model analysis.	Qualitative Research Method - Multi-criteria decision analysis	Winter, S. R., Thropp, J. E., & Rice, S. (2019).
Developing policy integrating sustainability: A case study into aviation.	Qualitative Research Method – Case Study	McManners, P. J. (2016).
Circular Supply Chain Case Studies in Aviation Sustainability.	Qualitative Research Method – Case Study	Mahama, A. J., & Iyer, S. S. (2023).
Social embeddedness for sustainability in the aviation sector.	Qualitative Research Method – Document review	Payán-Sánchez, B., Plaza-Úbeda, J. A., Pérez-Valls, M., & Carmona-Moreno, E. (2018).

continued on following page

Table 1. Continued

Article	Research Method	Resources
Framework for Integration of Health Monitoring Systems in Life Cycle Management for Aviation Sustainability and Cost Efficiency.	Quantitative Research Method	Kabashkin, I., Perekrestov, V., Tyncherov, T., Shoshin, L., & Susanin, V. (2024).
Green commercial aviation supply chain—A European path to environmental sustainability.	Qualitative Research Method - Multiple focus groups	Modarress Fathi, B., Ansari, A., & Ansari, A. (2023).
Sustainability assessment and engineering of emerging aircraft technologies—Challenges, methods and tools.	Qualitative Research Method - Descriptive analysis	Pinheiro Melo, S., Barke, A., Cerdas, F., Thies, C., Mennenga, M., Spengler, T. S., & Herrmann, C. (2020).
Global tourism, climate change and energy sustainability: assessing carbon reduction mitigating measures from the aviation industry.	Quantitative Research Method - Bibliometric Analysis	Leal Filho, W., Ng, A. W., Sharifi, A., Janová, J., Özuyar, P. G., Hemani, C., Heyes, G., Njau, D., & Rampasso, I. (2023).
Comprehensive Strategic Analysis for Sustainability: An Aviation Industry Case Study.	Quantitative Research Method and Qualitative Research Method -Case study	Heyes, G., Urquhart, C., Hooper, P., & Thomas, C. (2023).
Development taxonomy of human risk factors for corporate sustainability in aviation sector.	Qualitative Research Method – Document review	Yazgan, E. (2018).
A review of sustainability in aviation: A multidimensional perspective.	Qualitative Research Method – Document review	Guimarans, D., Arias, P., Tomasella, M., & Wu, C. L. (2019).
Sensitivity analysis of a hybrid MCDM model for sustainability assessment—an example from the aviation industry.	Qualitative Research Method - Multi-criteria decision analysis	Markatos, D. N., Malefaki, S., & Pantelakis, S. G. (2023).

Sustainability efforts in aviation are a serious issue on a global scale. Despite the crises it has experienced in certain periods (September 11 attacks, Asian crisis, Covid-19 pandemic, etc.), the sector is developing rapidly. The latest estimates show that demand for air transportation will increase by an average of 4.3% per year over the next 20 years. By the mid-2030s, at least 200,000 flights are expected to take off and land from all over the world. In addition, this growth is not limited to passenger traffic; cargo traffic is also expected to continue to grow along a similar curve. Sector stakeholders also plan to support an almost doubling of passenger and cargo numbers by 2036 (ICAO, 2024).

Considering the environmental problems of the aviation sector, it is known that environmental problems will increase with the expected development of the sector. Although the development of technology and the implementation of sustainability studies aim to reduce the environmental, economic and social problems that the sector will create, it is not possible to completely eliminate these problems. In this context, it is possible to say that sustainability studies to be carried out by international/national organizations and sustainability reporting in the aviation industry concern the whole world. It has always been accepted that studies to be carried out in the industry should be supported by academic studies, and it is possible to state that the need for studies using qualitative research methods in studies to be carried out in this field continues.

It is a precaution to understand the qualitative research process well in order to use qualitative research methods in sustainability studies in aviation. The qualitative research process consists of the following stages, respectively (Creswell, 2022);

- Realizing the problem
- Analyzing the problem and determining its details
- Choosing the approach to problem solving
- Designing the study
- Data collection procedures
- Classifying and analyzing data
- Interpreting data
- Reporting data
- Successful completion of the qualitative research process

In the literature, the sustainability studies examined in the aviation industry mostly use qualitative research methods, as shown in Table 1. The most frequently used qualitative research methods include situation analysis. Situation analysis is a methodological approach that involves in-depth examination of a limited system using multiple data collection methods to collect systematic information about how that system functions and works (Chmiliar, 2010). In research, it focuses on the questions of how and why, it is a method in which the researcher has no control over events and the event or phenomenon is studied within its own natural life framework (Subaşı and Okumuş, 2017). Another method used is document analysis. Analysis of written documents and documents is an information collection method used in qualitative research both on its own and to support information obtained through interviews and observations. The information collected in qualitative research can be in the form of observation notes, interview records, documents, pictures and other graphic presentations (such as drawings, tables). Coding, analyzing and interpreting the various types of information collected requires a systematic approach (Coffey and Atkinson, 1996; Wolcott, 1994).

Systematic literature review is another of these studies and is the research conducted by Afonso et al., (2023). They categorized the studies as shown in figure 2 with the aim of transforming aviation into a more sustainable sector.

Figure 2. Solution categories of sustainable aviation

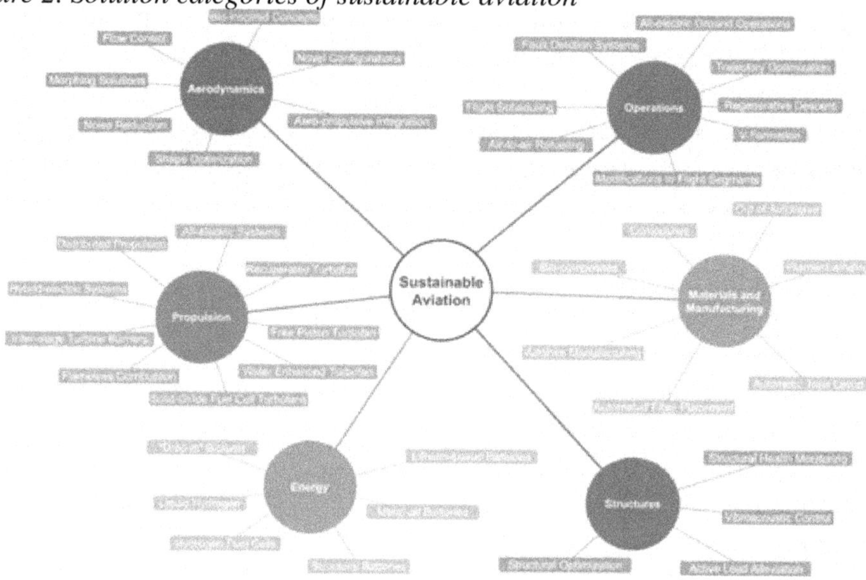

Kaynak: Afonso, F., Sohst, M., Diogo, C. M., Rodrigues, S. S., Ferreira, A., Ribeiro, I., Marques, R., Rego, F., Sohouli, A., Portugal, J., Policarpo, H., Soares, B., Ferreira, B., Fernandes, E.C., Lau, F. & Suleman, A. (2023). Strategies towards a more sustainable aviation: A systematic review. *Progress in Aerospace Sciences, 137*, 100878.

This study provides a comprehensive overview of the broad strategies for more sustainable aviation; flight operations, energy storage, propulsion systems, aerodynamics, structures, materials and manufacturing processes (Afonso et al., 2023).

Other methods used in sustainability studies in aviation are multi-criteria decision-making and focus group interviews. Multi-criteria decision-making is a research consisting of algorithms developed for people to make the most beneficial decisions (Dalbudak and Rençber, 2022). It can be defined as problems where the best alternative is selected from possible solution sets where more than one criteria is optimized (Turan, 2018).

Focus group interview, one of the qualitative research methods, is the use of the effect of group dynamics in an unstructured interview and discussion between a small group and a leader, to obtain in-depth information and produce ideas (Çokluk, Yılmaz, & Oğuz, 2011). The distinguishing point in focus groups is the presence of group interaction in the answers given to the researcher's questions. Interaction in focus groups provides a clear idea of how others think and talk. Interaction in focus groups is very important in terms of obtaining much more information than other methods or techniques (Işık & Semerci, 2019).

Other studies on sustainability in aviation using qualitative research methods are aimed at examining the sustainability reports of companies in the industry. It is important to evaluate the reports, their importance, accountability, what alternative research methods can be, and what responsibilities are given to companies, stakeholders, and employees in the sector, using qualitative research methods. While qualitative research methods help to examine and analyze the policies of international/national organizations and countries in aviation sustainability studies, it also helps to understand the decisions of companies in the industry, the perceptions of managers when analyzing external factors, and even the perceptions of how they direct the sustainability studies carried out in operations, employees, and passengers. Content analysis, which is used as a qualitative research method, is a method that can be used for a wide variety of data sources, including textual data, visual data and audio data (Stemler, 2015: 1). Sustainability studies in aviation, as a newly developing topic in the literature, can be applied with the use of content analysis, word cloud, code cloud and code distribution model analysis methods.

Quantitative research methodology has been and continues to be the preferred research methodology in aviation research. However, close examination of data collection techniques can raise questions about the actual approach taken, and the results are often a reduction of a qualitative process to a purely numerical value. This process often results in a loss of detail. Therefore, qualitative research is encouraged in research on sustainability in aviation (Ferroff et al., 2012). Qualitative research methods allow for an in-depth focus on the meanings that research participants attribute to a particular phenomenon. It can also be stated that it is best to create detailed and in-depth data collection when there is more than one source of information (Creswell and Poth, 2018).

CONCLUSION

Sustainability is an important issue in ensuring the continuity of humanity as a subject that concerns all of humanity and the future. Sustainability, which is shown as one of the solutions to global warming and climate change, constitutes an important agenda item for industries and sectors that are heavily dependent on the use of fossil fuels. Sustainability can be effective in eliminating environmental problems, especially emissions, originating from the aviation industry. However, in the aviation industry, which is quite expensive, focusing only on environmental problems in the context of sustainability is not the right approach. Sustainability studies to be carried out in economic and social, and even cultural and technological terms should be encouraged, implemented and followed. All studies to be carried out in the literature on sustainability in aviation should aim to close the gap in this area. However, it is very important to conduct studies using qualitative research methods in particular. Considering the usefulness of qualitative research methods in addressing social, economic and political problems, which has been proven in the literature, it can be stated that studies in this direction should be increased.

The success of sustainability is a complex issue that affects communities, industries, and many businesses, and requires the sector or business to cooperate with all stakeholders. To put it in a broader perspective, studies must be continued for every correction, change, or strategy to be developed for sustainability work in the aviation sector. These changes can be technological, operational, regulatory, or market-based or behavioral changes. In other words, new strategies can be developed in all studies to be conducted with stakeholders. In this study, it is suggested that qualitative research methods be used in academic research in the development of strategies for sustainability in aviation.

In recent years, in the context of global warming, the aviation sector has become one of the most visible sectors with all its stakeholders, considering the resources it uses and its impacts as well as the source of climate change. Therefore, sustainability in aviation has become a prominent issue in academia and the business world. Qualitative research methods in social sciences, which act as a bridge between academia and the business world in order to facilitate collaborations towards sustainable aviation, to share ideas and good practices, and to develop joint projects, can play a key role in achieving success.

Qualitative research methods are an approach that prioritizes the investigation and understanding of social phenomena within the environment they are connected to, with an understanding based on theory formation. It presents a modeling study that explains some previously unknown results in relation to each other based on the information collected. This situation requires the researcher to be flexible, to reshape the research process according to the information collected, and to follow an inductive approach both in the formation of the research design and in the analysis of the information collected. Therefore, this study reveals the importance of qualitative research methods in sustainability studies in aviation, sheds light on the deficiencies in this direction in the literature, and offers a different perspective for the qualitative studies to be conducted. In addition, with the acceleration of sustainability studies in aviation, the sustainability reports to be published by the companies in the industry require detailed examination. It is seen that different qualitative research methods can be used especially in the examination of sustainability reports in the aviation industry and that this research method can be a pioneer for detailed studies.

REFERENCES

Afonso, F., Sohst, M., Diogo, C. M., Rodrigues, S. S., Ferreira, A., Ribeiro, I., Marques, R., Rego, F., Sohouli, A., Portugal, J., Policarpo, H., Soares, B., Ferreira, B., Fernandes, E. C., Lau, F., & Suleman, A. (2023). Strategies towards a more sustainable aviation: A systematic review. *Progress in Aerospace Sciences*, 137, 100878. DOI: 10.1016/j.paerosci.2022.100878

Ağaçsapan, B., & Çabuk, A. (2019). Sustainable aviation: GIS for airport. *GSI Journals Serie B: Advancements in Business and Economics, 2*(1).

Ahmad, S., & Xu, B. (2021). A cognitive mapping approach to analyse stakeholders' perspectives on sustainable aviation fuels. *Transportation Research Part D, Transport and Environment*, 100, 103076. DOI: 10.1016/j.trd.2021.103076

Albelo Ph, D, J. L., Cuevas, H. M., Aguiar, M., Piccone, C., Petitt, K., & Villagomez, R. (2022). Defining Aviation "Skills" to Ensure Effective, Safe, and Efficient Evaluations: A Qualitative Study. *International Journal of Aviation Research*, 14(1), 1.

Almashhadani, M., & Almashhadani, H. A. (2023). The impact of sustainability reports to enhance company performance and its sustainability. *International Journal of Scientific and Management Research*, 149–163.

Alpman, E., & Göğüş, A. Y. (2017). Sustainable development indicators in aviation. *Journal of Sustainable Aviation Research*, 2(1), 1–11.

Ansell, P. J. (2023). Review of sustainable energy carriers for aviation: Benefits, challenges, and future viability. *Progress in Aerospace Sciences*, 141, 100919. DOI: 10.1016/j.paerosci.2023.100919

Aras, G., & Crowther, D. (2009). Making sustainable development sustainable. *Management Decision*, 47(6), 975–988. DOI: 10.1108/00251740910966686

Barclay, K., Voyer, M., Mazur, N., Payne, A. M., Mauli, S., Kinch, J., Fabinyi, M., & Smith, G. (2017). The importance of qualitative social research for effective fisheries management. *Fisheries Research*, 186, 426–438. DOI: 10.1016/j.fishres.2016.08.007

Barutçu, S. B., & Çolakoğlu, Ü. (2024). Sustainability Analysis in the Aviation Sector; Turkiye. *Journal of Travel and Tourism Research*, 24(24), 106–124.

Basiago, A. D. (1999). Economic, social, and environmental sustainability in development theory and urban planning practice. *The Environmentalist*, 19(2), 145–161. DOI: 10.1023/A:1006697118620

Baumgartner, R. J., & Rauter, R. (2017). Strategic perspectives of corporate sustainability management to develop a sustainable organization. *Journal of Cleaner Production*, 140, 81–92. DOI: 10.1016/j.jclepro.2016.04.146

Baumgärtner, S., & Quaas, M. (2010). What is sustainability economics? *Ecological Economics*, 69(3), 445–450. DOI: 10.1016/j.ecolecon.2009.11.019

Bexell, M., & Jönsson, K. (2017, January). Responsibility and the United Nations' sustainable development goals. [). Routledge.]. *Forum for Development Studies*, 44(1), 13–29. DOI: 10.1080/08039410.2016.1252424

Bossell, H. (1999). *Indicators for Sustainable Development: Theory, Method, Applications: A Report to the Balaton Group, International Institute for Sustainable Development*. IISD.

Bozlağan, R. (2005). Historical background of the idea of sustainable development. In *Journal of Social Policy Conferences* (No. 50, pp. 1011-1028). Istanbul University.

Britten, N. (1995). Qualitative Research: Qualitative Interviews in Medical Research. *BMJ (Clinical Research Ed.)*, 311(6999), 251–253. DOI: 10.1136/bmj.311.6999.251 PMID: 7627048

Bulak, M. E., Zengin, F. H., & Çiftçi, F. S. (2021). Global ölçekte faaliyet gösteren havalimanlarının göreceli sürdürülebilirlik analizi. *Pamukkale Üniversitesi Mühendislik Bilimleri Dergisi*, 27(4), 504–512.

Çalışkan, A. Ö. (2012). Sustainability Reporting. *Journal of Accounting and Taxation Studies*, 5(1), 41–68.

Chmiliar, l. (2010). Multiple-case designs. In A. J. Mills, G. Eurepas & E. Wiebe (Eds.), Encyclopedia of case study research (pp 582-583). USA: SAGE Publications.

Coffey, A., & Atkinson, P. (1996). *Making sense of qualitative data: Complementary research strategies*. Sage.

Çokluk, Ö., Yılmaz, K., & Oğuz, E. (2011). Nitel bir görüşme yöntemi: Odak grup görüşmesi. *Kuramsal Eğitimbilim*, 4(1), 95–107.

Colglazier, W. (2015). Sustainable development agenda: 2030. *Science*, 349(6252), 1048–1050. DOI: 10.1126/science.aad2333 PMID: 26339011

Creswell, J. W. (2002). *Educational research: Planning, conducting, and evaluating quantitative*. Prentice Hall Upper Saddle River.

Creswell, J. W., & Poth, C. N. (2018). *Qualitative inquiry & research design: Choosing among five approaches* (4th ed.). Sage.

Dalbudak, E., & Rençber, Ö. F. (2022). Çok kriterli karar verme yöntemleri üzerine literatür incelemesi. *Gaziantep Üniversitesi İktisadi ve İdari Bilimler Fakültesi Dergisi*, 4(1), 1–17. DOI: 10.55769/gauniibf.1068692

De Grosbois, D. (2016). Corporate social responsibility reporting in the cruise tourism industry: A performance evaluation using a new institutional theory based model. *Journal of Sustainable Tourism*, 24(2), 245–269. DOI: 10.1080/09669582.2015.1076827

Demir, Ö. ve Sezgin, E. E. (. (2014). Institutionalization and sustainability in family businesses: A research conducted in TRB1 region. *Electronic Turkish Studies*, 9(5), 707–725. DOI: 10.7827/TurkishStudies.6904

Demirci, S., Doğan, T. G., & Eroğlu, E. Havalimanlarında sürdürülebilirlik: Singapur Changi Havalimanı. *Düzce Üniversitesi Süs ve Tıbbi Bitkiler Botanik Bahçesi Dergisi, 2*(1), 14-26.

Demirkaya, G., & Kuşhan, M. (2023). Green airports. *Makina, 31*(178).

Diedrich, A., Upham, P., Levidow, L., & van den Hove, S. (2011). Framing environmental sustainability challenges for research and innovation in European policy agendas. *Environmental Science & Policy*, 14(8), 935–939. DOI: 10.1016/j.envsci.2011.07.012

Eid, A., Salah, M., Barakat, M., & Obrecht, M. (2022). Airport sustainability awareness: A theoretical framework. *Sustainability (Basel)*, 14(19), 11921. DOI: 10.3390/su141911921

Elhmoud, E. R., & Kutty, A. A. (2020). Sustainability assessment in aviation industry: A mini-review on the tools, models and methods of assessment.

Elkington, J. (2006). Governance for sustainability. *Corporate Governance*, 14(6), 522–529. DOI: 10.1111/j.1467-8683.2006.00527.x

Ertan, Y. (2018). Sustainability Reporting in Turkey (2005-2017). *Journal of Accounting and Taxation Studies*, 11(3), 463–478.

Ferroff, C. V., Mavin, T. J., Bates, P. R., & Murray, P. S. (2012). A case for social constructionism in aviation safety and human performance research. *Aeronautica*, 2(1), 1–12.

Frangapoulos, C., (2006), Exergy, energy system analysis, and optimization, in Encyklopedia of Life Supporting System.

García, J. L., & Sanz, J. M. (2018). Climate Change, Ethics and Sustainability: An innovative Approach. *Journal of Innovation & Knowledge*, 70-75.

Gemmer, E. (2019). Aviation sustainability. Honors Projects. 425. https://scholarworks.bgsu.edu/honorsprojects/425

Guimarans, D., Arias, P., Tomasella, M., & Wu, C. L. (2019). A review of sustainability in aviation: A multidimensional perspective. *Sustainable transportation and smart logistics*, 91-121.

Hartley, J. (2004). Case study research. In Cassell, C., & Symon, G. (Eds.), *Essential guide to qualitative methods in organizational research* (pp. 323–333). Sage. DOI: 10.4135/9781446280119.n26

Heyes, G., Urquhart, C., Hooper, P., & Thomas, C. (2023). Comprehensive Strategic Analysis for Sustainability: An Aviation Industry Case Study. *Sustainability (Basel)*, 15(11), 8806. DOI: 10.3390/su15118806

ICAO. (2024). Future of aviation. https://www.icao.int/Meetings/FutureOfAviation/Pages/default.aspx

Idowu, S. O., & Louche, C. (Eds.). (2011). *Theory and practice of corporate social responsibility* (p. 284). Springer.

International Civil Aviation Organization. (2011). Aviation & Sustainability. Determining the complex environmental, economic and social impacts that are defining aviation's future. *ICAO Journal*, 66(6), 1–36.

Işık, E., & Semerci, Ç. (2019). Nitel araştırmalarda veri üçgenlemesi olarak odak grup görüşmesi, bireysel görüşme ve gözlem. *Turkish Journal of Educational Studies*, 6(3), 53–66. DOI: 10.33907/turkjes.607997

Jeremić, V., & Sachs, J. D. (2014). The United Nations in the age of sustainable development. *The Economic and social review, 45*(2, Summer), 161-188.

Jia, X., Macário, R., & Buyle, S. (2023). Expanding horizons: A review of sustainability evaluation methodologies in the airport sector and beyond. *Sustainability (Basel)*, 15(15), 11584. DOI: 10.3390/su151511584

Kabashkin, I., Perekrestov, V., Tyncherov, T., Shoshin, L., & Susanin, V. (2024). Framework for Integration of Health Monitoring Systems in Life Cycle Management for Aviation Sustainability and Cost Efficiency. *Sustainability (Basel)*, 16(14), 6154. DOI: 10.3390/su16146154

Kanmaz, U. (2024). Qualitative Impact Analysis of Sustainable Aviation Fuels on Aviation Sustainability. [JSHSR]. *International Journal of Social and Humanities Sciences Research*, 11(107), 920–934.

Karagiannis, I., Vouros, P., Skouloudis, A., & Evangelinos, K. (2019). Sustainability reporting, materiality, and accountability assessment in the airport industry. *Business Strategy and the Environment*, 28(7), 1370–1405. DOI: 10.1002/bse.2321

Karakuş, G. (2023). A study on Turkish Airlines sustainability reports. Aerospace Research Letters (ASREL). *Journal*, 2(2), 86–113. DOI: 10.56753/ASREL.2023.2.4

Karam, H., Anwama, E., Davidson, I. E. A., Alfazari, H., Krykhtine, F., & Mora-Camino, F. (2022, September). A Methodology for Evaluating Aviation Sustainability Perspectives. In *International Seminar on Aeronautics and Energy* (pp. 277-289). Singapore: Springer Nature Singapore.

Karaman, A. S., Kilic, M., & Uyar, A. (2018). Sustainability reporting in the aviation industry: Worldwide evidence. *Sustainability Accounting. Management and Policy Journal*, 9(4), 362–391.

Kavacık, M., Zafer, S., Yıldız, A., & Karaman, D. (2012). Sustainable development in aviation industry and the case of Turkish airlines. *Journal of Animal and Veterinary Advances*, 9(3), 547–550.

Kılıç, M., Uyar, A., & Karaman, A. S. (2019). What impacts sustainability reporting in the global aviation industry? An institutional perspective. *Transport Policy*, 79, 54–65. DOI: 10.1016/j.tranpol.2019.04.017

Koç, S., & Durmaz, V. (2015). Airport corporate sustainability: An analysis of indicators reported in the sustainability practices. *Procedia: Social and Behavioral Sciences*, 181, 158–170. DOI: 10.1016/j.sbspro.2015.04.877

Korul, V. (2001). Management of Airports' Relationships with the Environment and Analysis of Environmental Pollution Practices at Airports Open to International Traffic in Turkey (Master's thesis, Anadolu University), Turkey.

Leal Filho, W., Ng, A. W., Sharifi, A., Janová, J., Özuyar, P. G., Hemani, C., Heyes, G., Njau, D., & Rampasso, I. (2023). Global tourism, climate change and energy sustainability: Assessing carbon reduction mitigating measures from the aviation industry. *Sustainability Science*, 18(2), 983–996. DOI: 10.1007/s11625-022-01207-x PMID: 36105893

Lee, D. S., Fahey, D. W., Skowron, A., Allen, M. R., Burkhardt, U., Chen, Q., Doherty, S. J., Freeman, S., Forster, P. M., Fuglestvedt, J., Gettelman, A., De León, R. R., Lim, L. L., Lund, M. T., Millar, R. J., Owen, B., Penner, J. E., Pitari, G., Prather, M. J., & Wilcox, L. J. (2021). The contribution of global aviation to anthropogenic climate forcing for 2000 to 2018. *Atmospheric Environment*, 244, 117834. DOI: 10.1016/j.atmosenv.2020.117834 PMID: 32895604

Levontin, P., Kulmala, S., Haapasaari, P., & Kuikka, S. (2011). Integration of biological, economic, and sociological knowledge by Bayesian belief networks: The interdisciplinary evaluation of potential management plans for Baltic salmon. *ICES Journal of Marine Science*, 68(3), 632–638. DOI: 10.1093/icesjms/fsr004

Litman, T. (2011). Sustainability and livability: Summary of definitions, goals, objectives and performance indicators. https://www.vtpi.org/sus_liv.pdf

Machnik-Słomka, J. (2017). Entrepreneurship of aviation industry in the perspective of sustainable development. *International Multidisciplinary Scientific GeoConference SGEM*, 17, 301–308.

Mahama, A. J., & Iyer, S. S. (2023). Circular Supply Chain Case Studies in Aviation Sustainability. In *Handbook of Research on Designing Sustainable Supply Chains to Achieve a Circular Economy* (pp. 351–376). IGI Global. DOI: 10.4018/978-1-6684-7664-2.ch017

Mak, B., & Chan, W. W. (2006). Environmental reporting of airlines in the Asia Pacific region. *Journal of Sustainable Tourism*, 14(6), 618–628. DOI: 10.2167/jost586.0

Man, M., & Bogeanu-Popa, M. M. (2020). Impact of non-financial information on sustainable reporting of organisations' performance: Case study on the companies listed on the bucharest stock exchange. *Sustainability (Basel)*, 12(6), 2179. DOI: 10.3390/su12062179

Markatos, D. N., Malefaki, S., & Pantelakis, S. G. (2023). Sensitivity analysis of a hybrid MCDM model for sustainability assessment—An example from the aviation industry. *Aerospace (Basel, Switzerland)*, 10(4), 385. DOI: 10.3390/aerospace10040385

Markatos, D. N., & Pantelakis, S. G. (2022). Assessment of the impact of material selection on aviation sustainability, from a circular economy perspective. *Aerospace (Basel, Switzerland)*, 9(2), 52. DOI: 10.3390/aerospace9020052

McManners, P. J. (2016). Developing policy integrating sustainability: A case study into aviation. *Environmental Science & Policy*, 57, 86–92. DOI: 10.1016/j.envsci.2015.11.016

Mertens, D. M., & Hesse-Biber, S. (2013). Mixed methods and credibility of evidence in evaluation. *New Directions for Evaluation*, 2013(138), 5–13. DOI: 10.1002/ev.20053

Miles, M. B., & Huberman, A. M. (1994). *Qualitative data analysis: An expanded source book* (2nd ed.). Sage.

Modarress Fathi, B., Ansari, A., & Ansari, A. (2023). Green commercial aviation supply chain—A European path to environmental sustainability. *Sustainability (Basel)*, 15(8), 6574. DOI: 10.3390/su15086574

Mousavi, S., & Bossink, B. A. (2017). Firms' capabilities for sustainable innovation: The case of biofuel for aviation. *Journal of Cleaner Production*, 167, 1263–1275. DOI: 10.1016/j.jclepro.2017.07.146

Mrazova, M. (2014). Sustainable development-the key for green aviation. *INCAS bulletin, 6*(1), 109.

Oto, N. (2011). Environmental sustainability and airports: The example of Esenboğa Airport. (Master's thesis, Ankara University), Turkey.

Payán-Sánchez, B., Plaza-Úbeda, J. A., Pérez-Valls, M., & Carmona-Moreno, E. (2018). Social embeddedness for sustainability in the aviation sector. *Corporate Social Responsibility and Environmental Management*, 25(4), 537–553. DOI: 10.1002/csr.1477

Payán-Sánchez, B., Plaza-Úbeda, J. A., Pérez-Valls, M., & Carmona-Moreno, E. (2018). Social embeddedness for sustainability in the aviation sector. *Corporate Social Responsibility and Environmental Management*, 25(4), 537–553. DOI: 10.1002/csr.1477

Pfister, T., Schweighofer, M., & Reichel, A. (2016). *Sustainability*. Routledge. DOI: 10.4324/9781315882888

Pinheiro Melo, S., Barke, A., Cerdas, F., Thies, C., Mennenga, M., Spengler, T. S., & Herrmann, C. (2020). Sustainability assessment and engineering of emerging aircraft technologies—Challenges, methods and tools. *Sustainability (Basel)*, 12(14), 5663. DOI: 10.3390/su12145663

Qiu, R., Hou, S., Chen, X., & Meng, Z. (2021). Green aviation industry sustainable development towards an integrated support system. *Business Strategy and the Environment*, 30(5), 2441–2452. DOI: 10.1002/bse.2756

Rawson, R., & Hooper, P. D. (2012). The importance of stakeholder participation to sustainable airport master planning in the UK. *Environmental Development*, 2, 36–47. DOI: 10.1016/j.envdev.2012.03.013

Rodoplu, H., & Erdoğan, T. (2022). Examining environmental sustainability practices on energy and climate change at airports. Detay Publishing, Editor: Kasım KİRACI, Number of editions: 1, Number of pages: 216, ISBN: 978-605-254-595-9, Section Pages: 41-55.4.

Rupcic, L., Pierrat, E., Saavedra-Rubio, K., Thonemann, N., Ogugua, C., & Laurent, A. (2023). Environmental impacts in the civil aviation sector: Current state and guidance. *Transportation Research Part D, Transport and Environment*, 119, 103717. DOI: 10.1016/j.trd.2023.103717

Rye, L., Blakey, S., & Wilson, C. W. (2010). Sustainability of supply or the planet: A review of potential drop-in alternative aviation fuels. *Energy & Environmental Science*, 3(1), 17–27. DOI: 10.1039/B918197K

Santa, S. L. B., Ribeiro, J. M. P., Mazon, G., Schneider, J., Barcelos, R. L., & de Andrade, J. B. S. O. (2020). A Green Airport model: Proposition based on social and environmental management systems. *Sustainable Cities and Society*, 59, 102160. DOI: 10.1016/j.scs.2020.102160

Sarrah, M., Ajmal, M. M., & Mertzanis, C. (2021). Identification of sustainability indicators in the civil aviation sector in Dubai: A stakeholders' perspective. *Social Responsibility Journal*, 17(5), 648–668. DOI: 10.1108/SRJ-06-2019-0203

Scoones, I. (2007). Sustainability. *Development in Practice*, 17(4-5), 589–596. DOI: 10.1080/09614520701469609

Selimoglu, S. K., & Caliskan, A. O. (2018). Sustainability reporting in airlines industry: Evidence from Turkey. *International Journal of Sustainable Aviation*, 4(2), 114–132. DOI: 10.1504/IJSA.2018.094224

Sharno, M. A., & Hiloidhari, M. (2024). Social sustainability of biojet fuel for net zero aviation. *Energy for Sustainable Development*, 79, 101419. DOI: 10.1016/j.esd.2024.101419

Sikdar, S. K. (2004). Science of sustainability. *Clean Technologies and Environmental Policy*, 7(1), 1–2. DOI: 10.1007/s10098-004-0261-z

Singh, J., Rana, S., Abdul Hamid, A. B., & Gupta, P. (2023). Who should hold the baton of aviation sustainability? *Social Responsibility Journal*, 19(7), 1161–1177. DOI: 10.1108/SRJ-05-2021-0181

Skouloudis, A., Evangelinos, K., & Moraitis, S. (2012). Accountability and stakeholder engagement in the airport industry: An assessment of airports' CSR reports. *Journal of Air Transport Management*, 18(1), 16–20. DOI: 10.1016/j.jairtraman.2011.06.001

Stemler, S.E. (2015). Content Analysis. *Emerging Trends in the Social and Behavioral Sciences,* 1-14.

Stiebe, M. (2023). Stakeholder Perceptions on Sustainability Challenges and Innovations in General Aviation. *Sustainability (Basel)*, 15(23), 16505. DOI: 10.3390/su152316505

Subaşı, M., & Okumuş, K. (2017). Bir araştırma yöntemi olarak durum çalışması. *Atatürk Üniversitesi Sosyal Bilimler Enstitüsü Dergisi*, 21(2), 419–426.

Sustainability Practices, P. W. C. (2011). *Turkish Business World Assessment Report*. PWC.

Topçu, G. (2023). Bibliometric analysis on the place of sustainability in aviation in international literature. [JSHSR]. *International Journal of Social and Humanities Sciences Research*, 10(99), 2419–2428.

Torum, O., & Yılmaz, A. K. (2009). Sustainability Management in Aviation: Research on Sustainability Practices for Airports in Turkey. *Journal of Aeronautics and Space Technologies*, 4(2), 47–58.

Trotter, R. T.II. (2012). Qualitative research sample design and sample size: Resolving and unresolved ıssues and inferential İmperatives. *Preventive Medicine*, 55(5), 398–400. DOI: 10.1016/j.ypmed.2012.07.003 PMID: 22800684

Turan, G. (2018). *Çok Kriterli Karar Verme. B. F. Yıldırım, & E. Önder içinde, Çok Kriterli Karar Verme Yöntemleri*. Dora Basım Yayınları.

Voyer, M., Gollan, N., Barclay, K., & Gladstone, W. (2015). 'It s part of me'; understanding the values, images and principles of coastal users and their influence on the social acceptability of MPAs. *Marine Policy*, 52, 93–102. DOI: 10.1016/j.marpol.2014.10.027

Wennersten, R., & Qie, S. (2018). United Nations sustainable development goals for 2030 and resource use. *Handbook of sustainability science and research*, 317-339.

Werbach, A. (2011). Strategy for sustainability. *Strategic Direction, 27*(10).

Winter, S. R., Thropp, J. E., & Rice, S. (2019). What factors predict a consumer's support of environmental sustainability in aviation? A multi-model analysis. *International Journal of Sustainable Aviation*, 5(3), 190–204. DOI: 10.1504/IJSA.2019.103502

Wolcott, H. F. (1994). *Transforming qualitative data: Description, analysis, and interpretation*. Sage.

Wu, P. J., & Yang, C. K. (2021). Sustainable development in aviation logistics: Successful drivers and business strategies. *Business Strategy and the Environment*, 30(8), 3763–3771. DOI: 10.1002/bse.2838

Yazgan, E. (2018). Development taxonomy of human risk factors for corporate sustainability in aviation sector. *Aircraft Engineering and Aerospace Technology*, 90(6), 1012–1022. DOI: 10.1108/AEAT-04-2018-0133

Zieba, M., & Johansson, E. (2022). Sustainability reporting in the airline industry: Current literature and future research avenues. *Transportation Research Part D, Transport and Environment*, 102, 103133. DOI: 10.1016/j.trd.2021.103133

Section 2
Qualitative Research Application in Air Transportation Management

Examples of qualitative research conducted in the field of air transport management

Chapter 8
Sustainable Operations in the Aviation Industry: A Qualitative Study on Behavioural Conformity and Contradictions

Armağan Macit
https://orcid.org/0000-0002-5694-8285
Aviation HVS, Turkey

Deniz Macit
https://orcid.org/0000-0002-7439-7202
Aviation HVS, Turkey

ABSTRACT

The aviation industry faces significant sustainability challenges. Efforts aim to reduce carbon emissions, conserve natural resources, and increase societal benefits by balancing environmental and operational efficiency. Personnel play a critical role in implementing sustainability through training, incentives, and the development of corporate cultures that encourage environmentally responsible behaviours. Conversely, insufficient resources and institutional pressures can undermine long-term goals and lead to unsustainable practices. Insights from personnel, particularly regarding processes such as waste management and energy efficiency, are vital for strategic improvements. This study investigates sustainability behaviours through semi-structured interviews with 18 aviation interns who observed various work groups. The findings reveal both sustainable and unsustainable practices, paving the way for the aviation industry.

DOI: 10.4018/979-8-3693-7403-0.ch008

1. INTRODUCTION

The aviation industry, due to its significant role in the global economy and its position as one of the fundamental elements of transportation, stands out for its environmental impacts. This industry conducts extensive operational processes, making the extent to which these processes align with sustainability principles a critical subject of inquiry. Sustainability is addressed in various dimensions, such as the preservation of natural resources, reduction of the carbon footprint, and enhancement of societal benefits, while emphasizing the importance of striking a balance between operational efficiency and environmental impacts in the aviation industry (ICAO, 2019).

Sustainability-aligned approaches include the utilization of techniques that enhance fuel efficiency, the transition to renewable energy sources, and the implementation of effective waste management policies. Such measures reduce environmental burdens while simultaneously offering cost-effective long-term solutions that support the industry's sustainability (ATAG, 2020). However, certain practices that do not adhere to sustainability principles also exist. These include the use of outdated aircraft fleets, inadequate maintenance and repair processes, excessive fuel consumption, and the use of environmentally harmful chemicals (Boeing, 2021). While such practices may yield short-term operational gains, they increase the risk of environmental degradation and economic losses in the long term.

In the aviation industry, sustainability-oriented practices implemented in processes such as check-in, boarding, and ramp operations are critically important for reducing environmental impacts and enhancing operational efficiency. In the check-in process, the use of digital solutions stands out as an application that significantly reduces paper waste (Heiets et al., 2022). Electronic tickets, digital boarding passes, and mobile check-in options minimize paper usage, thereby reducing resource consumption and accelerating processes. Additionally, the adoption of energy-efficient displays and devices has become widespread to improve energy efficiency during these processes. Such technological solutions not only enhance passenger experience but also contribute to achieving sustainability goals (Corlett, 2022).

Sustainability-focused behaviours in airport operations include various strategies to minimize the environmental impacts of operational processes. Specifically, during the boarding process, the use of energy-efficient gate management systems prevents unnecessary aircraft idling, thereby reducing fuel consumption (Uysal & Soğut, 2017). In ramp operations, converting ground support equipment (GSE) to electric or hybrid models significantly lowers carbon emissions and reduces noise pollution compared to traditional fossil fuel-based equipment (Alruwaili, 2022). Furthermore, the effective implementation of waste management and recycling practices minimizes waste generated during these processes, supporting environmental sustainability.

These sustainability-oriented behaviours are considered significant steps in the aviation industry to enhance operational efficiency and reduce environmental impacts.

In the aviation industry, personnel play a critical role in the adoption and implementation of sustainability principles in operational processes. Through training and continuous professional development programs, increasing the environmental awareness of personnel supports the effective implementation of sustainable practices (Alameeri et al., 2017). Employees' conscientious behaviours in areas such as energy efficiency, waste management, and carbon footprint reduction contribute to minimizing the environmental impacts of operational processes. Furthermore, the implementation of incentive systems aligned with sustainability goals motivates personnel in this direction, encouraging the widespread adoption of sustainable practices (Mızrak, 2023). Shaping organizational culture to support sustainability facilitates the alignment of personnel with these values and enhances the success of long-term sustainability strategies.

On the other hand, unsustainable behaviours among aviation personnel can negatively impact the environmental performance of operational processes. Insufficient training and a lack of awareness may lead to behaviours that result in energy wastage or improper waste management practices (Nguyen et al., 2023). Moreover, organizational pressures and short-term operational goals can cause personnel to overlook sustainability principles (Lynes & Andrachuk, 2008). For instance, cost-reduction strategies such as using outdated aircraft or opting for environmentally harmful chemicals may encourage personnel participation in such practices (Gibbs et al., 2017). These unsustainable behaviours can lead to both environmental degradation and long-term harm to the industry's economic sustainability.

In the aviation industry, the active participation and role of personnel are of great importance in ensuring the sustainability of operations. Personnel, being directly involved in operational processes, are in the best position to observe the environmental impacts of these processes and their alignment with sustainability goals (Buhusayen et al., 2021). With direct experience of the environmental consequences of decisions and actions taken in daily operations, personnel play a key role in identifying behaviours that align with or deviate from sustainability principles. For instance, in activities such as optimizing fuel consumption, monitoring waste management practices, or enhancing energy efficiency, the individual observations and feedback of personnel provide critical data for making strategic decisions aimed at improving the sustainability of operations (Mızrak, 2023).

At the same time, personnel are the best positioned to understand the causes of unsustainable behaviours and how these behaviours reflect in operational processes. Practices that contradict sustainability principles are often related to operational challenges, organizational pressures, or inadequate resources faced by personnel (Frank Alparslan, 2024). Personnel, by directly experiencing such situations and their causes, can develop solutions to address these misalignments. Therefore, feedback from personnel is crucial in achieving sustainability goals and can guide strategic decision-making aimed at improving operational processes (Kimathi, 2024). As a result, ensuring sustainability in the aviation industry requires making the best use of the knowledge and experience of personnel involved in operational processes.

This study aims to examine behaviours in aviation industry operational processes that align with or deviate from sustainability principles. In this context, the study seeks to benefit from the sustainability observations of aviation trainee students who have observed various business groups within the aviation industry. Semi-structured interviews were conducted with 18 trainee students working in different aviation business groups. These students were asked about the positive and negative aspects of sustainability-related behaviours implemented by their respective organizations. The interviews were analysed descriptively and interpretively, leading to inferences regarding behaviours that align or do not align with sustainability within different aviation business groups and operations.

2. CONCEPTUAL FRAMEWORK

Before examining sustainability in the context of aviation, it is crucial to clearly define concepts such as sustainability and operational sustainability to better understand the essence of the topic. Particularly, explaining sustainability principles and emphasizing how different industries integrate this concept is important for understanding operational sustainability in the dynamic aviation industry. In this context, this section of the study explains the concept of sustainability, addresses its classifications in various fields of application, and then highlights the importance of operational sustainability in aviation.

2.1. The Concept of Sustainability

With industrialization, production volumes increased rapidly, and as a result, concerns about the unchecked use of production resources emerged. In this context, the United Nations (UN) held the Environment and Human Conference in 1972 and published the Brundtland Report titled "Our Common Future" in 1987 (Brundtland & Khalid, 1987). The decisions made during the 1992 Rio Conference are considered some of the most significant decisions regarding sustainability and the environment (Tüyen, 2020: 92). Since 1978, the term "sustainability" has been used not only in technological papers and reports but also in policy documents. The concept of sustainability was also utilized in the 1989 G-20 Summit report (Kidd, 1992: 2).

Following the 1992 Rio Conference, the UN organized another conference in 2012 called Rio+20 (Giovannoni & Fabietti, 2013: 27).

Sustainability, as a policy concept, traces its origins to the 1987 Brundtland Report. This report, which adopted the concept of sustainability, is the one that brought it to its current widespread recognition (Kuhlman & Farrington, 2010: 3436). In the Brundtland Report, sustainability is defined as "meeting the needs of the present without compromising the ability of future generations to meet their own needs" (WCED, 1987: 43).

Due to the pervasive nature of the sustainability concept, various discourses have emerged over time, and it has often been associated with social responsibility, environmental management, or corporate sustainability, which are typically addressed separately (Giovannoni & Fabietti, 2013: 22). Sustainability is not a new concept; it has a long history and has evolved over time (Kidd, 1992). Sustainability is a normative concept about how humans should behave toward nature and how they are responsible to one another and future generations (Baumgärtner & Quaas, 2010: 445).

The concept of sustainability is generally considered to be multidimensional (Giovannoni & Fabietti, 2013: 21). Sustainability is defined as maintaining well-being over a long, even uncertain, period of time (Kuhlman & Farrington, 2010: 3441). The concept of sustainability is also defined by Pearce et al. (2013) as "the future generation having an environmental stock that is no less than the stock inherited from the previous generation."

In summary, when considered from a historical perspective, it can be seen that the concept of sustainability originated from environmental concerns. Over time, economic and social dimensions were added to this environmental aspect. Therefore, this concept now encompasses not only environmental management but also economic development, social responsibility, and corporate sustainability, appearing across many different disciplines.

2.2. Different Dimensions of Sustainability

Sustainability, an approach aimed at the conservation of natural resources, achieving balanced economic growth, and increasing societal welfare, is one of the prominent global priorities today. When expanded to include economic and socio-cultural dimensions, this concept can be addressed from a broad perspective, considering its fields of application and scope. In this section of the study, sustainability is examined from the perspective of business and management, as it is a focal point of the topic, while other aspects are briefly addressed to provide a general framework.

2.2.1. Sustainability from Environmental, Economic, and Social Perspectives

When examining the scope of sustainability types, it is observed that sustainability is fundamentally addressed as environmental, economic, and social sustainability. Sustainability is a concept in which these three core elements are examined in an integrated manner (CASEM, 2023). The multidimensionality of sustainability was also emphasized by Rio+20. In the Rio+20 outcome document, "The Future We Want" refers to the three dimensions of sustainable development: economic, social, and environmental (Giovannoni & Fabietti, 2013: 28). While there are strong connections between economic sustainability and environmental sustainability, there are also some overlaps between the three types of sustainability (Goodland, 1995: 2). One of the fundamental discourses regarding the concept of sustainability is the environmental discourse, which refers to the relationship between humans and nature (Giovannoni & Fabietti, 2013: 24).

Environmental sustainability aims to make global life-support systems indefinitely sustainable and is defined as "the preservation of natural capital" (Goodland, 1995: 10). Social sustainability is inextricably linked to both economic and environmental sustainability (Foot & Ross, 2017: 107). Environmental sustainability is also a prerequisite for social sustainability, and the primary goal of sustainable development is the reduction of poverty (Goodland, 1995: 4). Social sustainability is defined as "a condition that improves life within communities and a process within communities that can achieve this condition" (McKenzie, 2004). In another definition, it is expressed as "the necessary social conditions to support environmental sustainability, which is at the core of sustainable development" (Chiu, 2003: 19). The Brundtland Report defines social sustainability as "development that meets the needs of the present without compromising the ability of future generations to meet their own needs" (WCED, 1987: 40). Economic sustainability refers to resource efficiency and achieving sustainability in economic activities in the long term. This concept also aims to maintain a balance between economic growth and the use of natural resources (CASEM, 2023).

In summary, it is observed that the concept of sustainability forms a whole with its environmental, social, and economic aspects. In the sustainability concept, which is based on these three core focuses, all three dimensions are seen to support and complement each other. This integrated approach constitutes an indispensable paradigm for creating a sustainable future on a global scale.

2.2.2. Sustainability from a Business and Management Perspective

In businesses, the concept of sustainability refers to the ability to adapt to changes brought about by sustainability-related factors. In this context, sustainability involves adapting to legal regulations based on environmental, social, and economic trends (Altensis, 2023). Elkington and Rowlands (1999) explain sustainability from the "Triple Bottom Line" perspective. According to this, businesses should measure not only their financial performance but also their environmental and social performance. The types of sustainability considered from a business and management perspective can be expressed as corporate, operational, financial, and strategic sustainability.

Corporate sustainability is defined as the use of environmental, economic, and social elements in a business's activities and strategies to create long-term value (Tuna & Besler, 2015: 174). Gladwin et al. (1995) define corporate sustainability as "the consideration of sustainable growth at the business level." This dimension of sustainability reflects the application of sustainable development at the organizational level (Dalgıç et al., 2018: 20).

The concept of corporate sustainability is closely linked to financial sustainability, as it depends on access to financial resources (Can, 2019: 9). In terms of finance, sustainability refers to financial products that consider environmental factors throughout the processes and show sensitivity to the environment in investments (Volz et al., 2015). Financial sustainability is defined as ensuring economic, social, and environmental benefits for investors, shareholders, customers, employees, and society (YKB, 2022). Sustainable finance refers to the type of finance in which environmental, social, and governance (ESG) issues are taken into account while providing funding to investors (KuveytTürk, 2023). This concept expresses that the focus of sustainability in businesses is sustainable business capacity. In other words, it explains that funds must be viable and sufficient in the long term, and there must be efficiency in management (Can, 2019: 14).

Strategic sustainability, an essential element in maintaining financial sustainability, is defined as a set of strategies in which environmental, social, and corporate governance approaches are implemented under the sustainability umbrella (Enexion, 2023). This concept fundamentally refers to the preparation of a company's sustainability strategy and policies and the development of an action plan for the necessary steps in future periods (ESG, 2012). Integrating the three core principles of sustainability—economic, social, and environmental dimensions—into the core strategic and operational processes of businesses and including them in decision-making mechanisms is crucial for healthy corporate sustainability. A company's strategic sustainability can either be separate from the corporate strategy or incorporated into it (Altensis, 2023). For the sustainability phenomenon to take root at the operational and management levels of a company, it depends on strategically addressing this phenomenon (Yurdakul, 2024: 282). A sustainability strategy is expressed as a roadmap that determines the resources a company will use to achieve its sustainability goals, the managerial and operational steps it will take, and how it will adapt its current activities and business processes to these policies (Escarus, 2024).

The dynamics that drive a company towards sustainability encompass both the internal and external environment. The emergence of operational and strategic sustainability is possible by considering these environments together (Yurdakul, 2024: 302). While implementing sustainability-related strategies, companies must also ensure they align with their operations and business processes. Operational sustainability refers to a company or business operating by considering the basic principles of sustainability in its operations. Its main goal is to reduce environmental impacts by optimizing resource use in the long term while applying short-term business objectives. It can be described as a management approach that takes into account not only profitability but also environmental and social responsibilities.

In short, from a business and management perspective, sustainability is not limited to environmental responsibilities; it also appears in economic, social, and strategic contexts. Businesses develop sustainability strategies to achieve long-term value creation goals and attempt to increase their chances of success by integrating them into their operational processes. Corporate, financial, or operational sustainability practices, when approached in an integrated manner, allow businesses to enhance their competitive advantage. From this perspective, sustainability stands out as a contemporary necessity in business management.

3. THE AVIATION INDUSTRY AND SUSTAINABILITY

As briefly mentioned in the previous section, the types of sustainability are generally classified into environmental, economic, and social categories. Since each industry has its own unique dynamics, sustainability can also be addressed industrially. In this context, when sustainability is examined from the perspective of the aviation industry, it becomes clear that the industry holds a significant position at the intersection of these three general classifications. Operational sustainability in the aviation industry encompasses a wide range of processes, including fuel management, energy efficiency, flight planning, and waste management. From this perspective, for example, the reduction of carbon emissions can be considered environmental sustainability, cost optimization represents economic sustainability, and improving the welfare of those employed in the industry is social sustainability. When examined in this context, it can be said that achieving operational sustainability in the aviation industry will contribute not only to industryal success and growth but also to global sustainability goals. Due to the industry's critical role in sustainability, it is extremely important to structure operational processes with sustainable approaches.

The aviation industry, which holds a crucial place in today's advanced economies, contributes significantly to both economic and social development. In addition to its global contributions, the industry also has environmental impacts. Specifically, carbon dioxide and other greenhouse gas emissions raise significant concerns regarding sustainability. The industry's role in this regard is being increasingly questioned, and various regulations are being introduced (Barutçu & Çolakoğlu, 2024: 107).

Operational sustainability in the aviation industry is a critical area where general sustainability principles are translated into concrete industry-specific applications. Sustainable practices must be adopted in operational processes to minimize environmental impacts and enhance social welfare. In this framework, environmental, economic, and social sustainability practices are integrated into the operational processes of the industry. Thus, while fulfilling its environmental and social responsibilities, the industry also contributes to economic sustainability goals, ensuring industrial development and helping to build a more liveable future. The following section describes the operational sustainability practices in the industry.

3.1. Examples from Türkiye in Operational Sustainability Practices in the Aviation Industry

Aviation has taken on an important role in sustainability in order to continue its operations without compromising its global strategic significance. The environmental, economic, and social dimensions of sustainability are felt across all aviation operations. In the aviation industry, operational sustainability encompasses the entire goals of reducing environmental impacts, increasing economic efficiency, and maximizing societal benefits. Therefore, it is essential to focus on processes such as energy savings, emissions management, and waste reduction in aviation operations (Raimundo et al., 2023).

Each year, an increase in the amount of carbon dioxide in the Earth's atmosphere is observed. This is considered one of the main causes of global warming, climate change, and related phenomena. The transportation industry, in general, is known to be the second most harmful industry in terms of CO2 emissions, following the energy industry (Ritchie, 2024). When evaluating the aviation industry, issues such as carbon emissions resulting from high fuel and energy consumption, noise pollution created by aircraft, solid and liquid waste generated from operational processes, and environmental and water pollution are encountered. This shows that with the increase in aviation activities, the potential for environmental pollution also rises (Alpman and Göğüş, 2017).

In 2016, the International Civil Aviation Organization (ICAO) implemented the Carbon Offsetting and Reduction Scheme for International Aviation (CORSAIR). Under this plan, starting in 2019, the majority of airline operators began tracking and reporting their carbon footprints. In alignment with the United Nations Sustainable Development Goals, the International Air Transport Association (IATA) in 2021 accepted the commitment for member airlines to achieve net-zero carbon emissions by 2050. This commitment, which aims for zero carbon dioxide emissions, is considered a comprehensive decision that includes balancing methods and carbon capture technologies, known as "2050 Net Zero Carbon Emissions" (Barutçu and Çolakoğlu, 2024).

In line with the net-zero target, it is necessary to carry out development and improvement efforts in the industry. The measures that need to be implemented to achieve the net-zero goal for ensuring sustainability in the industry can be summarized as follows (Semtrio, 2024):

- Improving operational efficiency,
- Widespread use of Sustainable Aviation Fuels (SAF),
- Development of next-generation aircraft technologies,
- Effective management of non-offset table emissions.

Improvements made in factors affecting flight durations contribute to both enhancing operational efficiency and reducing emissions. The increase in operational efficiency enables a reduction in fuel consumption, thereby ensuring resource savings. Reduced fuel consumption also means a decrease in harmful emissions released into the environment. For example, optimizing air traffic management systems effectively can shorten flight durations. Additionally, flight routes optimized according to environmental factors such as wind speed and weather conditions allow for more efficient flights over shorter distances. The use of renewable energy sources in the provision of ground services is also an important factor in increasing the efficiency of operational processes. By reducing the environmental impacts of services performed at airports and ensuring sustainability in energy use, it contributes to operational efficiency (Semtrio, 2024).

The aviation industry has become a critical focal point for sustainability efforts due to its environmental impacts and high energy consumption. Today, airline operators and airports are striving to fulfil their responsibilities in this area by adopting environmentally sensitive and resource-efficient operational approaches. In Türkiye, the aviation industry has made significant progress in sustainability. Leading airlines operating in the country present their efforts to reduce environmental impacts through sustainability reports and programs. Furthermore, activity reports prepared by the General Directorate of State Airports Authority shed light on the environmental sustainability policies of airports (Barutçu & Çolakoğlu, 2024: 107). The following section discusses notable examples and strategies of sustainability practices in the aviation industry in Türkiye.

3.1.1. Examples of Operational Sustainability Practices in Airline Operators

Turkish Airlines (THY) has developed a sustainability program built around four focus areas (environment, economy, governance, society) to support sustainable development (http-1). Some examples of sustainability practices specific to THY are outlined below:

- ***Tomorrow On-Board***: In 2022, THY launched the "Tomorrow On-Board" sustainability vision to carry out sustainability management processes in line with a shared corporate understanding. This motto aims to guide the company's future sustainability practices and create a holistic communication strategy. With "Tomorrow On-Board," THY aims to consolidate its sustainability commitments and strategies under a single umbrella, conducting environmentally friendly and sustainable operations (THY, 2024: 33).
- ***CO2mission***: Through this initiative, THY offers passengers the opportunity to voluntarily offset the emissions caused by their flights. This program is defined as "a program where passengers can balance their flight-related carbon footprint by purchasing carbon credits from certified, climate change mitigation, and social development projects worldwide" (http-2).
- ***Carbon Reduction with Sustainable Aviation Fuel (SAF):*** Compared to fossil fuels, the use of SAF, which reduces carbon dioxide (CO2) emissions throughout the fuel's lifecycle, is part of THY's goal to achieve carbon neutrality by 2050 (http-3).

Turkish Airlines (THY) began its sustainability practices in 2009 with its sustainability vision. In 2014, the first sustainability report was created, and initial practices were launched in the same year. By 2017, these practices began to be implemented on flights. In 2020, the establishment of the corporate sustainability management department followed these developments. In 2021, the Sustainability Committee was formed. In 2022, the use of Sustainable Aviation Fuel (SAF) began, and in 2023, the carbon-neutral goal was announced, along with the establishment of the Türkiye Sustainable Aviation Platform (THY, 2024: 7).

In 2023, THY demonstrated its success in this field by receiving the "Most Sustainable Flag Carrier Airline" award from World Finance magazine. Through operational optimization projects carried out since 2008, THY reduced its carbon footprint, saving approximately 58,000 tons of fuel and preventing 181,000 tons of greenhouse gas emissions in 2022 (Irtak, 2023).

In summary, THY carries out its sustainability practices from a broad perspective focused on environment, economy, society, and governance. Since 2009, operating with a sustainability vision, THY has developed a strong roadmap supporting sustainable development by implementing innovative practices such as SAF usage, the carbon offset program (CO2mission), and the "Tomorrow On-Board" vision, aiming to reduce environmental impacts and achieve carbon-neutral goals.

Pegasus Airlines, within the framework of its corporate sustainability policy, has adopted the fundamental principle of defining its environmental, social, and corporate sustainability goals related to its activities and determining the necessary organizational responsibilities to achieve these goals. The company is committed to continuously improving its sustainability compliance programs and raising awareness among its stakeholders in this regard (http-4).

Operating one of the youngest and most fuel-efficient fleets in Türkiye and the world, Pegasus states in its 2023 sustainability report that the average age of its fleet, consisting of 110 aircraft, is 4.6 years. Through various operational initiatives applied with the operational team, the company continuously achieves significant efficiency and cost savings. The report indicates that these goals are a key source of strength for its sustainability efforts. Below are some examples of sustainability practices specific to Pegasus:

- ***PegasusEFB Project***: The Electronic Flight Bag (EFB) includes live flight maps, 24/7 live and real-time operational data, flight information, aircraft documentation, and many other elements, and it is fully integrated with all the company's other operational systems. Established in 2014, this project has enabled standard progress in flight operations management by embedding the application into all operational efficiency processes.
- ***Seat Cover Reuse Project***: The company inspects seat covers removed from aircraft and repairs those that are still usable. In this context, environmental impacts are reduced, and cost savings are achieved. The reduction in the carbon footprint also contributes to minimizing waste.

In its 2023 sustainability report, Pegasus expressed its sustainability policies under three main themes with the slogan "Our Route is a Sustainable Future": "Our Route is a Clean Future," "Our Route is an Equal and Multi-Vocal Future," and "Our Route is a Better Future Together." As a result of its many sustainability policies and practices, Pegasus received the "Sustainability-Supported Financing Deal of the Year" award at the Economic Sustainability Awards 2023.

3.1.2. Examples of Operational Sustainability Practices at Airports

Airports, operating 24/7, are facilities where a large number of people use the premises, resulting in significant consumption of resources such as energy and water, as well as the generation of large amounts of waste. This makes airports places where sustainability efforts are essential. Airport sustainability can be examined in three categories: social, environmental, and economic dimensions (Sreenath et al., 2021).

From a social perspective, airport sustainability includes factors such as minimizing noise pollution, promoting employee development, and supporting the community (Sreenath et al., 2021). People living near airport runways and terminal buildings, as well as the airport staff, are affected by noise pollution. In addition to the noise emissions created by aircraft, activities conducted within airport terminals also generate a certain level of noise emissions. To reduce the environmental harm caused by noise at airports, some restrictions have been introduced in the United States and European countries (Bose et al., 2019). ICAO has presented a four-step strategic roadmap to reduce noise emissions. These steps include reducing noise at the source, proper planning and management of airport land, selecting operational procedures with low noise emissions, and restricting aircraft operations during night-time (ICAO, 2024).

The value given to airport employees is seen as directly proportional to the value of the airport itself. In this context, employee development is considered an essential part of social sustainability (Miller, 2020). Training and career planning for employees, ensuring access to additional social benefits, and ensuring employee safety and security are evaluated as part of airports' social sustainability efforts. Airports are also thought to have a social responsibility to support the community against the environmental impacts they create. In areas where airports develop, positive effects on the economy and employment can be observed. The employment provided by airports indirectly creates an impact that extends to local businesses, which can be seen as an important effort in terms of social sustainability (Sreenath et al., 2021).

From an environmental perspective, airport sustainability includes parameters such as energy consumption, carbon emissions, water and waste management, and solid waste treatment (Sreenath et al., 2021). Airports' energy use is continuous and intensive. Energy consumption can also increase according to operational intensity. The use of fossil fuels in airport vehicles and the high consumption of electricity are some of the most common examples of energy usage. Airports must ensure energy savings at every stage, shifting towards green energy or renewable energy sources (Alba & Manana, 2016). Controlling carbon emissions at airports is crucial within the context of environmental sustainability. In addition to the carbon emissions generated by aircraft at airports, the carbon emissions produced by passengers and staff in enclosed spaces are also considered. To reduce this, various systems are being established at airports (Lamnatou & Chemisana, 2015). Airports consume significant amounts of water in operational processes and infrastructure services. Despite the increasing demand for air transport globally, given the scarcity of water resources, it is essential for airports to reduce their water consumption. Moreover, since most of the water is required to meet non-potable demands, this makes airports potential candidates for implementing sustainable practices (de Castro Carvalho et al., 2013). The waste generated in airport operations is important for environmental sustainability. Waste management is the responsibility of airports and other operating businesses. Certified waste contractors are engaged for waste management. These contractors are responsible for sorting and disposing of airport waste, especially hazardous waste (Santos et al., 2020).

From an economic perspective, airport sustainability refers to situations where businesses achieve high profitability, passenger satisfaction is high, and contributions to the national economy are significant (Sreenath et al., 2021). Airports directly and indirectly affect GDP. For this reason, developments in air transportation are observed as positive outcomes in the national economy (Sellner & Nagl, 2010). The services provided to passengers by airports impact passenger satisfaction and revenue. Consumers tend to prefer airports that offer higher quality services, demonstrating the effect of passenger satisfaction on airport revenues (Bezerra & Gomes, 2020). The adequacy of safety and security systems at airports is also seen as an important constraint in the choice of airports.

In general, it can be observed that airport sustainability practices are similar. However, airports aiming to increase their sustainability efforts and create a competitive advantage in this area also exhibit innovative practices. In this context, the sustainability practices carried out by the operator of Istanbul Airport, IGA, have been examined to gain insight into airport sustainability initiatives (IGA, 2024).

Istanbul Airport states its sustainability goals with the following statement: "We support the United Nations Sustainable Development Goals (SDGs) and, by adopting these principles as our guide, we work tirelessly with all our stakeholders to leave a more liveable world for future generations. As Istanbul Airport, we aim to be Carbon Neutral by 2050 and have been carrying out our activities in accordance with sustainability principles since our establishment."

In December 2020, Istanbul Airport achieved Carbon Accreditation by meeting the requirements of the ACI Airports Carbon Accreditation (ACA) Program and became part of the climate-friendly airports. In 2022, IGA, in line with its sustainability policy, successfully raised its certification from Level 1 to Level 3 in carbon footprint management and reduction efforts. Additionally, under the Energy Management Policy, the ISO 50001 Energy Management System is implemented to consciously manage energy consumption and ensure energy savings. Istanbul Airport takes measures to reduce its water footprint by measuring water consumption, setting targets, implementing water efficiency projects, and exploring alternative water sources for recycling. In line with its "Zero Waste" goal, it has developed a waste management system, including source separation and semi-automatic sorting systems, to recover waste, becoming the first airport in Türkiye to receive the Zero Waste Certificate.

Under the Biodiversity Action Plan launched in 2015, Istanbul Airport has been carrying out habitat and species protection efforts, as well as tree planting and ecosystem protection projects initiated in 2019 in collaboration with the General Directorate of Forestry. These efforts are carried out in detail with academics and stakeholders and continue to be monitored.

From a social sustainability perspective, IGA supports its employees' training and development processes by prioritizing employee experience, improving processes with digital tools, promoting equal opportunities, and tracking the proportion of women in employment and management positions. Furthermore, under the Social Investment Program, IGA develops projects in areas such as socio-economic support, entrepreneurship, agriculture and livestock, circular economy, basic healthcare, and cultural activities, aiming to benefit disadvantaged groups and prioritizing stakeholder collaboration.

4. METHODS

This study aims to examine the behaviours that are in compliance with and in violation of the principles of sustainability in operational processes in the aviation sector. In this context, it is aimed to benefit from the sustainability observations of the intern students who are receiving aviation education and who observe different business groups in the aviation sector. Data were collected by conducting semi-structured interviews with 18 students working in different business groups. 12 questions determined by the literature review were asked in the interview. Details of yes and no answers were requested for the questions. In addition, additional explanations were made to make the questions understandable. Thus, it was possible to obtain the results that the students drew from their observations within the framework of the questions, considering their entire internship period.

No demographic questions were asked in the study, only the company where the internship was conducted was asked to determine which group the company was in. In this study, when students' words were directly quoted, the code "IS" was used to refer to the intern students. The participating students were numbered between 1-18. It was determined that 3 of the 18 students (IS 1,2,3) did their internships in a private airport company, 2 (IS 4,5) in an airline food and beverage supply company, 1 (IS 6) in passenger hospitality lounges and 12 in airport ground services companies. It is known that 6 (IS 6-12) of the students who did their internships in ground services did their internships in the passenger services department, 4 (IS 13-16) in the lost and found department and 2 (IS 17,18) in the operations department. Therefore, when the 4 aviation business groups and ground services departments were considered separately, the results were evaluated in 6 groups. The descriptive analysis method and the interpretative analysis method were used together during the evaluation. All responses were evaluated by the author. The responses that attracted attention were shared by quoting. In the reporting phase of the process, separate analysis was made for each basic question. The findings were discussed within the scope of the created themes.

Qualitative research methods, which are increasingly gaining importance and are understood in academic studies, create an important area especially for making philosophical explanations and inferences (Roulston, 2010). Interviews are seen as an important data collection tool in qualitative research. In qualitative research, interviews can be classified as unstructured, semi-structured, and structured according to standardization processes such as preparation, organization, ordering, and direction of questions. In addition, the type of interview can be named differently according to technical elements such as whether the participants are individual or in groups, or whether the interview is conducted face-to-face, by phone or online (Creswell, 2018; Polat, 2022). In semi-structured interviews, it is preferred to keep the number of basic questions as low as possible and to obtain more detailed answers by generating new questions in the flow according to the answers given. Although there are different opinions about the preparation of questions, there is no single truth, but they should be understandable questions that basically serve the purpose of the research (Roulston, 2010; Polat, 2022). The dialogues obtained as a result of semi-structured interviews should be simplified and analysed. In the analysis phase, different methods such as description, classification, association, schematization, categorization, and interpretation can be used. In general, the analysis method needed is determined when the literature review is conducted and questions are created (Dömbekçi & Erişen).

Descriptive analysis is a type of qualitative analysis that involves summarizing and interpreting the obtained data according to previously determined themes. The aim of this type of analysis is to present the findings to the reader in a summarized and interpreted form (Yıldırım & Şimşek, 2003). Interpretive analysis is an analysis of the answers given by evaluating the data with the knowledge, experience and expertise of the researchers (İlerisoy, 2023).

5. FINDINGS

In the study, it was found that the 18 students who participated in semi-structured interviews were knowledgeable about the concept of sustainability and sustainability practices. Therefore, it can be said that the observations they made during their internships regarding sustainability are highly reliable for the respective stations.

- *1. Are there any rules related to the concept of sustainability adopted at the business where you did your internship?*

The three students who interned at a private airport operator mentioned that recycling and zero waste rules were adopted, that waste sorting and recycling bins were placed in various areas of the airport, that materials used for tea and coffee were selected to be recyclable, and that there were contracted businesses for waste recycling. Two students who interned at a food and beverage business stated that all expired products were routinely identified and sent for recycling. One student who interned at a passenger lounge mentioned that passengers were given thermoses to reduce plastic waste, that waste was collected in recycling bins, and that tablets had replaced paper in all operational processes. Two students who interned in the operations department of ground services businesses emphasized that no rules related to sustainability were adopted. Among the six students who interned in the passenger services department, only one mentioned that waste bins were available at the counter area for sorting, but that the staff did not pay attention to it, while five others stated that no rules related to sustainability had been adopted. Of the four students who interned in the lost property department, two said sustainability was not adopted, while the other two mentioned that their daily work plans were communicated via message to avoid paper waste, and that the empty spaces on the used papers were used as drafts, in an effort to reduce paper consumption.

IS 2 who interned at the airport operator stated, *"On the first day of our internship, it was explained throughout the entire airport tour that waste bins were placed, and the importance of sorting recyclables was emphasized. In the break room, all materials used for beverages like tea and coffee were chosen to be recyclable."* This suggests that the airport businesses have adopted the zero-waste principle and are trying to adapt all staff to this goal.

- *2. Does the business where you interned provide training to its staff on the concept of sustainability?*

Students who interned at the airport operator mentioned that staff received training on sustainability, but no training was organized for the interns. Students who interned at food and beverage businesses and passenger lounges stated that neither the staff nor the interns received any sustainability training. Twelve students who interned in ground services, including the operations, passenger services, and lost property departments, mentioned that the staff received face-to-face sustainability training, followed by annual online refresher training. Additionally, the students interning in ground services noted that sustainability topics were addressed during the training.

IS 8 who interned in the passenger services department of a ground services company said, *"I know that the staff received face-to-face training on topics like sustainability, quality systems, occupational health and safety, etc., when they started their work. Also, at certain times every year, these refresher training sessions are sent online to the staff's email. They repeat the sustainability training by answering the questions and watching the videos. Furthermore, in the first two days of our internship, we were also briefed on sustainability policies and what we are expected to do in this regard."* As can be understood from this, airport ground services companies prioritize staff training on sustainability, while no findings on sustainability training were found in other business groups.

- *3. Has the staff at the business where you interned adopted the zero waste system?*

Among the students who interned at the airport operator, one mentioned that the staff did not prioritize the zero waste issue, while another pointed out that they did care and were diligent about it. Students who interned at the food and beverage business stated that the staff had adopted the zero waste system and that everyone was careful about sorting and recycling. A student interning at a passenger lounge mentioned that the staff had no awareness of this issue. In ground services, students reported that the staff working in operations did not adopt the zero waste system, while in passenger services and lost property departments, it completely varied by person. Some staff were very attentive to the zero waste system, while others had no awareness of it. They also pointed out that attention to this issue decreased during periods of operational intensity.

IS 6 who interned in the passenger services department of a ground services company stated, *"The staff's awareness of zero waste varies from person to person. Some are very careful about it, while others completely ignore it. However, in situations of operational intensity, even the staff who pay attention might exhibit some flexibility in their behaviour."* This suggests that factors such as time constraints and intensity in operational departments can negatively impact the zero waste system.

- *4. Does the staff at the business where you interned have awareness of their carbon footprint?*

All 18 students who interned at different business groups answered "no" to this question. They mentioned that they had never heard such a discussion or that no processes related to this were being implemented. This suggests that carbon footprint awareness is not yet fully established in aviation operational processes.

IS 14 who interned in the lost property department of a ground services company stated, *"I haven't seen any discussion, effort, or actions taken regarding this issue."* This indicates that awareness of this issue is low.

- *5. Is the staff at the business where you interned aware of the carbon emissions created by the aviation industry?*

Sixteen students answered "yes" to this question, indicating awareness, while one student who interned in ground services and one who interned at a passenger lounge answered "no," stating they were not aware. In their responses, the students mentioned that the staff was aware that aircraft contribute to environmental pollution and noise pollution, and they were also aware that new-generation aircraft are being developed to address these issues.

IS 5 who interned at a food and beverage business stated, *"Yes, they are aware. Since they load catering onto the aircraft, all the staff are familiar with the types of aircraft. They know that new-generation aircraft cause less environmental and noise pollution, and some aircraft have information about this on the cabin panels."* This indicates that there is a high level of awareness regarding the environmental and noise pollution caused by aircraft.

- *6. Has the business where you interned adopted digitalization?*

Students who interned at the airport operator stated that the business had adopted digitalization and that many new digital systems were in use. Students who interned at the food and beverage business also mentioned that digitalization was adopted and that every process was carried out digitally, even noting that operational disruptions occurred during internet outages. A student who interned at a passenger lounge mentioned that the business had developed digital programs for areas like fast track, where digitalization was implemented. Students who interned in the operations department of ground services reported that digitalization had been adopted, with many digital programs used for operational processes, and that documents to be taken to aircraft were not printed but digitally transferred. Students who interned in the lost property department stated that, thanks to digitalization, the rate of lost luggage was low, and they used the World Tracer system. Students who interned in the passenger services department mentioned that digitalization was adopted, with all processes managed by computer systems, and that kiosk and self-bagging services had developed.

IS 9 who interned in the passenger services department of a ground services company stated, *"Yes, digitalization has been adopted. Each airline has different check-in programs, and those without are served through the ground services program. The same programs are also used for boarding, and there are kiosk machines, with their numbers continuously increasing, and equipment for self-bagging services is being introduced. I can say that there is constant development in this area."* This suggests that digitalization is at the heart of aviation operations. With the impact of digitalization, significant support is provided for sustainability.

- *7. Has digitalization been adopted by the staff at the business where you interned?*

All students answered "yes" to this question, indicating that digitalization has been adopted. They stated that everyone had been trained on digital programs and that they were proficient in all the programs they actively used in their work. However, students who interned in ground services pointed out that not everyone had the same level of proficiency; more experienced staff could perform different tasks in the programs and correct many mistakes, while less experienced staff could only carry out more standard tasks.

IS 10 who interned in the passenger services department of a ground services company stated, *"All the staff receive training on the programs, and since they directly use them in their work, they have to learn and use them. But of course, not everyone uses them in the same way. Supervisors, shift managers, or people who have worked at the same company for many years are proficient in all features of the programs, but inexperienced staff can only perform standard tasks."* This shows that, to varying degrees, the process of digitalization has been adopted in aviation operations. This can be considered an important effort for sustainability.

- *8. Does the business where you interned have any projects related to sustainability?*

One student who interned at the airport operator stated that there were no sustainability-related projects, while the other two students mentioned that they conducted a survey process with sustainability-related questions for passengers. Students who interned at food and beverage businesses and passenger lounges reported that they were not aware of any sustainability-related projects. Among the students who interned at ground services businesses, two stated that there were no projects, while others mentioned that there was a project, and the advertisements for this project were also shared in the media. One student additionally mentioned that, within the scope of social sustainability, the business provided educational scholarships to female students. This student stated that they became aware of the scholarship program during their internship and applied for it.

IS 17 who interned in the operations department of a ground services company said, *"The business has started a project about a clean future, they are constantly sharing it on social media, there are posters in the offices, and they also explain this project and the activities carried out in the company's magazine. The business is offering a scholarship, specifically for female students, and I found out about it from the office magazine. Since I met the requirements, I applied."*

This indicates that the ground services business is striving to create a competitive advantage by promoting its sustainability activities through written media such as social media, magazines, and posters. The scholarship program can be considered evidence of the business's activities in the realm of social sustainability.

- *9. Does the business where you interned have posters, advertisements, or announcements related to sustainability activities?*

Students who interned at the airport operator stated that they did not come across any posters, advertisements, or announcements, but they mentioned that the business's website had a sustainability section, which they reviewed. Students who interned at food and beverage businesses stated that they only saw sustainability-related information on some aircraft. A student who interned at a passenger lounge stated that they did not see any posters or advertisements. Students who interned at ground services businesses mentioned that there were sustainability-related announcements on social media ads and in the company's magazine.

IS 1 who interned at the airport operator stated, *"I did not come across any advertisements or announcements, but I saw it on the website. When I looked for some information on the website, there was a sustainability section, and annual sustainability reports were available. When I downloaded the report, I saw that it contained information about the activities carried out by the business."*

This situation can be considered evidence that businesses are attempting to report their sustainability activities.

- *10. What actions has the staff at the business where you interned taken to contribute to sustainability?*

Among the students who interned at the airport operator, one mentioned the use of paper cups instead of plastic cups, while another mentioned that information was provided through mobile apps instead of paper. Students who interned at food and beverage businesses pointed out the importance of recycling practices as an example. The use of tablets instead of paper for passenger greeting was considered an example of sustainable behaviour at the passenger lounge. Students who interned in the operations department of ground services mentioned that A4 paper was used double-sided only when necessary. Students who interned in passenger services evaluated behaviours such as not printing boarding passes for passengers who checked in via mobile or online, and directing passengers to kiosks, as positive actions for sustainability. Students who interned in the lost property department considered sending daily plans via message to staff instead of printing them as an example.

IS 4 who interned at a food and beverage business stated, *"Especially the tracking of any product nearing or past its expiration date is done very well. This is obviously related to food characteristics, but the recycling of the large amount of waste produced is entirely related to sustainability. There are agreements with some businesses that pick up the sorted waste and support recycling. Also, leftover food from shelters is sent to animals every day. I think this is an important social responsibility. The staff knows this, nothing goes to waste."*

In this case, it can be said that the staff working in aviation operations exhibit behaviours that align with sustainability. Additionally, it can be stated that food and beverage businesses display sustainability-aligned behaviours in terms of recycling and social sustainability.

- *11. What are the contradictory behaviours of the staff at the business where you interned in relation to sustainability?*

Two students who interned at the airport operator mentioned observing some staff members throwing their waste into bins without paying attention to waste sorting. One student who interned at a food and beverage business pointed out the excessive use of paper cups, while another mentioned the surplus of receipts printed by digital machines. A student who interned at a passenger lounge considered placing end-of-day reports and slips into envelopes made from A4 paper as a behaviour that contradicts sustainability. Among the 12 students who interned in ground services, 8 pointed out excessive use of paper cups. Additionally, students who interned in passenger services mentioned printing documents listing the names of transit passengers for each flight, and the excessive use of paper caused by kiosks tearing some boarding passes, both of which were seen as unsustainable behaviours.

IS 12 who interned in the passenger services department of a ground services company said, *"There is an excessive use of paper cups; maybe it's better than plastic, but it is still a waste. Also, when you put paper of a certain weight into the paper tray of the kiosk machines, it causes problems, leading to paper waste. Furthermore, for arrivals, the staff prints a separate copy for each flight to separate transit passengers. There is a lot of trash from one check-in counter, and sometimes there are misprinted baggage tags as well."*

This indicates that there are unsustainable behaviours among the staff working in aviation operations, and that excessive material waste occurs in some operational processes.

- *12. Was there any specific situation during your internship that was either aligned with or contradictory to sustainability? If so, what was it? If not, what recommendations would you make to the business where you interned regarding sustainability?*

All students who interned at various businesses stated that they did not observe any particularly special situations related to sustainability. Therefore, their recommendations for the businesses where they interned are as follows:

IS 2 who interned at the airport operator said, *"Staff should use their own cups for drinks like tea and coffee, and to encourage this, cups with the company logo could be given as gifts."*

IS 3 who interned at the airport operator recommended, *"Sustainability training could be increased, and the staff could be informed about what to pay attention to at each stage of operations, and this workflow could be shared with the staff."*

IS 4 who interned at a food and beverage business said, *"Businesses could implement a reward and punishment system related to sustainability. Employees who demonstrate sustainable behaviour could be given bonuses, while those who don't could be given warnings, reprimands, or salary deductions."*

IS 6 who interned at a passenger lounge said, *"The staff who put the most effort into sustainability could be incentivized with extra breaks or similar rewards."*

IS 17 who interned in the operations department of ground services recommended, *"To eliminate paper usage, all processes should be made accessible via mobile phones. This would be both more sustainable and save time. All digital applications could be converted into mobile apps."*

IS 1 who interned in the passenger services department of ground services suggested, *"The staff's attention to sustainability should be drawn. Large posters could be placed in the airport, raising awareness among both staff and passengers, and at least for passengers flying, public service announcements could help increase awareness."*

IS 15 who interned in the lost property department of ground services proposed, *"The amount of personal waste generated by the staff could be tracked, which would help increase their sense of responsibility. Additionally, training should be increased, and the effect of training on this topic should be assessed."*

6. DISCUSSION AND CONCLUSION

This study aimed to examine operational sustainability practices in the aviation industry, the differences between these practices across business groups, and the behaviours of business personnel that align with or contradict sustainability principles. The data obtained through semi-structured interviews were focused on for analysis. The findings of the study revealed that the level of awareness regarding sustainability varied depending on the type of business and operational intensity. The aviation industry's adoption and implementation of sustainability principles were presented through the themes derived from the findings. The themes identified regarding behavioural alignment and contradictions in sustainable aviation operations were: sustainability and education, carbon footprint and operational awareness, the role of digitalization, behaviours conflicting with sustainability, and positive sustainability practices.

The findings indicate that sustainability education is a critical necessity for the aviation industry. While ground services businesses support sustainability awareness through training and refresher programs, such training was found to be lacking in other business groups. For example, students who interned in food and beverage businesses emphasized that no such training was provided. This highlights the need for standardizing sustainability education across the aviation industry.

The study found that personnel in aviation businesses generally have low awareness of their carbon footprint. Specifically, the lack of knowledge about the impact of carbon emissions on business processes suggests that such awareness projects should be implemented on a larger scale. In this context, strategies and training programs to reduce carbon emissions should be developed.

The study also identified that digitalization plays an important role in sustainability. The adoption of digitalization in airport operations and ground services enhances the efficiency of operational processes and contributes to achieving sustainability goals. The development of mobile applications and the reduction of paper usage highlight the need for the industry to increase its investments in digitalization. However, the varying levels of digitalization adoption among staff indicate that the process needs to be implemented more uniformly.

It was observed that some behaviours of the staff were not aligned with sustainability principles. For example, the improper use of waste bins and excessive paper consumption in operational processes are factors negatively affecting sustainability goals. This situation necessitates the implementation of stricter policies to raise sustainability awareness.

Some businesses' recycling practices and social responsibility projects left a remarkably positive impression. For instance, waste management and food donations to animal shelters in food and beverage businesses demonstrate that sustainability efforts can be effective. Additionally, proposed reward and punishment mechanisms to encourage sustainable behaviours and the diversification of training programs could increase staff alignment with sustainability principles.

In summary, it has been observed that sustainability practices in the aviation industry vary significantly across business groups and departments. These differences necessitate the more consistent application of sustainability policies across the industry. Widespread digitalization, education programs, increased carbon footprint awareness, and diversification of social sustainability projects are steps that need to be taken to achieve sustainability goals in the industry.

Future research could provide a more comprehensive analysis by comparing sustainability practices in different aviation regions, potentially leading to more inclusive global results. In particular, the long-term effects of digitalization and the use of eco-friendly technologies on operational processes should be examined in more detail in the future. Such studies will play a guiding role in the aviation industry's sustainability journey.

REFERENCES

Air Transport Action Group [ATAG]. (2020). *Aviation: Benefits beyond borders.* https://aviationbenefits.org/

Akman Dömbekci, H., & Erişen, M. A. (2022). Nitel araştırmalarda görüşme tekniği. *Anadolu Üniversitesi Sosyal Bilimler Dergisi*, 22(Özel Sayı 2), 141-160.

Alameeri, A., Ajmal, M. M., Hussain, M., & Helo, P. T. (2017). Sustainability practices in the aviation industry: A study of UAE-based airlines. *International Journal of Sustainable Society*, 9(2), 119–147. DOI: 10.1504/IJSSOC.2017.086818

Alba, S. O., & Manana, M. (2016). Energy research in airports: A review. *Energies*, 9(5), 349. DOI: 10.3390/en9050349

Alparslan, F. (2024). Sustainability in aviation and safety culture. In *Contemporary sustainable organisational practices: A roadmap for transformation* (pp. 149–170). Springer Nature Switzerland. DOI: 10.1007/978-3-031-62474-2_11

Alpman, E., & Göğüş, A. Y. (2017). Havacılıkta sürdürülebilir gelişme göstergeleri. *Sürdürülebilir Havacılık Araştırmaları Dergisi*, 2(1), 1–11.

Alruwaili, M. (2022). *Sustainable airports: Green energy solutions to achieve carbon neutrality* (Doctoral dissertation, Cardiff University).

Altensis. (2023). Kurumsal sürdürülebilirlik. https://www.altensis.com/hizmetler/kurumsal-surdurebilirlik/surdurulebilirlik-stratejisi/

Barutçu, S. B., & Çolakoğlu, Ü. (2024). Havayolu taşımacılığında sürdürülebilirlik analizi; Türkiye. *Journal of Travel and Tourism Research*, 24(24), 106–124.

Baumgärtner, S., & Quaas, M. (2010). What is sustainability economics? *Ecological Economics*, 69(3), 445–450. DOI: 10.1016/j.ecolecon.2009.11.019

Bezerra, G. C. L., & Gomes, C. F. (2016). Measuring airport service quality: A multidimensional approach. *Journal of Air Transport Management*, 53, 85–93. DOI: 10.1016/j.jairtraman.2016.02.001

Boeing. (2021). *Sustainability report.* https://www.boeing.com/principles/environment/

Bose, T., Adhvaryu, B., & Chakraborti, S. (2019). Airport noise and urban planning: A case of Ahmedabad, India. *Population*, 3, 5–577.

Brundtland, G. H., & Khalid, M. (1987). Our common future. Oxford University Press, Oxford, GB.

Buhusayen, B., Seet, P. S., & Coetzer, A. (2021). Front-line management during radical organisational change: Social exchange and paradox interpretations. [h]. *Sustainability (Basel)*, 13(2), 893. DOI: 10.3390/su13020893

Can, İ. (2019). *Finansal sürdürülebilirlik göstergeleri ve değerlendirilmesi: Borsa İstanbul sürdürülebilirlik endeksi şirketleri üzerine bir araştırma (Yüksek Lisans Tezi)*. Marmara Üniversitesi.

CASEM. (2023). Sürdürülebilirlik nedir? https://www.casem.com.tr/surdurulebilirlik-nedir

Chiu, R. L. (2003). Social sustainability, sustainable development and housing development: The experience of Hong Kong. In *Housing and social change: East-west perspectives* (pp. 221–243). Routledge.

Corlett, B. M. (2022). *Improvement of airline check-in and boarding processes* (Doctoral dissertation, Stellenbosch University). Stellenbosch University. https://scholar.sun.ac.za

Creswell, J. W. (2018). *Nitel araştırma yöntemleri: Beş yaklaşıma göre nitel araştırma ve araştırma deseni (M. Bütün & S. Demir, Çev* (Kitabevi, S., Ed.).

Dalgıç Turhan, G., Özen, T., & Albayrak, R. S. (2018). Kurumsal sürdürülebilirlik kavramı, stratejik önemi ve sürdürülebilirlik performansı ölçümü. *Ege Stratejik Araştırmalar Dergisi*, 9(1), 17–37. DOI: 10.18354/esam.304155

de Castro Carvalho, I., Calijuri, M. L., Assemany, P. P., Silva, M. D. F. M., Neto, R. F. M., da Fonseca Santiago, A., & de Souza, M. H. B. (2013). Sustainable airport environments: A review of water conservation practices in airports. *Resources, Conservation and Recycling*, 74, 27–36. DOI: 10.1016/j.resconrec.2013.02.016

dos Santos, A. J., Mancini, S. D., Roveda, J. A. F., Ewbank, H., & Roveda, S. R. M. M. (2020). A fuzzy assessment method to airport waste management: A case study of Congonhas Airport, Brazil. *Journal of Air Transport Management*, 87, 101838. DOI: 10.1016/j.jairtraman.2020.101838

Elkington, J., & Rowlands, I. H. (1999). Cannibals with forks: The triple bottom line of 21st century business. *Alternatives Journal*, 25(4), 42–44.

Enexion. (2023). Stratejik sürdürülebilirlik. https://www.enexion.com.tr/stratejik-surdurulebilirlik-8/

Escarus. (2024). Sürdürülebilirlik stratejisi geliştirme. https://escarus.com/surdurulebilirlik-stratejisi-gelistirme/

ESG. (2012). Sürdürülebilirlik stratejisi. https://www.esgturkey.com/surdurulebilirlik-stratejisi/

Foot, D. K., & Ross, S. (2017). Social sustainability. In Sloan, G. (Ed.), *Teaching business sustainability* (pp. 107–125). Springer. DOI: 10.4324/9781351281805-7

Gibbs, L., Slevitch, L., & Washburn, I. (2017). Competency-based training in aviation: The impact on flight attendant performance and passenger satisfaction. *Journal of Aviation/Aerospace Education Research*, 26(2), 55–80. DOI: 10.15394/jaaer.2017.1716

Giovannoni, E., & Fabietti, G. (2013). What is sustainability? A review of the concept and its applications. In *Integrated reporting: Concepts and cases that redefine corporate accountability* (pp. 21–40). Springer. DOI: 10.1007/978-3-319-02168-3_2

Gladwin, T. N., Kennelly, J. J., & Krause, T. (1995). Shifting paradigms for sustainable development: Implications for management theory and research. *Academy of Management Review*, 20(4), 874–907. DOI: 10.2307/258959

Goodland, R. (1995). The concept of environmental sustainability. [REMOVED HYPERLINK FIELD]. *Annual Review of Ecology and Systematics*, 26(1), 1–24. DOI: 10.1146/annurev.es.26.110195.000245

Heiets, I., La, J., Zhou, W., Xu, S., Wang, X., & Xu, Y. (2022). Digital transformation of the airline industry. *Research in Transportation Economics*, 92, 101186. DOI: 10.1016/j.retrec.2022.101186

http-1. Türk Hava Yolları (THY). Sürdürülebilirlik. https://investor.turkishairlines.com/tr/kurumsal-yonetim/surdurulebilirlik

http-2. Türk Hava Yolları (THY), CO2mission. https://www.turkishairlines.com/tr-int/co2mission/

http-3. Türk Hava Yolları (THY), Sürdürülebilir havacılık yakıtı (SAF) nedir? https://www.turkishairlines.com/tr-int/surdurulebilir-hava-yakiti/

http-4. Pegasus Yatırımcı İlişkileri. Sürdürülebilirlik politikası. https://www.pegasusyatirimciiliskileri.com/medium/image/kurumsal-surdurulebilirlik-politikasi_1023/view.aspx

İGA. (2024). *Sürdürülebilirlik*. İGA Havalimanı. Erişim adresi: https://www.igairport.aero/surdurulebilirlik/

İlerisoy, M. (2023). Yorumlayıcı fenomenolojik analiz. *İstanbul Ticaret Üniversitesi Sosyal Bilimler Dergisi*, 22(46), 509-527.

International Civil Aviation Organization (ICAO). (2019). *Environmental protection.* https://www.icao.int/environmental-protection/Pages/default.aspx

International Civil Aviation Organization (ICAO). (2024). *Aircraft noise: Balanced approach to aircraft noise management.* https://www.icao.int/environmental-protection/Pages/noise.aspx

Irtak, K. (2023). THY'ye "En sürdürülebilir bayrak taşıyıcı hava yolu" ödülü. Anadolu Ajansı. https://www.aa.com.tr/tr/gundem/thyye-en-surdurulebilir-bayrak-tasiyici-hava-yolu-odulu/2936491

Kidd, C. V. (1992). The evolution of sustainability. *Journal of Agricultural & Environmental Ethics*, 5(1), 1–26. DOI: 10.1007/BF01965413

Kimathi, T. W. (2024). Leadership, corporate social responsibility, and sustainability in airlines: Case study for Kenya Airways.

Kuhlman, T., & Farrington, J. (2010). What is sustainability? [REMOVED HYPERLINK FIELD]. *Sustainability (Basel)*, 2(11), 3436–3448. DOI: 10.3390/su2113436

KuveytTürk. (2023). Sürdürülebilir finans ve çevresel etkileri. https://www.kuveytturk.com.tr/blog/surdurulebilirlik/surdurulebilir-finans-ve-cevresel-etkileri

Lamnatou, C., & Chemisana, D. (2015). A critical analysis of factors affecting photovoltaic-green roof performance. *Renewable & Sustainable Energy Reviews*, 43, 264–280. DOI: 10.1016/j.rser.2014.11.048

Lynes, J. K., & Andrachuk, M. (2008). Motivations for corporate social and environmental responsibility: A case study of Scandinavian Airlines. *Journal of International Management*, 14(4), 377–390. DOI: 10.1016/j.intman.2007.09.004

McKenzie, S. (2004). Social sustainability: Towards some definitions. Hawke Research Institute, Working Paper Series No. 27, University of South Australia.

Miller, C. (2020). Adult learning theory and its importance in employee training. *BizLibrary*. https://www.bizlibrary.com

Mızrak, F. (2023). Strategies for effective human resource management in the aviation industry: A case-based analysis. *Beykoz Akademi Dergisi*, 11(2), 82–109. DOI: 10.14514/beykozad.1346605

Nguyen, T. L. P., Nguyen, T. T. H., & Ključnikov, A. (2023). The influence of socially responsible human resource management on green behaviours in the aviation industry. *Journal of Competitiveness*.

Pearce, D., Markandya, A., & Barbier, E. (2013). *Blueprint 1: For a green economy*. Routledge. [REMOVED HYPERLINK FIELD] DOI: 10.4324/9781315070223

Pegasus Hava Taşımacılığı, A. Ş. (2023). 2023 sürdürülebilirlik raporu. https://www.pegasusyatirimciiliskileri.com/medium/image/pegasus-2023-surdurulebilirlik-raporu_1442/view.aspx

Polat, A. (2022). Nitel araştırmalarda yarı-yapılandırılmış görüşme soruları: Soru form ve türleri, nitelikler ve sıralama. *Anadolu Üniversitesi Sosyal Bilimler Dergisi*, 22(Özel Sayı 2), 161-182.

Raimundo, R. J., Baltazar, M. E., & Cruz, S. P. (2023). Sustainability in the airports ecosystem: A literature review. *Sustainability (Basel)*, 15(16), 12325. DOI: 10.3390/su151612325

Ritchie, H., Rosado, P., & Roser, M. (2024). Breakdown of carbon dioxide, methane and nitrous oxide emissions by industry. *Our World in Data*. Erişim adresi: https://ourworldindata.org/emissions-by-industry

Roulston, K. (2010). Reflective interviewing: A guide to theory and practice. *Sage (Atlanta, Ga.)*. Advance online publication. DOI: 10.4135/9781446288009

Sellner, R., & Nagl, P. (2010). Air accessibility and growth: The economic effects of a capacity expansion at Vienna International Airport. *Journal of Air Transport Management*, 16(6), 325–329. DOI: 10.1016/j.jairtraman.2010.04.003

Semtrio (2024). Havacılıkta sürdürülebilirlik: Sektörde yeşil dönüşüm. https://www.semtrio.com/blog/havacilikta-surdurulebilirlik-sektorde-yesil-donusum

Sreenath, S., Sudhakar, K., & Yusop, A. F. (2021). Sustainability at airports: Technologies and best practices from ASEAN countries. *Journal of Environmental Management*, 299, 113639. DOI: 10.1016/j.jenvman.2021.113639 PMID: 34479146

Tuna, Ö., & Besler, S. (2015). Kurumsal sürdürülebilirlik anlayışı ve uygulamaları: Örnek olay çalışması. *Journal of Faculty of Political Science*, 52, 171–200.

Türk Hava Yolları (THY). (2024). 2023 Sürdürülebilirlik raporu. https://cdn.turkishairlines.com/m/38a092fe993bc5a0/original/2023-Surdurulebilirlik-Raporu.pdff

Türk Hava Yolları (THY). (2024). Sürdürülebilirlik yolcuğumuz. https://cdn.turkishairlines.com/m/3fe0ae64a7afc39e/original/PowerPoint-Presentation.pdf

Tüyen, Z. (2020). İşletmelerde sürdürülebilirlik kavramı ve sürdürebilirliği etkileyen etmenler. İstanbul Ticaret Üniversitesi Sosyal Bilimler Dergisi, 19(37), 91-117.

UNESCO. (1978). *Intergovernmental Conference on Environmental Education: Final Report*. UNESCO and UNEP. https://unesdoc.unesco.org/images/0003/000327/032763eo.pdf

Uysal, M. P., & Sogut, M. Z. (2017). An integrated research for architecture-based energy management in sustainable airports. *Energy*, 140, 1387–1397. DOI: 10.1016/j.energy.2017.05.199

Volz, U., Böhnke, J., Knierim, L., Richert, K., Roeber, G. M., & Eidt, V. (2015). *Financing the green transformation: How to make green finance work in Indonesia*. Springer. DOI: 10.1057/9781137486127

WCED. (1987). *Our common future*. Oxford University Press.

Yapı Kredi Bankası (YKB). (2022). Finansal sürdürülebilirlik nedir? https://www.yapikredi.com.tr/blog/surdurulebilirlik/finansal-okuryazarlik/detay/finansal-surdurulebilirlik-nedir

Yıldırım, A., & Şimşek, H. (2003). *Sosyal bilimlerde nitel araştırma yöntemleri*. Seçkin.

Yurdakul, M. (2024). Şirketlerde sürdürülebilirliği stratejik olarak gündeme getiren dinamikler. *Yönetim Bilimleri Dergisi*, 22(52), 282–307. DOI: 10.35408/comuybd.1351915

Chapter 9
Challenges and Solution Approaches for Unmanned Aerial Vehicles in Air Traffic Control:
A Case Study

Arif Tuncal
https://orcid.org/0000-0003-4343-6261
International Science and Technology University, Warsaw, Poland

ABSTRACT

The aim of the study was to explore the challenges in managing Unmanned Aerial Vehicles (UAVs) in Air Traffic Control (ATC). It is a qualitative case study involving semi-structured interviews with 36 air traffic controllers. Through coding, four main findings emerged: training, procedures, technology, and legal regulations. The findings highlighted the necessity for enhanced training for both UAV operators and air traffic controllers, updated procedures, the establishment of Unmanned Traffic Management (UAM), technological enhancements for communication and surveillance, and comprehensive new legal regulations. It is recommended that similar studies be continued in the future, considering the rapid advancements in UAV technology.

DOI: 10.4018/979-8-3693-7403-0.ch009

INTRODUCTION

The historical process and development of air traffic control (ATC) highlights significant stages from the inception of aviation with the Wright brothers' first flight in 1903 to the present day. During World War I, the rapid advancement in aviation technology and increased aircraft traffic raised safety concerns, prompting industry leaders to establish safety standards. In 1929, ATC services were first provided, with aircraft directed using signal flags from a wheelbarrow serving as a makeshift control tower, coordination being limited to simple visual signals. The world's first radio-equipped control room was established in 1930 at Cleveland's Municipal Airport, pioneering ATC towers. A significant milestone occurred in 1936 with the establishment of ATC centers, marking the inception of airlines tracking their flights between airports along their routes to enhance safety measures (Hartman, 2014). In the subsequent years, the increasing air traffic and the need to prevent aircraft accidents necessitated the development of more systematic and organized ATC systems. In the 1950s and 1960s, the rise of jet aircraft and international air traffic required the development of more sophisticated and integrated ATC systems (Hunt & Zellweger, 1987). Today, ATC is supported by advanced automation systems, GPS technology, and data connections. ATC reaches a new dimension by including unmanned aerial vehicles (UAVs) into airspace. Integration of UAVs into ATC systems, marking the beginning of a new evolutionary process, is progressively increasing (Fasano et al., 2023; Hatfield et al., 2020; Matus & Materna, 2021; Shrestha, Oh & Kim, 2021). One of the fundamental reasons behind this increase is the growing prevalence and diversification of UAV technology. Initially used for military purposes, UAVs are now employed for various civilian purposes. Efforts continue for their use in urban air mobility for purposes such as air taxis. It is projected that the production of UAVs, which was 2 million units in 2021, will increase to 6.5 million units by 2030 in response to this increasing demand (International Civil Aviation Organization [ICAO], 2024). In the United States, the leading country in the market, there are currently 781.781 registered UAVs (Federal Aviation Administration [FAA], 2024).

With the increase in the number of UAVs, incidents involving UAVs flying without control have increased, raising serious concerns for aviation safety. In March 2013, an Alitalia pilot reported seeing a small UAV while approaching John F. Kennedy International Airport. In March 2014, the crew of a Boeing 777 at Vancouver International Airport noticed a UAV 30 meters away from their aircraft. On March 22, 2024, US Airways Flight 4650 narrowly avoided a collision with a UAV while landing at Tallahassee Regional Airport. On May 30, 2014, an ATR 72 approaching London Southend Airport encountered a UAV flying 25 meters away. In July 2014, an Airbus A320 landing at Heathrow Airport nearly collided with a UAV passing within 20 feet of its left wing. In July 2015, a Lufthansa aircraft narrowly avoided a collision with a UAV while landing at Warsaw Chopin Airport. In September 2015, a Boeing 737 taking-off from London Stansted Airport narrowly avoided a collision with a UAV flying 23 meters away at 150 meters altitude. Similarly, in October 2015, a Dornier 328 taking-off from Manchester Airport encountered a UAV flying 15 meters away from the left wingtip at 900 meters altitude. In August 2016, a rescue helicopter carrying an accident victim to Klagenfurt Hospital had to maneuver to avoid a UAV. On April 17, 2016, an Airbus A320 landing at London Heathrow Airport collided with a UAV belonging to the Metropolitan Police, but no severe damage occurred. In September 2017, a civilian UAV collided with a Black Hawk helicopter in New York; the helicopter landed safely. On October 12, 2017, a Beechcraft King Air A100 belonging to Skyjet Aviation approaching Jean Lesage Airport in Quebec City collided with a UAV; the aircraft landed safely with minor wing damage. In January 2017, a UAV operator at Hangzhou Xiaoshan International Airport was arrested for filming aircraft during a landing at close range. On March 25, 2018, an Air New Zealand Boeing 777-200 landing at Auckland Airport passed within 5 meters of a UAV during the final approach. On December 13, 2018, the nose of Aeromexico Flight 773 landing at Tijuana International Airport was damaged due to a possible UAV collision. On December 19, 2018, all take-offs and landings at Gatwick Airport were suspended due to a UAV incident. On August 10, 2021, a Cessna 172 approaching Buttonville Municipal Airport collided with a UAV belonging to the York Regional Police, causing severe damage to the aircraft, but it landed safely (Chiper et al., 2022; Wikipedia, 2024).

Such incidents necessitate the development of new strategies and technologies in civil aviation and ATC. However, there may be some things that could be improved in this process. UAV flights, which were initially planned in segregated airspace and restricted to specific areas (Simpson, Stoker & Brennan, 2009), have shifted towards unsegregated airspace due to increasing demand (Taraglio, Nanni & Taurino, 2012). UAV integration into airspace poses challenges such as safe management of UAVs under high traffic demands in ATC (D'Amato, Notaro & Mattei, 2018; Diiulio, Militello & Klein, 2020; Guan et al., 2020; Ho et al., 2019; Labib et al., 2019; Matolak, 2015; Mueller, Kopardekar & Goodrich, 2017; Persiani & Bagassi, 2013; Sandor, 2017; Sarim et al., 2019; Simpson & Stoker, 2006; Tuncal & Uslu, 2021; Wang et al., 2022; Zeng, Zhang & Lim, 2016). The approach of finding temporary solutions to current problems is not sustainable, and managing a large number of UAV flights in unsegregated airspace will pose serious challenges (Ribeiro et al., 2017). In this case, air traffic controllers must acquire new skills to monitor and manage the increasing UAV traffic alongside conventional aircraft (Gui et al., 2020). UAVs differ from conventional aircraft in terms of flight-control systems, autopilots, ground-control stations, communication systems, and navigation modes (Aber et al., 2019). These differences may result in a more complex environment for air traffic controllers. Hence, the experiences and perspectives of air traffic controllers are invaluable in this distinct process.

The challenges faced by air traffic controllers play a critical role in identifying the measures and technologies needed for integrating UAVs into ATC. However, there is a literature gap between theoretical research and practical applications. Qualitative studies provide in-depth and nuanced information about experiences and perceptions (Stiles, 1993; Vishnevsky & Beanlands, 2004).

The aim of the study is to identify the challenges in the ATC dimension of UAV flights based on the experiences and perceptions of air traffic controllers. As a qualitative method, the case study is preferred for its ability to provide an in-depth and detailed examination of these challenges. The findings of the study contribute significantly to understanding the critical factors related to the integration of UAVs and improve existing practices in this field. UAV integration is a crucial process that shapes the future of the aviation sector, and the study provides information to facilitate this process to be more effective and safer. The findings contribute to a more profound comprehension of the operational practices of air traffic controllers and the integration of UAVs into civilian airspace, thereby providing a foundational basis for future research and informing policy development. The research question guiding the study is:

What are the challenges of managing UAVs in ATC, and how can their integration into civilian airspace be improved?

METHODOLOGY

Research Design

The research employed a qualitative methodology, emphasizing the collection and analysis of qualitative data. This approach enables an in-depth exploration of participants' experiences, attitudes, and perspectives (Austin & Sutton, 2014; Fossey et al., 2002; Ohman, 2005; Smith, 1996).

A case study design was used in line with the qualitative research approach. A case study is a research method that allows for an in-depth examination of a specific event, situation, or phenomenon (Crowe et al., 2011; Gilgun, 1994). The study explored the operational challenges and solutions related to UAV flights from the perspective of air traffic controllers.

Qualitative research is concerned with understanding phenomena by gathering data through methods such as interviews, observations, and document analysis (Sargeant, 2012). A semi-structured interview technique was used for the purpose of data collection. This method allows for in-depth and flexible discussions on a specific topic by directly engaging with participants. Semi-structured interviews allow participants to express their thoughts and provide detailed information relevant to the research objectives in a natural and unstructured manner (Adams, 2015; Barriball & While, 1994; Carruthers, 1990; Longhurst, 2009). The objective of the interviews was to gain a more comprehensive understanding of the experiences, perspectives, and perceptions of the participants (DeJonckheere & Vaughn, 2019; Doody & Noonan, 2013).

Case

The case focused on UAV flights within the ATC system, examining the challenges faced by controllers due to the increasing prevalence of UAVs in unsegregated airspace. Key challenges included managing UAVs alongside conventional aircraft, adapting to differences in flight control systems and communication technologies, and acquiring new skills to ensure safe and efficient traffic management.

Since the early 2000s, there has been a notable increase in the utilization of UAVs. This exponential and unregulated expansion has resulted in a number of significant incidents and accidents in the field of aviation in recent years. The accessibility of UAVs of varying sizes has increased considerably in recent times. In addition to their use for hobbies and entertainment, UAVs are increasingly being considered for use in the transportation of goods, in commercial applications, and in the provision of air taxi services. The market for UAVs is expected to become a significant contributor to the global economy in the near future. Furthermore, this development is affecting the aviation sector, particularly with regard to urban air mobility, where there are considerable societal efforts aimed at integrating UAVs (Aydin, 2019; Chamata & Winterton, 2018; Çınar & Tuncal, 2023; Eibfeldt et al., 2020; Macias et al., 2019; MacSween, 2003; Macsween-George, 2003; Tan et al., 2021; Tepylo, Straubinger & Laliberte, 2023).

However, the experiences and perceptions of air traffic controllers, who are responsible for air traffic management, are crucial. They face real challenges when managing airspace, and it is vital to identify and solve these problems. Information obtained from air traffic controllers through purposive sampling regarding the integration of UAV flights, along with systematic analysis within the research limitations, is important in this case.

Within this context, it is clear that more empirical research is essential to further understand the complex dynamics of UAV flights within the ATC system. The research is required for the development of effective strategies for safe and efficient airspace management and the support of UAV development.

Research Participants

The primary data source of the study is air traffic controllers selected through a purposeful sampling method. The participants were intentionally chosen from three different ATC units providing distinct air traffic services, as well as from various experience groups. This diversity was included to capture a comprehensive range of perspectives related to different service environments and experience levels. The data collection process concluded when data saturation was achieved by obtaining similar responses from 36 participants, indicating fulfillment of the designated objective and sufficiency of the collected data for analysis (Denny & Weckesser, 2022; Grady, 1998). Additionally, the sample size was found to be acceptable for the study according to the literature (Bertaux, 1981; Charmaz, 2006; Green & Thorogood, 2018).

Table 1 provides demographic information of participants. The majority of participants are male (64%). Participants with 6-10 years of experience constitute the lowest proportion at 8%, while those with 0-5 years of experience account for 14%. Participants with 11-20 and over 20 years of experience equally contribute to the study at 39%. Undergraduates have the highest participation rate (69%). At the unit level, the highest participation is observed in Aerodrome Control Towers (TWR) at 47%, followed by 31% in Approach Control Unit (APP) and 22% in Area Control Centre (ACC).

A particularly noteworthy point in the demographic information is that 89% of participants provide ATC services to UAVs. This aspect is crucial for comprehending the case under research in the study.

Table 1. Demographic information

		n
Gender	Female	13
	Male	23
Experience	0-5 years	5
	6-10 years	3
	11-20 years	14
	Over 20 years	14
Education level	Undergraduate	25
	Postgraduate	11
Unit	Aerodrome Control Towers (TWR)	17
	Approach Control Unit (APP)	11

continued on following page

Table 1. Continued

		n
	Area Control Centre (ACC)	8
Providing ATC service to UAVs	Yes	32
	No	4
Total		36

Data Collection

A semi-structured set of questions was developed to be used in the interviews. This set included four open-ended "Why" and "How" questions to provide flexibility during interviews and to explore the challenges of UAV flights in the real air traffic environment within a limited framework, as shown in Table 2. The content, accuracy, and language usage of the questions were reviewed three times by experts. Field experts reviewed the questions to ensure they were appropriate, clear, and aligned with the objectives of the study. Necessary adjustments were made to the questions based on the feedback from the experts. The content, language usage, and overall structure of the form were reviewed, and the final version was provided. Alongside the four main questions asked of each participant, follow-up questions were employed to gain more detailed insights. Observations were conducted in addition to interviews with participants, involving participation in the operational process, to enhance the trustworthiness of the study (Creswell & Miller, 2000). The observations were conducted briefly during the provision of ATC services to UAVs at the ATC-OC (Air Traffic Control Operation Center). Short field notes were taken after the observations, and as the findings from the analysis of the interview recordings encompassed the field notes, the observational findings were not presented separately.

The interviews with air traffic controllers working in APP and ACC were conducted face-to-face, while those in TWR were conducted online. The interviews were audio-recorded, with a total recording time of 545 minutes.

Table 2. Semi-structured interview questions

How do air traffic controllers currently handle unmanned aerial vehicle flights within airspace? **Why is the integration of unmanned aerial vehicles into airspace considered challenging from an operational perspective?** **Why are there more challenges in managing UAVs in traffic management compared to managing conventional traffic?** **How can the safe and effective integration of UAVs into airspace be ensured?**

Data Analysis

The data analysis involved using the content analysis technique, which consists of several steps, after transcribing the interview recordings into written format. Initially, the data was thoroughly examined for accuracy. Subsequently, grouping was conducted to identify potential codes. After combining the codes under 12 categories and naming the segments, four themes were created, as shown in Table 3. The generated codes and themes were then reviewed by another researcher, and consensus was reached on the identified codes and themes. Additionally, the study results were confirmed by presenting them to a group of participants to ensure the validity of the study (Johnson, 1997; Morse et al., 2002). Quotations were coded for each participant according to the sequence of their interviews as "ATC".

Table 3. Research themes

1.	Training	1.1. UAV operators training 1.2. ATC training 1.3. Experience sharing and observation
2.	Procedures	2.1. Segregated airspace 2.2. Centralized system 2.3. Procedural deficiencies & development
3.	Technology	3.1. Surveillance/ communication 3.2. UAV level capabilities 3.3. Collision avoidance
4.	Legal regulation	4.1. Operational regulation 4.2. Safety 4.3. Licensing

RESULTS AND DISCUSSIONS

The research themes emerged after coding, with results and discussions provided under each sub-heading. This structure allows for a comprehensive analysis of the findings in relation to each aspect of the study.

Training

Significant findings were obtained when considering the coding related to training in the study. Participants indicated insufficient awareness among UAV operators regarding traffic flow in airspace (ATC25, ATC32) and a general lack of training (ATC8, ATC6, ATC22). It was highlighted that air traffic controllers also required training in technical knowledge and flight principles related to UAVs (ATC4, ATC19, ATC36). Some participants explicitly expressed the lack of education within ATC (ATC34) and the necessity for UAV technical knowledge and flight principles training for ATC (ATC19). Additionally, it was mentioned that UAV operators should receive training in aviation rules and communication (ATC24, ATC31), and such training could address their deficiencies in aviation knowledge and rules (ATC5, ATC25). Some participants remarked that the competency of UAV operators is debatable (ATC13), highlighting the importance of experience sharing (ATC14). Furthermore, it was noted that observation of operations between air traffic controllers and UAV operators would be beneficial (ATC10). The sub-coding and examples of participant quotations are presented in Table 4.

Table 4. Training examples of participant quotations

Sub-coding	Examples of participant quotations
UAV operators training	**ATC24:** *"UAV operators need training on air traffic rules and communication. It is really crucial for them to understand how to navigate the airspace properly."* **ATC25:** *"UAV operators often lack awareness of the air traffic flow in the airspace. This can lead to potential conflicts and safety issues."*
ATC training	**ATC4:** *"ATC really needs more training. It is essential for them to stay updated with the latest practices."* **ATC19**: *"ATC should receive training on UAV technical knowledge and flight principles. This kind of education is crucial for effective air traffic management."*
Experience sharing and observation	**ATC10:** *"ATC should be given the chance to observe UAV flights at ground control stations. The operational processes of UAVs are pretty complex, and it is really important for ATC to understand those processes"* **ATC14:** *"I think there is not enough knowledge transfer between ATC and UAV operators. This is because both sides do not fully understand the challenges they face in operational processes. "*

These findings suggest the need for reassessing both UAV operator and ATC training, emphasizing the importance of training supported by mutual experience and observation. Air traffic controllers highlight continuous training to ensure proficiency in their duties (Pavlinovic, Juricic & Antulov-Fantulin, 2017; Richardson, 1979; Schultheis, 2014). Air traffic controllers balance training with real-world performance for safe air traffic flow (Çınar & Tuncal, 2024). Air traffic controllers should be trained for new technology by ensuring they are ready for advancements, overcoming any challenges, and equipping them with the skills (Smith & Douglas, 1998). It is essential to note that UAVs represent a form of this new technology, possessing distinct features and operational characteristics compared to conventional manned aircraft. Knowledge of UAV features and operational requirements is crucial for air traffic controllers to manage UAVs in controlled airspace effectively (Liu, Zhang & Guan, 2018; Nyboe et al., 2023; Park, Lee & Im, 2014; Theunissen et al.,

2005). Therefore, integrating UAV-specific training modules into the curriculum of air traffic controller training programs is essential. On the other hand, the training of UAV operators, despite being a relatively new concept in aviation, is deemed a crucial necessity. Annex 1 by the International Civil Aviation Organization (ICAO) highlights the importance of UAV operators possessing knowledge in specific areas related to their licensing. These areas encompass fundamental aviation subjects such as air law, navigation, and radiotelephony (ICAO, 2022). Furthermore, it is imperative that UAV operators highlight training not solely for routine operations but also for unforeseen and emergency scenarios, showcasing essential competencies throughout the selection process (Beneigh, 2021). In conclusion, emphasizing the comprehensive training of UAV operators and air traffic controllers in UAV operations is vital. This training is essential not only for their own proficiency but also for minimizing challenges that air traffic controllers may encounter. Additionally, promoting seamless cooperation in shared airspace operations requires mutual experience sharing and observation. Mutual experience sharing and observation result in more positive evaluations of one another's task capabilities (Wittenbaum & Bowman, 2004), leading to enhanced understanding and increased cohesion between ATC and UAV operators (Dunn & Holt, 2004).

Procedure

Significant findings were obtained when considering the coding related to procedure issues in the study. Participants emphasized the need for specialized airspace models for UAVs (ATC1, ATC20) and suggested that UAVs flying in remote areas away from conventional flight routes, highlighting potential risks and delays (ATC18, ATC23, ATC30, ATC33). Furthermore, they advocated for conducting UAV flights in segregated airspace (ATC12, ATC15, ATC17) and organizing them in a manner that did not pose problems for other air traffic (ATC19). Participants also pointed out procedural deficiencies, stating that there were shortcomings in ATC procedures for UAV flights (ATC34) and suggesting the development of new procedures to address these deficiencies (ATC1). Additionally, they highlighted the importance of establishing a centralized management structure for coordinating between UAVs and conventional aircraft (ATC1, ATC2, ATC27, ATC28). Suggestions such as creating a joint management center for UAV operations (ATC1) and coordinating UAVs in the same area through a single center (ATC26) stood out in this regard. Restricting UAV flights in specific areas (ATC17), defining UAV flight corridors that did not interfere with conventional aircraft routes (ATC7, ATC11, ATC24), establishing flight procedures in case of radio loss (ATC9), and ensuring direct communication between the established center and ATC facilitated safer and more organized operations (ATC29). The sub-coding and examples of participant quotations are presented in Table 5.

Table 5. Procedure examples of participant quotations

Sub-coding	Participant quotations
Segregated airspace	**ATC18**: *"UAV flights should happen well away from conventional aircraft routes. Keeping them in separate areas ensures safety and minimizes potential conflicts."* **ATC20**: *"UAVs need to keep flying within designated airspace. It is important to establish clear no-fly zones to avoid interference with manned aircraft."* **ATC24**: *"We need to establish dedicated flight corridors for UAVs that won't interfere with conventional aircraft routes. It is all about keeping the airspace organized and safe for everyone."*
Centralized system	**ATC1:** *"Creating a unified management center for UAV operations would help ensure smoother coordination across different regions."* **ATC2:** *" Unmanned Traffic Management (UTM) hub should also function as a collaboration point within the ATM structure, helping all parties work together more effectively."* **ATC29:** *"We should set up a central hub for direct communication and coordination. It will streamline the process and improve response times."*
Procedural deficiencies & development	**ATC1:** *"Developing proper procedures for UAV operations is essential. It is the only way to handle the increasing traffic safely."* **ATC9:** *"In case of a UAV radio failure, we need a clear path set up in advance. It helps avoid confusion and ensures a safe resolution."* **ATC34:** *"There are gaps in the current ATC procedures when it comes to UAV flights. We really need to address those to ensure smoother operations."*

These findings indicate the importance of discussing segregated airspace and a centralized system for UAVs, as well as addressing procedural deficiencies and development. Current ATC procedures may not fully address the unique operational characteristics and requirements of UAV flights, particularly in shared airspace scenarios. ATC procedures are crucial for managing large numbers of manned and unmanned aircraft with minimal human intervention, especially in controlled airspace with UAVs (Chin et al., 2022; Meyer et al., 2018). By updating ATC procedures to incorporate UAV operations, airspace authorities can establish clear guidelines for UAV operations, minimizing disruptions to conventional air traffic and optimizing airspace utilization. Segregated airspace and corridors, dedicated specifically to UAV operations, offer a solution to mitigate potential conflicts and enhance safety by providing a controlled environment for UAV flights. Initially, the focus was on having separate airspace for UAVs to ensure safety and prevent interference with conventional aircraft (Simpson, Stoker & Brennan, 2009). However, the current trend is shifting towards integrating UAVs into the same airspace (Taraglio, Nanni & Taurino, 2012). This shift emphasizes the need for updated procedures and advanced management systems like Unmanned Traffic Management (UTM) to ensure safe and efficient operations. UTM encompasses systems and technologies designed to manage UAV operations in airspace, ensuring their safe integration with manned aircraft and other airspace users (Alkadi & Shoufan, 2021; Carraminana et al., 2021). UTM plays a crucial role in coordinating UAV flights, providing real-time information, and managing airspace congestion. However, effective UTM implementation requires seamless interaction and collaboration with conventional ATM systems to ensure airspace safety and efficiency. Integration between UTM and ATM systems is essential to facilitate the exchange of data, coordination of flights, and resolution of potential conflicts between UAVs and manned aircraft (Fasano et al., 2023). The integration of UTM with updated ATC procedures is essential to address the evolving dynamics of airspace management, ensuring the safe coexistence of both UAVs and manned aircraft while optimizing the use of airspace resources.

Technology

Significant findings were obtained when considering the coding related to technological issues in the study. Participants noted a major deficiency in UAVs due to the absence of collision avoidance systems like TCAS (ATC3, ATC35). Additionally, serious issues arose in UAV surveillance and detection, such as radar losses (ATC4, ATC19). The lack of bidirectional communication capability between UAVs and ATC created operational challenges and communication issues. The problem expressed as ATC-UAV communication problem complicated the safe and effective management of UAVs (ATC4, ATC32). Furthermore, unreliable altitude readings in UAVs triggering TCAS alarms led to altitude-related issues (ATC18, ATC21). Technical deficiencies in UAVs, particularly in communication and surveillance areas, were notably evident (ATC25, ATC35). The sub-coding and examples of participant quotations are presented in Table 6.

Table 6. Technology examples of participant quotations

Sub-coding	Participant quotations
Surveillance/ communication	**ATC4:** *"There is a surveillance issue, we are experiencing radar detection loss."* **ATC32:** *"UAVs do not have two-way communication with ATC, which is an operational challenge."*
UAV level capabilities	**ATC18:** *"Conventional aircraft are facing TCAS alerts due to level busts, and this is becoming a recurring problem."* **ATC21:** *"There is an issue with the reliability of UAV altitudes, and it can trigger the TCAS system unnecessarily."*
Collision avoidance	**ATC35:** *"It is important for UAVs to be equipped with a system like TCAS to prevent collisions. This would help them navigate safely in congested airspace."*

These findings emphasize the critical necessity for technological improvements concerning surveillance/communication capabilities, level capabilities, and collision avoidance systems of UAVs. Integrating collision avoidance systems, like TCAS, into UAVs is crucial for enhancing safety and avoiding mid-air collisions (Xu, 2013; Zhang et al., 2018). A visual input-based see-and-avoid system for UAVs serves as the main collision avoidance system (Zsedrovits et al., 2011). However, the see-and-avoid concept in civil aviation has significant limitations that hinder its effectiveness in preventing midair collisions (Morris, 2005). Therefore, equipping UAVs with TCAS systems similar to those in conventional aircraft is essential for safety (Bolz & Nowacki, 2023). Problems related to UAV surveillance and detection, such as radar losses, reveal vulnerabilities that could lead to operational failures, highlighting the importance of robust and reliable surveillance systems. Efficient ATC systems rely on effective surveillance and detection mechanisms (Gui et al., 2020; Sridhar & Chatterji, 1997). Hence, there is a crucial need for an effective UAS detection and tracking method (Ganti & Kim, 2016). The lack of bidirectional communication capabilities between UAVs and ATC complicates operational management and poses significant communication challenges, indicating the urgent requirement for an enhanced communication infrastructure (Matus & Materna, 2021). In addition to conventional voice communication between ATC and UAV operators, communication can also be established between ATC and UAVs via datalink. Datalink is a technology that facilitates data communication between aircraft and ATC (Sharples et al., 2007). This technology is particularly significant for UAVs, as conventional voice communication can sometimes be inadequate, and datalink communication offers a safer and more effective alternative. Moreover, improving altitude measurement systems in UAVs is a crucial priority to guarantee precise altitude data and mitigate altitude-related incidents. ADS-B technology could be perceived as a solution in this regard. ADS-B utilizes satellite-based navigation and positioning data to relay aircraft location and altitude information to air traffic controllers and neighboring aircraft (Ali, 2017). Ensuring safe and efficient airspace management necessitates technological advancements and the establishment of effective communication and surveillance systems between UAVs and ATC.

Legal Regulation

Significant findings were obtained when considering the coding related to technological issues in the study. Participants emphasized the need for international determination of separation minima between UAVs and conventional aircraft (ATC20). It was indicated that there were generally deficiencies in the regulations, particularly regarding flight safety (ATC24). Regulations were suggested to be established regarding UAVs' compliance with General Air Traffic (GAT) rules in

GAT airspace (ATC12), and detailed regulations concerning airspace usage were deemed necessary (ATC6). The lack of licensing for UAV operators was perceived as a significant issue (ATC16), and it was emphasized that responsibilities needed to be clearly defined in case of potential accidents and incidents (ATC8). Furthermore, it was suggested that a system should be established to take necessary preventive and regulatory measures in the event of rule violations (ATC17). The sub-coding and examples of participant quotations are presented in Table 7.

Table 7. Legal regulation examples of participant quotations

Sub-coding	Participant quotations
Operational regulation	**ATC6:** *"We need regulations on airspace usage, specifically for separation procedures between UAVs and between UAVs and conventional aircraft."* **ATC12:** *"There needs to be regulation ensuring that UAVs in GAT airspace comply with GAT rules."*
Safety	**ATC17:** *"To enhance safety, a system needs to be established for preventive and corrective measures in case of rule violations."* **ATC24**: *"There are gaps in regulations when it comes to flight safety."*
Licensing	**ATC8:** *"The fact that UAV operators aren't currently licensed is a problem; we need to define accountability in case of incidents or accidents."* **ATC16:** *"Current regulations need to be updated to allow for licensing of UAV operators, just like in conventional aviation."*

These findings highlight the urgent need to update existing legal regulations that govern UAV operations to address shortcomings in operation, safety, and licensing. This ensures that regulations stay relevant and effective in managing the complexities of UAV operations. One key reason for updating legal regulations is the rapid advancements in UAV technology (Abro et al., 2022; Cracknell, 2017; Nakamura & Kajikawa, 2017). As UAV capabilities evolve, current regulations may become outdated or insufficient to address new challenges. Without timely updates, regulatory gaps could compromise airspace safety and efficiency. Moreover, new legislation is necessary to fill existing gaps and tackle emerging issues specific to UAV operations. Initially, UAVs were regulated by conventional aviation laws, which required training, compliance with rules, and registering operators and their drones to ensure traceability (Cracknell, 2017). However, these conventional laws may not fully cover the unique characteristics and needs of UAVs, leading to ambiguity or inadequate oversight (Srivastava, 2020). Creating new laws specifically for UAVs can provide clearer regulations on various issues (Cracknell, 2017; Kochan, Kruk & Rutkowska, 2018), such as responsibility frameworks, airspace access, licensing, flight rules, separation, and safety. Authorities can improve regulatory compliance and promote industry standardization by establishing a dedicated legal regulation for UAV operations.

CONCLUSION

The integration of UAVs into conventional airspace poses significant challenges for ATC. These challenges arise from the unique characteristics of UAVs. In the study, a comprehensive analysis of the challenges faced by ATC in managing UAV flights is presented. The analysis revealed four primary categories: training, operational procedures, technological advancements, and legal regulations.

Firstly, there is a clear need for enhanced training for both UAV operators and air traffic controllers. This training should focus on improving awareness of airspace traffic flow among UAV operators and providing air traffic controllers with the necessary technical knowledge to manage UAVs effectively. Integrating specialized UAV training modules into existing training programs is essential for adequately preparing personnel for UAV operations. Secondly, updating ATC procedures and integrating UTM systems is crucial. The analysis shows that initially, UAVs were restricted to flying in limited areas. However, now UAV flights are expanding to encompass all airspace, highlighting the importance of the UTM system. By updating procedures and incorporating UTM systems, airspace authorities can establish clear guidelines for UAV operations, reducing disruptions to conventional air traffic while maximizing airspace utilization. Furthermore, technological enhancements are crucial for enhancing communication and surveillance between UAVs and ATC. Integration of collision avoidance systems, improvement of surveillance and detection capabilities, establishment of bidirectional communication channels, and enhancement of altitude measurement systems are necessary for safe and efficient airspace management. Lastly, there is an urgent need to establish a comprehensive legal regulation governing UAV operations. Current regulations often fail to ensure flight safety and comply with General Air Traffic (GAT) rules. Establishing licensing requirements for UAV operators at the national level, clarifying responsibilities in the event of accidents and incidents, and implementing systems to address rule violations are essential steps. Developing specific legal regulations tailored to UAV operations is crucial for providing more precise guidelines on airspace access, licensing, flight rules, separation, and safety.

The study highlights the necessary actions across training, procedures, technology, and legal frameworks to ensure the safe and efficient integration of UAVs into airspace. Implementing these recommendations will not only support the growth of the aviation sector but also improve flight safety within ATC. Given the rapid advancements in UAV technology, future research is recommended to explore integration challenges, particularly for simultaneous operations of UAVs and conventional aircraft.

REFERENCES

Aber, J., Marzolff, I., Ries, J., & Aber, S. (2019). *Small-format aerial photography and UAS imagery: Principles, techniques and geoscience applications*. Academic Press.

Abro, G., Zulkifli, S., Masood, R., Asirvadam, V., & Laouti, A. (2022). Comprehensive Review of UAV Detection, Security, and Communication Advancements to Prevent Threats. *Drones (Basel)*, 6(10), 284. DOI: 10.3390/drones6100284

Adams, W. C. (2015). Conducting semi-structured interviews. *Handbook of practical program evaluation*, 492-505. DOI: 10.1002/9781119171386.ch19

Ali, B. (2017). *Automatic Dependent Surveillance Broadcast (ADS-B)*. Routledge., DOI: 10.4324/9781315566382-4

Alkadi, R., & Shoufan, A. (2021). Unmanned Aerial Vehicles Traffic Management Solution Using Crowd-Sensing and Blockchain. *IEEE Transactions on Network and Service Management*, 20(1), 201–215. DOI: 10.1109/TNSM.2022.3201817

Austin, Z., & Sutton, J. (2014). Qualitative research: Getting started. *The Canadian Journal of Hospital Pharmacy*, 67(6), 436–440. DOI: 10.4212/cjhp.v67i6.1406 PMID: 25548401

Aydin, B. (2019). Public acceptance of drones: Knowledge, attitudes, and practice. *Technology in Society*, 59, 101180. DOI: 10.1016/j.techsoc.2019.101180

Barriball, K., & While, A. (1994). Collecting data using a semi-structured interview: A discussion paper. *Journal of Advanced Nursing*, 19(2), 328–335. DOI: 10.1111/j.1365-2648.1994.tb01088.x PMID: 8188965

Beneigh, T. (2021). Unmanned Aircraft System Operations. Barnhart, R. K., Marshall, D. M., & Shappee, E. (Eds.). (2021). *Introduction to unmanned aircraft systems*. Crc Press.

Bertaux, D. (1981). *From the life-history approach to the transformation of sociological practice*. Biography and society: The life history approach in the social sciences. California: SAGE Publications.

Bolz, K., & Nowacki, G. (2023). Air transport safety in UAV operational conditions. *Journal of Civil Engineering and Transport*, 5(1), 9–23. Advance online publication. DOI: 10.24136/tren.2023.001

Carraminana, D., Campana, I., Bergesio, L., Bernardos, A., & Besada, J. (2021). Sensors and Communication Simulation for Unmanned Traffic Management. *Sensors (Basel)*, 21(3), 927. DOI: 10.3390/s21030927 PMID: 33573192

Carruthers, J. (1990). A Rationale for the Use of Semi-Structured Interviews. *Journal of Educational Administration*, 28(1), 63–68. DOI: 10.1108/09578239010006046

Chamata, J., & Winterton, J. (2018). A conceptual framework for the acceptance of drones. *The International Technology Management Review*, 7(1), 34–46. DOI: 10.2991/itmr.7.1.4

Charmaz, K. (2006). *Constructing Grounded Theory: A Practical Guide Through Qualitative Analysis*. SAGE.

Chin, C., Gopalakrishnan, K., Balakrishnan, H., Egorov, M., & Evans, A. (2022). Protocol-Based Congestion Management for Advanced Air Mobility. *Journal of Air Transportation (Reston, Va.)*, 31(1), 35–44. DOI: 10.2514/1.D0298

Chiper, F. L., Martian, A., Vladeanu, C., Marghescu, I., Craciunescu, R., & Fratu, O. (2022). Drone detection and defense systems: Survey and a software-defined radio-based solution. *Sensors (Basel)*, 22(4), 1453. DOI: 10.3390/s22041453 PMID: 35214355

Çınar, E., & Tuncal, A. (2023). The Future of UAVs in Urban Air Mobility: Public Perception and Concerns. *Türkiye İnsansız Hava Araçları Dergisi*, 5(2), 50–58. DOI: 10.51534/tiha.1381175

Çınar, E., & Tuncal, A. (2024). A comprehensive analysis of competency and training perspectives among air traffic controllers. *Aviation*, 28(2), 54–63. DOI: 10.3846/aviation.2024.21501

Cracknell, A. (2017). UAVs: Regulations and law enforcement. *International Journal of Remote Sensing*, 38(8-10), 3054–3067. DOI: 10.1080/01431161.2017.1302115

Creswell, J. W., & Miller, D. L. (2000) Determining validity in qualitative inquiry. *The ory into Practice,* 39(3), 124–130.

Crowe, S., Cresswell, K., Robertson, A., Huby, G., Avery, A., & Sheikh, A. (2011). The case study approach. *BMC Medical Research Methodology*, 11(1), 100–100. DOI: 10.1186/1471-2288-11-100 PMID: 21707982

D'Amato, E., Notaro, I., & Mattei, M. (2018, June). Distributed collision avoidance for unmanned aerial vehicles integration in the civil airspace. In *2018 International Conference on Unmanned Aircraft Systems (ICUAS)* (pp. 94-102). IEEE. DOI: 10.1109/ICUAS.2018.8453432

DeJonckheere, M., & Vaughn, L. (2019). Semistructured interviewing in primary care research: A balance of relationship and rigour. *Family Medicine and Community Health*, 7(2), e000057. Advance online publication. DOI: 10.1136/fmch-2018-000057 PMID: 32148704

Denny, E., & Weckesser, A. (2022). How to do qualitative research? *BJOG*, 129(7), 1166–1167. DOI: 10.1111/1471-0528.17150 PMID: 35430773

Diiulio, J., Militello, L., & Klein, D. (2020). UAS integration in congested terminal airspace: Challenges posed to pilots. *Journal of Unmanned Vehicle Systems*, 8(2), 79–88. DOI: 10.1139/juvs-2019-0015

Doody, O., & Noonan, M. (2013). Preparing and conducting interviews to collect data. *Nurse Researcher*, 20(5), 28–32. DOI: 10.7748/nr2013.05.20.5.28.e327 PMID: 23687846

Dunn, J., & Holt, N. (2004). A Qualitative Investigation of a Personal-Disclosure Mutual-Sharing Team Building Activity. *The Sport Psychologist*, 18(4), 363–380. DOI: 10.1123/tsp.18.4.363

Eibfeldt, H., Vogelpohl, V., Stolz, M., Papenfub, A., Biella, M., Belz, J., & Kügler, D. (2020). The acceptance of civil drones in Germany. *CEAS Aeronautical Journal*, 11(3), 665–676. DOI: 10.1007/s13272-020-00447-w

Fasano, G., Crespillo, O., Sabatini, R., Roy, A., & Ogan, R. (2023). From the Editors of the Special Issue on Urban Air Mobility and UAS Airspace Integration: Vision, Challenges, and Enabling Avionics Technologies. *IEEE Aerospace and Electronic Systems Magazine*, 38(5), 4–5. Advance online publication. DOI: 10.1109/MAES.2023.3266992

Federal Aviation Administration [FAA] (2024). *Drones by the Numbers (as of 2/29/24).* Retrieved May 18, 2024, from https://www.faa.gov/uas

Fossey, E., Harvey, C., McDermott, F., & Davidson, L. (2002). Understanding and Evaluating Qualitative Research. *The Australian and New Zealand Journal of Psychiatry*, 36(6), 717–732. DOI: 10.1046/j.1440-1614.2002.01100.x PMID: 12406114

Ganti, S., & Kim, Y. (2016). Implementation of detection and tracking mechanism for small UAS. *2016 International Conference on Unmanned Aircraft Systems (ICUAS),* 1254-1260. DOI: 10.1109/ICUAS.2016.7502513

Gilgun, J. (1994). A Case for Case Studies in Social Work Research. *Social Work*, 39, 371–380. DOI: 10.1093/sw/39.4.371

Grady, M. P. (1998). Qualitative and action research: A practitioner handbook. Phi Delta Kappa International. *Archives of Sexual Behavior*, 41, 1319–1320.

Green, J., & Thorogood, N. (2018). *Qualitative Methods for Health Research*. SAGE.

Guan, X., Lyu, R., Shi, H., & Chen, J. (2020). A survey of safety separation management and collision avoidance approaches of civil UAS operating in integration national airspace system. *Chinese Journal of Aeronautics*, 33(11), 2851–2863. DOI: 10.1016/j.cja.2020.05.009

Gui, G., Zhou, Z., Wang, J., Liu, F., & Sun, J. (2020). Machine Learning Aided Air Traffic Flow Analysis Based on Aviation Big Data. *IEEE Transactions on Vehicular Technology*, 69(5), 4817–4826. DOI: 10.1109/TVT.2020.2981959

Hartmann, J. H. (2014). *Feasibility study of Air Traffic Control Towers around the globe: International research regarding the local influences providing an optimal structural design for air traffic control towers around the globe in an economical perspective*, Master thesis, TU Delft, Delft University of Technology.

Hatfield, M., Cahill, C., Webley, P., Garron, J., & Beltran, R. (2020). Integration of Unmanned Aircraft Systems into the National Airspace System-Efforts by the University of Alaska to Support the FAA/NASA UAS Traffic Management Program. *Remote Sensing (Basel)*, 12(19), 3112. DOI: 10.3390/rs12193112

Ho, F., Geraldes, R., Gonçalves, A., Rigault, B., Sportich, B., Kubo, D., Cavazza, M., & Prendinger, H. (2019). Decentralized Multi-Agent Path Finding for UAV Traffic Management. *IEEE Transactions on Intelligent Transportation Systems*, 23(2), 997–1008. DOI: 10.1109/TITS.2020.3019397

Hunt, V., & Zellweger, A. (1987). The FAA's Advanced Automation System: Strategies for Future Air Traffic Control Systems. *Computer*, 20(2), 19–32. DOI: 10.1109/MC.1987.1663475

International Civil Aviation Organization [ICAO]. (2022). *Annex 1, Personnel Licensing* (Thirteenth Edition). Montréal, Quebec, Canada.

International Civil Aviation Organization [ICAO]. (2024). *Increased use of unmanned aircraft systems (UAS)*. Retrieved May 18, 2024, from https://www.icao.int/annual-report-2021/Pages/emerging-and-cross-cutting-aviation-issues-increased-use-of-unmanned-aircraft-systems-uas.aspx

Johnson, R. (1997). Examining the Validity Structure of Qualitative Research. *Education 3-13*, 118, 282.

Kochan, A., Kruk, A., & Rutkowska, P. (2018). Prospects for unmanned aerial vehicles market development in Poland. *WUT Journal of Transportation Engineering*, 123, 41–53. DOI: 10.5604/01.3001.0013.7352

Labib, N., Danoy, G., Musial, J., Brust, M., & Bouvry, P. (2019). Internet of Unmanned Aerial Vehicles—A Multilayer Low-Altitude Airspace Model for Distributed UAV Traffic Management. *Sensors (Basel)*, 19(21), 4779. DOI: 10.3390/s19214779 PMID: 31684133

Liu, Y., Zhang, X., & Guan, X. (2018). Framework Design of Unmanned Aerial Systems (UAS) Integration in the National Airspace System (NAS). *2018 13th World Congress on Intelligent Control and Automation (WCICA)*, 1323-1327. DOI: 10.1109/WCICA.2018.8630592

Longhurst, R. (2009). Interviews: In-depth, semi-structured. *International encyclopedia of human geography*, 580-584. DOI: 10.1016/B978-008044910-4.00458-2

Macias, M., Barrado, C., Pastor, E., & Royo, P. (2019). The future of drones and their public acceptance. In *2019 IEEE/AIAA 38th Digital Avionics Systems Conference (DASC)* (pp. 1-8). IEEE. DOI: 10.1109/DASC43569.2019.9081623

MacSween, S. (2003). A public opinion survey-unmanned aerial vehicles for cargo, commercial, and passenger transportation. In *2nd AIAA" Unmanned Unlimited" Conf. and Workshop & Exhibit* (p. 6519). DOI: 10.2514/6.2003-6519

Macsween-George, S. L. (2003). Will the public accept UAVs for cargo and passenger transportation? In *2003 IEEE Aerospace Conference Proceedings (Cat. No. 03TH8652)* (Vol. 1, pp. 1-367). IEEE. DOI: 10.1109/AERO.2003.1235066

Matolak, D. (2015). Unmanned aerial vehicles: Communications challenges and future aerial networking. *2015 International Conference on Computing, Networking and Communications (ICNC)*, 567-572. DOI: 10.1109/ICCNC.2015.7069407

Matus, P., & Materna, M. (2021). Approaches of chosen air navigation services providers to UAV integration into air traffic control systems. *Práce a štúdie - Vydanie 10*. DOI: 10.26552/pas.Z.2021.2.26

Meyer, D., Wypych, T., Petrovic, V., Strawson, J., Kamat, S., & Kuester, F. (2018). *An air traffic control simulator for test and development of airspace management schemes. In 2018 IEEE Aerospace Conference, 1-8.*, DOI: 10.1109/AERO.2018.8396575

Morris, C. (2005). Midair collisions: Limitations of the see-and-avoid concept in civil aviation. *Aviation, Space, and Environmental Medicine*, 76(4), 357–365. PMID: 15828635

Morse, J., Barrett, M., Mayan, M., Olson, K., & Spiers, J. (2002). Verification Strategies for Establishing Reliability and Validity in Qualitative Research. *International Journal of Qualitative Methods*, 1(2), 13–22. DOI: 10.1177/160940690200100202

Mueller, E. R., Kopardekar, P. H., & Goodrich, K. H. (2017). Enabling airspace integration for high-density on-demand mobility operations. In *17th AIAA Aviation Technology, Integration, and Operations Conference* (p. 3086). DOI: 10.2514/6.2017-3086

Nakamura, H., & Kajikawa, Y. (2017). Regulation and innovation: How should small unmanned aerial vehicles be regulated? *Technological Forecasting and Social Change*, 128, 262–274. DOI: 10.1016/j.techfore.2017.06.015

Nyboe, F., Malle, N., Hoang, V., & Ebeid, E. (2023). Open-Source Hardware/Software Architecture for Autonomous Powerline-Aware Drone Navigation and Recharging. *2023 International Conference on Unmanned Aircraft Systems (ICUAS)*, 1233-1240. DOI: 10.1109/ICUAS57906.2023.10155937

Ohman, A. (2005). Qualitative methodology for rehabilitation research. *Journal of Rehabilitation Medicine*, 37(5), 273–280. DOI: 10.1080/16501970510040056 PMID: 16208859

Park, W., Lee, K., & Im, K. (2014). A Proposal on the Aviation Rules of the Military UAV in the National Airspace System. *Journal of the Korean Society for Aviation and Aeronautics*, 22(1), 22–31. DOI: 10.12985/ksaa.2014.22.1.022

Pavlinovic, M., Juricic, B., & Antulov-Fantulin, B. (2017). Air traffic controllers' practical part of basic training on computer based simulation device. *2017 40th International Convention on Information and Communication Technology, Electronics and Microelectronics (MIPRO)*, 920-925. DOI: 10.23919/MIPRO.2017.7973554

Persiani, C., & Bagassi, S. (2013). Route planner for unmanned aerial system insertion in civil non-segregated airspace. *Proceedings of the Institution of Mechanical Engineers. Part G, Journal of Aerospace Engineering*, 227(4), 687–702. DOI: 10.1177/0954410012439975

Ribeiro, L., Giles, S., Katkin, R., Topiwala, T., & Minnix, M. (2017). Challenges and opportunities to integrate UAS in the National Airspace System. 2017 Integrated Communications,*Navigation and Surveillance Conference (ICNS)*. DOI: 10.1109/ICNSURV.2017.8011942

Richardson, H. (1979). The FAA Approach to Air Traffic Controller Training. *Journal of Educational Technology Systems*, 7(4), 373–381. DOI: 10.2190/BVNB-23HF-KQ8C-D6GL

Sandor, Z. (2017). Challenges Caused by the Unmanned Aerial Vehicle in the Air Traffic Management. *Periodica Polytechnica Transportation Engineering*, 47(2), 96–105. DOI: 10.3311/PPtr.11204

Sargeant, J. (2012). Qualitative Research Part II: Participants, Analysis, and Quality Assurance. *Journal of Graduate Medical Education*, 4(1), 1–3. DOI: 10.4300/JGME-D-11-00307.1 PMID: 23451297

Sarim, M., Radmanesh, M., Dechering, M., Kumar, M., Pragada, R., & Cohen, K. (2019). Distributed Detect-and-Avoid for Multiple Unmanned Aerial Vehicles in National Air Space. *Journal of Dynamic Systems, Measurement, and Control*, 141(7), 071014. Advance online publication. DOI: 10.1115/1.4043190

Schultheis, S. (2014). Integrating advanced technology into air traffic controller training. In *14th AIAA Aviation Technology, Integration, and Operations Conference* (p. 2421). DOI: 10.2514/6.2014-2421

Sharples, S., Stedmon, A., Cox, G., Nicholls, A., Shuttleworth, T., & Wilson, J. (2007). Flightdeck and air traffic control collaboration evaluation (FACE): Evaluating aviation communication in the laboratory and field. *Applied Ergonomics*, 38(4), 399–407. DOI: 10.1016/j.apergo.2007.01.012 PMID: 17499573

Shrestha, R., Oh, I., & Kim, S. (2021). A survey on operation concept, advancements, and challenging issues of urban air traffic management. *Frontiers in Future Transportation*, 2, 1. DOI: 10.3389/ffutr.2021.626935

Simpson, A., Stoker, J., & Brennan, V. (2009). *UAS Safety in Non-segregated Airspace*. INTECH Open Access Publisher. DOI: 10.5772/6492

Simpson, A. J., & Stoker, J. (2006). Safety challenges in flying UAVS (unmanned aerial vehicles) in non-segregated airspace. In *The First Institution of Engineering and Technology International Conference on System Safety, 2006.* (pp. 8-pp). IET.

Smith, A., & Douglas, A. (1998). Introducing new technology to the air traffic controller: Implications for skill acquisition and training. *17th DASC. AIAA/IEEE/SAE. Digital Avionics Systems Conference. Proceedings (Cat. No.98CH36267)*, 2, F54/1-F54/8 vol.2. DOI: 10.1109/DASC.1998.739833

Smith, J. (1996). Qualitative methodology: Analyzing participants' perspectives. *Current Opinion in Psychiatry*, 9(6), 417–421. DOI: 10.1097/00001504-199611000-00010

Sridhar, B., & Chatterji, G. (1997). *Computationally efficient conflict detection methods for air traffic management*. Proceedings of the 1997 American Control Conference (Cat. No.97CH36041), 2, 1126-1130 vol.2. DOI: 10.1109/ACC.1997.609708

Srivastava, S., Gupta, S., Dikshit, O., & Nair, S. (2020). A review of UAV regulations and policies in India. *Proceedings of UASG 2019: Unmanned Aerial System in Geomatics 1*, 315-325. DOI: 10.1007/978-3-030-37393-1_27

Stiles, W. (1993). Quality control in qualitative research. *Clinical Psychology Review*, 13(6), 593–618. DOI: 10.1016/0272-7358(93)90048-Q

Tan, L. K. L., Lim, B. C., Park, G., Low, K. H., & Yeo, V. C. S. (2021). Public acceptance of drone applications in a highly urbanized environment. *Technology in Society*, 64, 101462. DOI: 10.1016/j.techsoc.2020.101462

Taraglio, S., Nanni, V., & Taurino, D. (2012). Hardware on-Board Implementation of a Possible Solution to Introduce UAVs into Non Segregated Areas. In *ICINCO (1)* (pp. 177-182).

Tepylo, N., Straubinger, A., & Laliberte, J. (2023). Public perception of advanced aviation technologies: A review and roadmap to acceptance. *Progress in Aerospace Sciences*, 138, 100899. DOI: 10.1016/j.paerosci.2023.100899

Theunissen, E., Goossens, A., Bleeker, O., & Koeners, J. (2005). UAV mission management functions to support integration in a strategic and tactical ATC and C2 environment. In *AIAA modeling and simulation technologies conference and exhibit* (p. 6310). DOI: 10.2514/6.2005-6310

Tuncal, A., & Uslu, S. (2021). Kentsel Hava Hareketliliği Kavramının Gelişiminde İki Önemli Faktör: ATM ve Toplum. *Karamanoğlu Mehmetbey Üniversitesi Sosyal ve Ekonomik Araştırmalar Dergisi*, 23(41), 564–577.

Vishnevsky, T., & Beanlands, H. (2004). Qualitative research. *Nephrology Nursing Journal : Journal of the American Nephrology Nurses' Association*, 31-2, 234–238. DOI: 10.4324/9780203299975-17 PMID: 15114810

Wang, H., Dattel, A., Mummert, E., & Haris, S. (2022). Assessing Air Traffic Controllers' Stress and Performance with Uav Integration in Future Air Traffic Management. *Proceedings of the Human Factors and Ergonomics Society Annual Meeting*, 66(1), 38–38. DOI: 10.1177/1071181322661332

Wikipedia. (2024). List of unmanned aerial vehicle-related incidents. In *Wikipedia, The Free Encyclopedia*. Retrieved May 18, 2024, from https://en.wikipedia.org/wiki/List_of_unmanned_aerial_vehicle-related_incidents

Wittenbaum, G., & Bowman, J. (2004). A social validation explanation for mutual enhancement. *Journal of Experimental Social Psychology*, 40(2), 169–184. DOI: 10.1016/S0022-1031(03)00091-X

Xu, Y. (2013). TCAS/ADS-B integrated surveillance and collision avoidance system. In *Conference of the 2nd International Conference on Computer Science and Electronics Engineering (ICCSEE 2013)* (pp. 666-669). Atlantis Press. DOI: 10.2991/iccsee.2013.169

Zeng, Y., Zhang, R., & Lim, T. (2016). Wireless communications with unmanned aerial vehicles: Opportunities and challenges. *IEEE Communications Magazine*, 54(5), 36–42. DOI: 10.1109/MCOM.2016.7470933

Zhang, Z., Zhang, J., Wang, P., & Chen, L. (2018). Research on Operation of UAVs in Non-isolated Airspace. *Computers, Materials & Continua*, 57(1), 151–166. DOI: 10.32604/cmc.2018.02890

Zsedrovits, T., Zarándy, Á., Vanek, B., Péni, T., Bokor, J., & Roska, T. (2011). Visual Detection and Implementation Aspects of a UAV See and Avoid System. *2011 20th European Conference on Circuit Theory and Design (ECCTD)*, 472-475. DOI: 10.1109/ECCTD.2011.6043389

Chapter 10
Every Air Travel Begins With a Customer Journey:
Applying Service Design Principles in Airport Planning and Operations

Elad Harison
https://orcid.org/0000-0002-8356-5626
Shenkar College of Engineering and Design, Israel

Yael Lahav
Shenkar College of Engineering and Design, Israel & Bar Ilan University, Israel

ABSTRACT

This qualitative study analyzes and enhances the air travel experience by integrating observations and interviews with passengers and airport staff in Europe and the U.S. Seven observers conducted semi-structured interviews with 64 participants, mapping the entire passenger journey. Four passenger personas—Business Commuter, Family Traveler, Leisure Traveler, and Senior Traveler—were identified. The study provides a framework for assessing passenger experiences, offering guidelines for airport service improvements. By focusing on touchpoint analysis, pain points, and persona development, airports can enhance satisfaction through better service design. Emerging technologies such as biometric systems and AI may transform airport operations, with a holistic approach integrating customer journey mapping and service blueprinting. The concept of "dynamic personas" that adapt to real-time data is explored, as well as strategies for seamless touchpoints and collaboration between airlines, security, and airport operators.

DOI: 10.4018/979-8-3693-7403-0.ch010

CHAPTER SUMMARY

This study employs a qualitative approach to comprehensively analyze and improve the air travel experience, integrating observations and interviews with passengers and ground staff members. Seven observers were positioned across airports in Europe and the U.S., conducting semi-structured interviews with 64 passengers and staff members. The study documented the entire passenger journey, from pre-travel planning to final destination arrival, using customer journey mapping, touchpoint analysis, and persona creation. Four distinct passenger personas were identified: Business Commuter, Family Traveler, Leisure Traveler, and Senior Traveler. The methodology of the study constructed a framework for assessing passenger experiences and offers practical guidelines for service improvements in airports. By analyzing the effectiveness of each touchpoint and understanding the diverse needs of different passenger groups, airports can enhance overall satisfaction and create a more positive travel experience. This comprehensive approach provides valuable insights into factors influencing passenger satisfaction and highlights areas for potential improvements in airport operations, including analysis of touchpoints, pain points, and best practices. To complement the empirical methodology used in this study, a comprehensive overview of service design methodologies and their application in airport planning and operations is presented. The study explores the evolving landscape of customer experience in the aviation industry, focusing on employing qualitative research methods for optimizing the passenger journey through a multi-touchpoint analysis. By applying key service design concepts, such as customer journey mapping, touchpoint and pain point analyses, persona development, and human-centric design principles, airports can create more seamless and satisfying experiences for passengers while improving operational efficiency.

This study also examines the integration of emerging technologies, including biometric systems, artificial intelligence, and mobile applications, in transforming traditional airport operations and services to enhance various touchpoints throughout the airport experience of passengers. Further, the study emphasizes the importance of understanding the entire passenger journey, from pre-flight planning to post-flight experiences, and proposes an integrated framework that combines customer journey mapping with service blueprinting techniques. This holistic approach allows for a more comprehensive view of the airport service ecosystem, considering both passenger perspectives and operational processes. Additionally, the evolution of persona development in airport service design is based on the concept of "dynamic personas" that adapt to real-time data and changing passenger needs.

Frontstage actions play an essential role in shaping passenger perceptions and satisfaction, forming strategies for effective service recovery and the integration of technology in customer service roles. They also examine the application of human-centered design principles in various aspects of airport operations, including security design and wayfinding systems.

Identification and management of pain points across different airport types and sizes offer a classification scheme based on information, performance, and hospitality themes. They highlight the use of the concept of "seamless touchpoints" and the potential for coordinated efforts between airlines, security agencies, and airport operators to mitigate pain points and enhance overall passenger satisfaction.

By applying service design methodologies in airports within the realm of qualitative research methods, the contribution of this study to the growing body of literature on airport service design is significant. In particular, the study provides a comprehensive analysis of key service design methodologies and practical applications that can serve both managers and academics in the field of civil aviation in forming new approaches to improve airport operations and passenger experiences.

1. INTRODUCTION

Service design has emerged as a major discipline in contemporary customer and user experience-driven sectors. It includes the planning and organization of people, infrastructure, communication, and material components of a service to improve its quality and the customers' experience. Essentially, service design is about creating services that are useful, usable, efficient, effective, and desirable. It involves analyzing the needs of customers, service providers, and other stakeholders. Following these notions, designing, prototyping, and implementing solutions to meet these needs are carried out (Becker and Jaakkola, 2020; Tueanrat et al., 2021).

Service design is a multifaceted discipline that combines elements of design thinking, user experience, and business strategy to create services that meet user needs while achieving organizational goals. By focusing on the customer journey, touchpoints, pain points, and human-centric design principles, service designers can create experiences that are not only functional but also positively engaging and memorable (Vink et al., 2021).

One of the major methods applied in service design is the customer journey. This refers to the complete sequence of experiences a customer goes through when interacting with a company or with a brand. Understanding the customer journey provides insights into customer behavior, needs, and pain points at each stage of their interaction with a service.

The customer journey typically consists of several stages:

1. Awareness: When the customer first becomes aware of a need or a service.
2. Consideration: When the customer actively researches and compares options.

3. Purchase: The actual transaction or commitment to the service.
4. Service Delivery: The period during which the customer receives and uses the service.
5. Post-Service: The period after service delivery, including follow-ups and potential repurchases.

(Tueanrat et al., 2021; Dalkıran, 2023)

Mapping the customer journey involves creating a visual representation of these stages, often in the form of a timeline (West et al., 2020). This map includes not only the customer's actions but also their thoughts, feelings, and motivations at each stage (Alwidian, 2022). It helps identify opportunities for improvement and innovation in the service delivery process. By mapping this journey, service providers can identify areas where customers might feel dissatisfied, confused, frustrated, or unsupported, and design adequate interventions to improve the overall experience of customers (West et al., 2022).

Complementary to the customer journey, are touchpoint and pain point analyses. Touchpoints are the individual transactions or moments of contact between the customer and the service provider throughout the customer journey (Lundin and Kindström, 2024). These can be physical (such as a reception desk), digital (such as a website or a mobile application), or human (such as a phone call with customer service or an interaction with an airport employee) (Vannucci and Pantano, 2020; Weidig et al., 2024). Touchpoints represent the concrete manifestations of the service in the customer's experience. Therefore, effective service design ensures that each touchpoint is not only functional but also consistent with the overall experience and service promise (Gao et al., 2022). To achieve this goal, careful consideration of both the frontend (what the customer sees and experiences) and the backend (the systems and processes that support the service delivery) is required.

Pain points are the specific problems, frustrations, or inconveniences that customers experience during their journey. Identifying and addressing pain points is a key objective of service design. Common pain points include long waiting times, confusing processes, lack of information, or poor customer service. Once identified, these pain points become opportunities for innovation and improvement. For example, in an airline service, a common pain point might be the stress and uncertainty associated with flight delays. A service design solution may involve creating a proactive communication system that keeps passengers informed about their flight status and provides clear instructions and support in case of delays (Nordqvist, 2022; Holz et al., 2024).

Another method used in service design is the creation of personas that represent different user types within a targeted demographic, attitude and/or behavior set that may use a site, brand, product or service in a similar way. Personas help teams understand and empathize with the different types of customers they serve (Bradley et al., 2021). A well-crafted persona includes collection of demographic information (age, occupation, location, etc.), behaviors and preferences of each user type, as well as their pain points and frustrations (Jansen et al., 2022). For instance, a public transportation service to and from an airport might develop personas – fictional characters that represent typical passengers - such as "Commuter Carl," a frequent user of the system who prioritizes reliability and efficiency, and "Tourist Tina," an occasional user who values clear information and ease of use. These personas would then guide decision-making throughout the service design process.

Human-centric design emphasizes the importance of considering human needs, capabilities, and behaviors in every step of the design process (Sun et al., 2022). This approach ensures that the resulting service is not only functional but also intuitive, accessible, and satisfying for the end-users. Human-centric design is based on empathy (i.e. understanding the users' needs, desires, and challenges), iteration (i.e. continuously refining designs based on user feedback and testing), inclusivity (i.e. ensuring the service is accessible and usable for people with diverse abilities and backgrounds) and co-creation (i.e. involving users and stakeholders in the design process) (Singh et al., 2022). For example, a human-centric approach to designing an airline service portal would involve extensive user research to understand needs and pain points of passengers, prototyping and testing of different interface designs, and ongoing refinement based on user feedback.

Design thinking is a problem-solving approach that is highly relevant to service design. It emphasizes understanding users, challenging assumptions, redefining problems, and creating innovative solutions to prototype and test (Foster, 2021). The design thinking process typically involves five stages: First, empathizing by understand the users and their needs. Second, articulating the problem to be solved. Third, ideating by generating a wide range of creative ideas. Fourth, prototyping through quick, low-fidelity representations of ideas. Finally, testing the prototypes with users and gathering feedback. This approach helps teams move beyond obvious solutions to innovate in ways that truly address user needs (Dam and Siang, 2021; Rijaluddin and Safitri, 2023).

Additionally, design thinking in service design includes service ecology analysis that refers to the complex system of stakeholders, processes, and resources that collectively deliver a service. Understanding this ecology is crucial for effective service design, as changes in one part of the system can have ripple effects throughout it. Additionally, stakeholder mapping is used as a key tool for visualizing and analyzing the service ecology. It involves identifying all parties involved in or affected by the service, including not just customers and frontline staff, but also backend support, partners, regulators, and even competitors.

Prototyping in service design involves creating scaled-down versions of a service to test ideas and gather user feedback before full implementation. Unlike product prototyping, service prototyping often involves role-playing, simulations, or limited real-world trials. Some of the methods of service prototyping include desktop walkthrough (using small props and figures to act out service scenarios), service roleplay (team members act out different parts of the service experience), experience prototyping (creating a physical environment that simulates the service) and digital prototyping (using digital tools to simulate digital touchpoints).

These tools and methodologies, from personas and journey maps to prototyping techniques, provide a robust toolkit for overcoming complex service design challenges. As services continue to play an increasingly important role in the global economy and in particular in air travel, the principles of service design will become more important for organizations seeking to differentiate themselves and create value for their customers. Ultimately, successful service design requires a deep understanding of human needs and behaviors, combined with the creativity to envision new solutions and the practical skills to implement them.

This chapter explores the application of service design methodologies in airport planning and operations, addressing the increasing challenges in managing passenger expectations and operational efficiency in the aviation industry. This research aims to optimize the overall design of the customer journey through a comprehensive analysis of key touchpoints and pain points. Moreover, it contributes to the growing body of literature on airport service design by providing a multi-touchpoint analysis of the passenger journey. The results highlight the potential for service design methodologies to significantly improve airport operations and passenger experiences. Doing so, airports can create a more seamless and satisfying journey for travelers.

2. LITERATURE REVIEW

The aviation industry has been experiencing significant changes and challenges in recent years, particularly in the wake of the global COVID-19 pandemic. Airports, as critical nodes in the air transportation network, have been at the forefront of adapting to new passenger expectations, safety requirements, and operational constraints. In this context, service design has emerged as an important approach to enhance the airport experience for travelers while optimizing operational efficiency.

The COVID-19 pandemic has highlighted the need for resilient service design in airports. Kuo et al. (2022) examined the impact of the pandemic on airport networks and passenger flows, highlighting the need for service design approaches that consider the entire travel ecosystem, including connections to other transportation modes. Cankurtaran and Beverland (2020) discussed how service design can be applied to respond to crises, providing valuable lessons for airport service designers in preparing for future disruptions.

The concept of customer journey mapping has gained significant traction in airport service design research over the past few years. Usman et al. (2022) emphasize the importance of understanding the entire passenger journey, from pre-flight planning to post-flight experiences. Their study demonstrates how a comprehensive customer journey map can help airport managers identify critical touchpoints and areas for improvement throughout the passenger experience. Building on this work, Wongyai et al. (2024) propose an integrated framework that combines customer journey mapping with service blueprinting techniques. This approach allows for a more holistic view of the airport service ecosystem, considering both the passenger perspective and the operational processes that support each stage of the journey. The authors argue that this integrated approach can lead to more effective service innovations and process optimizations.

Recent research has also focused on the impact of technology on the customer journey. Dalkıran (2023) explores the role of digital touchpoints in shaping the airport experience, highlighting how mobile apps, self-service kiosks, and biometric systems are transforming traditional journey maps. The findings of Dalkıran's research suggest that while these technologies can streamline processes and reduce waiting times, they also introduce new challenges in terms of user adoption and potential points of friction.

Lee-Anant and Monpanthong (2021) highlighted the growing importance of technology integration in airport service design. Their study examined factors influencing airport technology selections at various touchpoints, emphasizing the need for a balanced approach that considers both passengers needs and operational efficiency. Their study also incorporated persona development to better understand the diverse needs of different passenger groups, ranging from tech-savvy frequent flyers to less experienced travelers.

Chutiphongdech and Vongsaroj (2022) analyzed the technical efficiency and productivity of regional and local airports in Thailand. Their research identified key touchpoints that influence overall airport efficiency and passenger satisfaction, providing insights into areas where service design interventions could yield significant improvements.

The development and use of personas have become increasingly sophisticated in airport service design research. Chonsalasin et al. (2021) present a data-driven approach to creating airport passenger personas, utilizing a combination of demographic data, behavioral patterns, and psychographic characteristics. Their study demonstrates how these nuanced personas can inform targeted service improvements and marketing strategies.

Expanding on the concept of personas, Kachniewska (2021) introduce the idea of "dynamic personas" that evolve based on real-time data and changing passenger needs. This approach recognizes that travelers may shift between different persona categories depending on factors such as trip purpose, time constraints, or unforeseen circumstances. The authors argue that this flexible persona framework allows for more adaptive and personalized service design strategies.

Interestingly, recent research has also begun to consider non-passenger personas in airport service design. Puls and Wittmer (2021) explore the needs and experiences of airport employees, airline staff, and other stakeholders who interact with the airport environment on a regular basis. Their work highlights the importance of designing services that cater to the needs of these diverse user groups, ultimately contributing to a more efficient airport ecosystem.

The concept of frontstage actions, which encompass the visible interactions between airport staff and passengers, has been a focus of several recent studies. Story et al. (2020) address the various stakeholders involved in the provision of airport services as "actors" and analyze the impact of frontline employee behavior on passenger satisfaction. Their analysis emphasizes the role of emotional intelligence and cultural competence in service delivery and suggests that investing in employee training and empowerment can significantly enhance the overall airport experience. In a related study, Lipkin and Heinonen (2022) examine the effectiveness of various frontstage service recovery strategies in airport settings. Their study identifies best practices for addressing passenger complaints and resolving service failures, highlighting the importance of proactive communication and personalized solutions. Another emphasis of their study refers to the critical role of well-designed frontstage actions in maintaining positive passenger perceptions, even in challenging situations. The integration of technology into frontstage actions has also been a topic of recent interest. Schepers and Streukens (2022) investigate the use of service robots and AI-powered chatbots in airport customer service roles. While their findings suggest that these technologies can enhance efficiency and consistency in service delivery, they also note the importance of maintaining a human touch in certain high-stakes or emotionally charged interactions.

Human-centered design (HCD) approaches have become increasingly prevalent in airport service design research. Almarri et al. (2021) emphasized the importance of taking a holistic and systematic design approach to address the evolving challenges facing the aviation industry and airports. They present a comprehensive framework for applying HCD principles to airport security design, emphasizing the need to consider diverse user needs, accessibility requirements, and cultural factors. Their work demonstrates how HCD can lead to more inclusive and user-friendly airport environments.

Cankurtaran and Beverland (2020) explore the application of service design principles in responding to crises, using the COVID-19 pandemic as a case study. While not specifically focused on airports, their research provides valuable insights into how human-centered design approaches can be applied to a broad range of service environments, which is particularly relevant to the dynamic nature of airport operations. Their study is particularly significant for understanding how airports have redesigned processes and spaces to accommodate new health and safety requirements while maintaining a focus on passenger comfort and efficiency. Moreover, their work also highlights the adaptability of HCD principles in responding to rapidly changing circumstances and user needs. Based on these principles, Qing et al. (2021) and Sun et al. (2022) explore the application of HCD in airport wayfinding systems. Their research combines eye-tracking studies with qualitative user feedback to optimize signage, digital displays, and architectural cues that guide passengers through complex terminal spaces. They conclude that a human-centered approach to wayfinding can significantly reduce stress and improve the overall passenger experience.

Identifying and addressing pain points in the airport experience has been a central theme in recent service design research. A comprehensive analysis of passenger pain points across different airport types and sizes may reveal common challenges, such as security screening bottlenecks, baggage claim delays, and confusing terminal layouts. Pain points are valuable in providing insights for prioritizing service improvements and resource allocation. Holz et al. (2024) explore the concept of eliminating customer experience pain points in complex customer journeys, specifically within the airport context. The importance of the subject which their paper refers to results from the fact that emotional aspects of airport pain points, such as the effects of passenger stress, anxiety, and frustration, significantly impact their overall satisfaction. Their research is based on incident analysis and problem-centered interviews, as well as on analysis of online airport reviews. The study identified key pain points across the airport journey and proposed a classification scheme based on information, performance, and hospitality themes. Halpern and Mwesiumo (2021) discuss the impact of service failures on passenger satisfaction and the likelihood of promoting an airport online. Their study highlights the importance of effectively managing major pain points to minimize negative experiences and their subsequent effects on airport reputation.

The concept of touchpoints has also evolved in recent airport service design literature. Wang et al. (2021) discuss the idea of "seamless touchpoints," which can be applied to create a more cohesive and integrated passenger experience across different service areas. The implications of their research suggest that coordinating touchpoints between airlines, security agencies, and airport operators can lessen the magnitude of pain points and lead to improved overall satisfaction. In recent years, touchpoint design in airports has been experiencing a paradigm shift, evolving from traditional, often fragmented approaches to more seamless and integrated experiences that leverage emerging technologies. This transformation is driven by the dual need to enhance operational efficiency and elevate passenger satisfaction in an increasingly competitive aviation industry. Additionally, airports are under pressure to create new streams of revenue, and innovative touchpoint designs that offer new ways to achieve these goals. The adoption of cutting-edge technologies like biometric systems, the Internet of Things (IoT), Artificial Intelligence (AI), Machine Learning (ML), and mobile technologies are playing major roles in this ongoing evolution (Mota et al., 2021; Polat et al., 2024).

One of the most notable trends in airport touchpoint design is the integration of biometric technologies. The implementation of facial recognition, fingerprint scanning and other biometric systems is revolutionizing passenger processing at critical touchpoints, such as check-in, security screening, and boarding. Biometric systems not only enhance security by providing accurate and reliable identification but also significantly reduce processing times, leading to smoother and more efficient passenger experience. By eliminating the need for traditional identity verification methods, biometric technologies are reducing the time passengers spend waiting in queues, thereby increasing overall satisfaction and reducing stress levels associated with air travel (Rubio-Andrada et al., 2023). The concept of "single token travel" has emerged as a significant advancement in the application of biometric technologies within airports. In this model, a passenger's biometric data, typically captured at the initial touchpoint, and serves as their single form of identification throughout the entire airport journey. This approach simplifies the travel process by eliminating the need for multiple document checks, such as presenting a passport or boarding pass at various stages. Consequently, the adoption of single token systems can result in substantial time and passenger angst reductions in the total journey time through the airport, as highlighted in a study by Heggde et al. (2022). The seamless nature of "single token travel" not only enhances passenger convenience but also minimizes bottlenecks at critical touchpoints.

In addition to biometric technologies, the integration of the Internet of Things (IoT) devices and sensors is transforming the design of airport touchpoints. IoT technologies enable real-time monitoring of passenger flows, queue lengths, and the utilization levels of airport facilities. By collecting and analyzing data from these sensors, airports can optimize resource allocation, manage potential bottlenecks proactively, and improve overall operational efficiency. For instance, IoT sensors can detect overcrowding in certain areas and automatically redirect passengers to less congested routes, thereby reducing waiting times and enhancing the passenger experience (Florido-Benítez, 2022). The deployment of IoT technologies is also crucial for maintaining safety and security within the airport environment, as these systems can monitor environmental conditions, detect potential hazards, and alert authorities to any unusual activities. In this respect, IoT technologies may present substantial potential improvement and have become a significant part in the toolbox of service designers.

Complementary domains that benefits from the inputs of IoT devices are Artificial Intelligence (AI) and Machine Learning (ML). These technologies are increasingly being utilized to create more personalized and engaging experiences for passengers by analyzing vast amounts of data to predict passenger preferences and behaviors, allowing airports to offer tailored services and recommendations. For example, AI-powered chatbots and virtual assistants are being deployed to provide real-time information and support to passengers throughout their journey. These AI tools can answer common questions, provide updates on flight statuses, and even offer personalized suggestions for dining or shopping based on a passenger's preferences. Further, AI and ML are being used to analyze feedback from passengers and other data sources, such as property management systems, to continually redesign services to improve the passenger experience. For instance, AI algorithms can assess passenger feedback on noise levels, cleanliness, and service quality, enabling airports to make data-driven decisions that enhance overall satisfaction (Geske et al., 2024; Liu et al., 2022; Krishnan et al., 2024).

The concept of "smart airports" is gaining momentum as touchpoints are increasingly being designed to integrate into their services digital and physical experiences seamlessly. Smart airports leverage technologies like Augmented Reality (AR) and Virtual Reality (VR) to enhance various aspects of the customer journey of passengers. For example, AR and VR can be used to improve wayfinding by providing passengers with interactive, real-time navigation assistance, helping them to find their gates, restrooms, and other facilities more easily. Additionally, these technologies can offer immersive entertainment experiences, such as virtual tours of cultural landmarks, which passengers can enjoy while waiting for their flights. Moreover, AR and VR can be used to showcase retail offerings in innovative ways, such as virtual fitting rooms or interactive product demonstrations, thereby enhancing the shopping experience and potentially increasing sales in airport stores (Birrell et al., 2022; Monge-Zamorano, 2023).

Another area where technological innovation is making a significant impact on service design and, in particular, on touchpoint design is mobile technology. Mobile airport apps have become a central hub for passenger interactions, offering a wide range of features that enhance convenience and engagement. These apps often integrate real-time flight information, indoor navigation, mobile ordering for food and retail, and digital boarding passes. By consolidating these features into a single platform, airports can provide a more cohesive and streamlined experience for passengers. For example, indoor navigation systems within mobile apps can guide passengers to their destinations within the airport, reducing the likelihood of getting lost and minimizing stress. Similarly, mobile ordering allows passengers to pre-order food or shop for retail items, which they can then pick up at designated locations, thereby saving time and reducing queues in airport stores. Moreover, mobile technologies are facilitating more personalized and proactive communication between airports and passengers. Push notifications can be sent to the devices of passengers to inform them about important updates, such as gate changes or boarding times, ensuring that they stay informed throughout their journey. Additionally, mobile apps can collect data on the preferences and behaviors of passengers, allowing airports to offer customized promotions and services. For instance, passengers who frequently travel through a particular airport may receive personalized offers for discounts at their favorite stores or restaurants. This level of personalization not only enhances the passenger experience but also opens up new opportunities for airports to generate revenue through targeted marketing and advertising (Sia et al., 2023). While these technologies are available and are gradually adopted by airports worldwide as technological means that can simplify passenger experience, our study suggests that by applying service design methods and understanding the customer journey in-depth. Thereby, airports can better understand passenger needs and preferences and optimize their experience either by implementing advanced technologies or by other means.

3. METHODOLOGY

This study utilized a qualitative approach to thoroughly understand and enhance the air travel experience, integrating qualitative observations with detailed analysis of flow and satisfaction metrics. The research aimed at developing a comprehensive toolkit for enhancing airport services. The major goal of this methodology was to develop a comprehensive set of tools that can be applied across various airports worldwide. The methodology focuses on a capturing and analyzing qualitative data to identify areas for improvement in airport services, such as key touchpoints, pain points and best practices that can benefit the overall travel experience.

The basis of this research and its implementation in airports relies on precise observation and data collection processes. By positioning observers in strategic points across multiple airports in Europe and the U.S., the full scale and spectrum of the passenger experience were recorded.

The research included seven observers, positioned in the following airports: Ben-Gurion Airport (Israel), JFK Airport (U.S.), Newark Airport (U.S.), Orlando International Airport (U.S.), Kastrup (Denmark), Charles de Gaulle Airport (France) and Naples International Airport (Italy). Semi-structured interviews were selected as the main research method, since the flexibility of this method and format allow for the exploration of specific topics in greater depth, while still providing consistency in the collection of data. The observers were instructed to approach passengers politely, explain the purpose of the questionnaire and ensure that the responses were recorded with accuracy and impartiality. Observers were tasked with documenting each phase of the passengers' experience, beginning from their departure from home until they reached their final destination. Additionally, they were instructed to report any unique experiences or exceptional service points that could provide valuable insights for improving airport operations. Observers indicated in detail how travelers planned their trip to the airport, including the mode of transportation used (e.g., taxi, personal vehicle, public transportation, etc.). The ease of securing transportation was then evaluated, while noting any issues or delays that passengers encountered during this initial stage of their journey. The experiences of passengers that involved airport parking and drop-off areas were also assessed for the sake of documenting and studying the customer journey of passengers that arrive to the airport by car. Another aspect focused on the ability of different passengers to navigate from the airport entrance to the check-in counters was considered by the observers who recorded the clarity of signage and instructions, the duration of the check-in process, and any other challenges that passengers encountered during their check-in phase. Other aspects that were documented included the security screening and the efficiency of this process, the waiting time for screening, the helpfulness of the security personnel, and the difficulties and the challenges that passengers confronted during this stage of their journey. The pre-boarding experience was also addressed by documenting the activities of passengers before boarding, including shopping, dining, and resting. Other focal points were the availability of seating areas and facilities, and the ease of finding flight information. This approach enables to construct a holistic view of the passenger journey, allowing for identification of key elements that influence passenger satisfaction and experience. In total 64 interviews were conducted between February and July 2024 during daytime and night, according to the availability of observers. The questionnaires were manually filled in and were qualified as valid, with each observer interviewing eight to ten passengers and ground staff members (see Appendix A for the list of interviewed passengers). The analysis of the questionnaires was carried out by theming their contents according to the different stages of the customer journey of passengers and the responses were accordingly associated to each stage. Then, commonalities and differences in the experience of passengers across airports were identified.

On the basis of these observations and observers' reports, a detailed qualitative analysis of the customer journeys as recorded, covering all stages from the pre-travel stage to the final arrival at the destination, was carried out. The customer journey was divided into specific segments to facilitate detailed data collection and analysis. The organization of the boarding process was also assessed by the observers, with particular attention to any delays or issues through it. The interaction between passengers and ground staff during this phase was also documented. The two last phases to be observed were the arrival at the destination airport and the transportation to the final destination. Observers evaluated the ease with which passengers navigated the destination airport, including signage for baggage claim procedures and departure areas. Any delays or issues with baggage claim were also recorded, with the final part of the journey involving the arrangements for the transportation of passengers from the airport to their final destination. The ease of finding and using transportation options and any problems encountered during this part of the journey were also recorded.

In this study, various service design tools were implemented to analyze and to gain insights on the passenger experience. First, customer journey mapping was used to visualize the entire travel experience from the passenger's perspective. It helped in identifying critical occurrences and potential pain points that could impact overall satisfaction of passengers. Second, touchpoint analysis was applied to examine each interaction between the passenger and airport services. This included the check-in counters, security checkpoints, boarding gates, baggage claim, and ground transportation. Observers documented the efficiency, clarity of instructions, and quality of interactions at each touchpoint. Finally, by creating personas for different types of passengers, the varied needs and requirements of different passenger groups were classified. Passengers were divided into four major and distinct personas: Business Commuter, Family Traveler, Leisure Traveler, and Senior Traveler. Each persona was characterized by attributes such as age, gender, occupation, frequency of travel, primary needs, and typical challenges faced by this group of passengers. This approach facilitated a more tailored analysis of different traveler types, ensuring and enabling recommendations that could be effectively and precisely targeted to address the specific needs of diverse passenger groups.

Finally, the combination of customer journey analysis, service design tools, observational data, and semi-structured interviews allows for a comprehensive understanding of the passenger experience at airports. The findings from this study offer valuable insights into the factors that influence passenger satisfaction and highlight areas where improvements can be made to enhance the overall travel experience.

This methodology not only provides a robust framework for assessing passenger experiences but also provides practical guidelines for airports aspiring to implement service improvements. By understanding the customer journey in detail and analyzing the effectiveness of each touchpoint, airports can better meet the needs of their diverse passenger base, ultimately leading to higher levels of satisfaction and a more positive travel experience.

4. RESULTS

The results of the qualitative content analysis reveal key insights into the experience of passengers across various airports, focusing on specific touchpoints, pain points, and best practices in customer journeys that affect both their customer journey and overall satisfaction. The findings refer to the different stages of the passenger journey, and are focused on common themes, regional differences, and areas for improvement that are reflected in the different stages that passengers experience (see the classification of touchpoints by airport in Table 1 and by persona type in Table 2).

Leaving home is the very first stage of the customer journey. This stage, which involves the modes of transportation to the airport, has shown significant regional variations across regions and airports, as the analysis of the interviews with respondents indicate. In Israel, New York, and Orlando, finding a taxi was an easy task, while other public transportation options were generally limited or unavailable. As noted by one of the passengers interviewed at the JFK airport: "getting a taxi to JFK was straightforward and quick, which was a relief given my tight schedule. I wouldn't think about taking an airport bus or a shuttle, given their long time of travel." (male, 35, Business Commuter). A passenger leaving from Ben-Gurion airport said that "we have an early morning flight and we had to arrive at 3 a.m. There are no buses or trains at this hour, so we had to pay for a taxi." (male, 73, senior traveler) In Italy, taxis were found to be more convenient than public transportation, though notably expensive, leading to passenger dissatisfaction. As noted by one of the interviewed passengers: "Naples is such a beautiful city, but getting to the airport was frustrating because of the taxi fares—they're ridiculously high… I just wish public transport were a more viable option here." (female, 28, leisure traveler). Another passenger indicated that "getting to Naples Airport was frustrating because the taxi fares are so high." (male, 68, senior traveler). In terms of economic value gained, these results suggest that the inconvenient pricing of taxis exceeds the value gained from their availability. In contrast, Denmark stood out for the ease of access to its public transportation, with passengers arriving to the airport by using its efficient public transport system. One of the interviewed passengers mentioned that "traveling to Kastrup with my family was very easy thanks to the fantastic public transportation system. We took the train straight to the airport, and it was both affordable and convenient." (male, 36, family traveler).

The second stage begins upon the arrival at the airport, and includes the check-in process and the navigation within the airport itself. At all the airports that are included in this study, the check-in counters were easily found by passengers, with clear signage and directions contributing to a smooth arrival process, with the exception of France's Charles de Gaulle Airport. This airport's signage was unclear and passengers experienced difficulties finding ground staff members that could assist them with directions. As one of the passengers expressed: "The airport's layout is confusing, and the signage didn't help. We spent a lot of time looking for the check-in counter, but once we found it, the staff were very helpful." (female, 40, family traveler). Another passenger mentioned that "the airport could be great if they improved the signs. It was hard to find my way around, and I wasted a lot of time asking for directions." (female, 32, leisure traveler). In general, the check-in process was efficient at all the airports for passengers without checked luggage. Passengers with luggage confronted long queues that affected their experience for the worse. The observations of long queues suggest that airports should rethink their queue management processes for the check-in of passengers with luggage to reduce their waiting times. Additionally, respondents indicated that the interaction with the airport ground staff at the check-in counters was positive, as the staff was helpful in resolving issues and answering their questions. The passenger quoted above added that "the airport staff members were kind and helpful, but the airport itself is confusing." (female, 32, leisure traveler).

The next stage, i.e. security checks, varied by airport, with Israel's Ben-Gurion Airport, France's Charles de Gaulle Airport, Italy's Naples International Airport and Denmark's Kastrup Airport providing a quick and smooth process, especially for passengers without luggage. In JFK Airport, Newark Airport and Orlando International Airport, passengers faced no significant waiting times, regardless of whether they had luggage or not, due to the use of efficient self-service kiosks for both issuing boarding passes to passengers and collecting their luggage for their flights. However, in the U.S. airports that were studied, passengers had to take off their shoes and go through security screening, which created long queues, though they were well managed and waiting times were reasonable. As a JFK passenger indicated: "JFK had a smooth check-in process thanks to the kiosks, but the security lines were longer than I'd hoped. I also didn't enjoy taking off my shoes during screening. It felt unnecessary." (male, 46, business commuter).

After the security checks, the pre-boarding experience was assessed. Interestingly, this experience across airports varied significantly, particularly in terms of shopping, dining, and seating areas. Some passengers confronted long queues in duty-free shops and food outlets, while others had minimal waiting times. A common pain point across all the airports that were examined in this study was the availability of seating areas and seats. It was found that airports lacked sufficient seating and in some cases tables were not cleared quick enough, making it difficult for passengers to find clean spaces in sitting areas. For example, one passenger at the Charles de Gaulle airport complained that "... in the seating areas it is almost impossible to find a spot. This is why I'm standing here near the gate because there aren't enough clean tables or chairs." (male, 24, leisure traveler). A passenger at the Ben-Gurion airport said that "the only negative thing that I experience is finding a clean seating area while waiting for my flight." (female, 66, senior traveler). Nevertheless, the flight information was easy to access, with clear signage throughout all airports, ensuring that passengers were well-informed. Delays also played a role in shaping passenger experiences. While most airports reported no significant delays, travelers in Israel experienced a cumulative 70-minute delay in flight departures, and in Orlando a 30-minute delay occurred. A passenger at the Orlando airport shared his experience as follows: "the 30-minute delay is a bit inconvenient, but the clear flight information keeps me informed. The airport staff are really helpful, which makes this situation better." (male, 51, business commuter).

The following stage, the boarding process, was identified as a significant domain for improvement across airports. In most cases, passengers boarded the plane based on their arrival at the gate rather than their seating order, leading to delays and confusion while boarding the aircraft. The current process of boarding caused frustration, as passengers struggled to find overhead space for luggage close to their seats. As one passenger at the Newark airport said " I had to struggle to find overhead space for my bag." (male, 50, business commuter). At the JFK airport, a passenger complained: "I couldn't find space for my bag in the overhead compartments." (male, 72, senior traveler). Streamlining this process involves boarding passengers from the back to the front of the plane, hence minimizing delays and conflicts between passengers, and improving the boarding experience. It is important to note that interactions with ground staff members during this stage were generally positive, being described as helpful and accommodating. For example, a passenger at the JFK airport indicated that "the boarding seems chaotic and could be better organized." (male,46, business commuter). Another passenger at the Newark airport described this stage as follows: "the boarding is a mess. It seems like everyone rushes to the gate at once." (male, 43, business commuter).

Upon arriving at the destination airport, most passengers found the signage somewhat confusing but manageable with the assistance of ground staff members, who were readily available and provided clear directions. Baggage claim had no significant delays, with luggage retrieval times being fast and taking no more than five minutes in most cases. An issue reported by passengers arriving to Charles de Gaulle was the lack of suitable restroom facilities near the baggage claim areas. Some airports lacked hangers for bags in the restrooms, causing inconvenience for passengers with luggage.

The last stage of the customer journey involved the transportation to the final destination. This stage significantly varied from one airport to another. Denmark stood out for its efficient public transportation system, with passengers easily navigating using clear signage and support from ground staff members. The setting at the ticket vending machine was praised, as attendants helped passengers purchase tickets quickly, shortening queue times and ensuring a smooth transit experience. The train stations were conveniently located and designed with ample seating and luggage space, contributing to a stress-free journey. However, in France, the signage of the terminal transportation was not clear enough, leading to passengers requesting assistance from ground staff members. Nonetheless, the transportation itself was comfortable and efficient once passengers located it.

Table 1. Satisfaction from touchpoints: low satisfaction in black, medium satisfaction in gray and high satisfaction in white.

Touchpoints/ Airport	Ben-Gurion (Israel)	JFK (U.S.)	Newark (U.S.)	Orlando International (U.S.)	Naples International (Italy)	Kastrup (Denmark)	Charles de Gaulle (France)
Check-In							
Security							
Pre-Boarding							
Boarding							
Baggage Claim							
Transportation							

Table 2. Passenger experience by traveler persona (each "+" represents a positive experience allowing visualization of the rating for each touchpoint by different personas).

Touchpoint	Business	Family	Leisure	Senior
Check-In	++++	++++	+++	+++
Security	++++	++++	+++	++++
Pre-Boarding	++	+++	+++	++
Boarding	++	++	+	+
Baggage Claim	++++	++++	++++	++++
Transportation	++++	++++	++++	++++

5. DISCUSSION

The results of the qualitative study provide valuable insights into the passenger experience across multiple international airports, highlighting critical touchpoints, pain points, and opportunities for service design improvements. This discussion analyzes the findings in relation to the existing literature and explores their implications for airport management and service design practices.

One of the most important findings from this study is the substantial variability in efficiency and service quality across different airports and touchpoints. This finding aligns with previous research highlighting the heterogeneity in airport service quality (Bakır et al., 2022; Da Rocha et al., 2022). For instance, the efficient public transportation system and helpful ground staff at the Kastrup Airport contrasted sharply with the unclear signage and disorganized boarding processes at the Charles de Gaulle Airport in Paris. This variability highlights the need for applying standardized best practices and continuous improvement initiatives that utilize airport service design methodologies. The observed differences between airports in the degrees of efficiency and service quality can be attributed to various factors, including local infrastructure, cultural contexts, and management practices. However, they also represent opportunities for knowledge sharing and benchmarking among airports. As Strokosch and Osborne (2023) argue, service improvements in complex environments such as airports often require a systematic approach that considers both the technical and the social factors within them. Our findings support this view and suggest that airports could benefit from adopting a more holistic approach towards service design that includes all the touchpoints that the passenger journey involves. Further, airports benefit from knowledge exchanges on implementation of service design methods and practices to enhance their operations by applying lessons from other airports worldwide.

Another major finding from this study suggests that clear signage and effective communication are crucial elements in enhancing passenger experience across all airports. This finding validates the results of prior studies how information accessibility assists in reducing passenger stress and improving satisfaction (Mainardes et al., 2021; Ma et al., 2022). The challenges faced by passengers at Charles de Gaulle Airport due to unclear signage exemplify the negative impacts of poor communication on the overall travel experience. The importance of clear communication extends beyond static signage to include real-time information systems and staff interactions. The results indicate that passengers significantly value the availability of helpful ground staff, particularly in situations where other forms of communication are unclear or fail them. Therefore, a multi-faceted approach to airport communication that combines clear visual cues, digital information systems, and well-trained staff is necessary to create a seamless information ecosystem for passengers.

Long queues, particularly in luggage-related processes, emerged as a major source of passenger dissatisfaction across multiple airports. This finding aligns with previous studies on the negative impact of waiting times on perceived service quality in airports (Tueanrat et al., 2021; Zdanowicz et al., 2022). The contrast between efficient self-service kiosks in some U.S. airports and long queues for luggage check-in at others highlights the importance of widespread adoption of technological solutions to address these challenges. However, the results of this study also indicate that technology alone is not a sufficient condition for resolving queuing issues. The experience at Ben-Gurion Airport, where security checks were smooth and quick despite the minimal use of self-service technology, indicates that well-designed processes and trained staff can also reduce waiting times. This result supports the argument made by Bellos and Kavadias (2021) that service design improvements often require a combination of technological and human-centered solutions. Our findings also affirm that staff helpfulness and friendliness have a significant effect on the overall passenger satisfaction. This finding is consistent with prior studies that emphasize the importance of human interactions in service environments (see for example Strokosch & Osborne, 2023). Additionally, our results suggest that the nature of staff interactions with passengers may need to evolve vis-à-vis the ongoing adoption of self-service technologies by airports. For instance, the successful implementation of self-service kiosks for both issuing boarding passes and collecting luggage at some U.S. airports signifies a shift in the role of staff from routine transactions to more complex problem-solving and customer support. This result aligns with the concept of "high-touch, high-tech" service design, where technology handles routine tasks, freeing up staff to provide more personalized and value-added services (Lee-Anant & Monpanthong, 2021).

The use of personas in our study (Business Commuter, Family Traveler, Leisure Traveler, and Senior Traveler) provided valuable insights into the ways in which different passenger groups experience various services and touchpoints. This approach aligns with recent research on the importance of understanding diverse passenger needs in airport service design (Chonsalasin et al., 2021; Kachniewska, 2021). For instance, our findings suggest that while some touchpoints, such as check-in and baggage claim, were generally rated positively across all personas, others, like pre-boarding and boarding, showed more variation. This result highlights the need to design services that can accommodate a range of passenger needs and preferences. Particularly, the lower satisfaction ratings for boarding among Leisure and Senior Travelers suggest that these groups may benefit from additional support or streamlined processes during this phase of the journey.

While our study did not explicitly focus on technological innovation, the findings point to several areas where emerging technologies could significantly enhance the experience of passengers. The success of self-service kiosks in some airports suggests that there is potential for further automation and digitalization of various touchpoints, as indicated also in other studies about the use of technologies, such as biometrics, IoT and AI, in airport operations (Mota et al., 2021; Polat et al., 2024). However, our results also caution against viewing technology as a universal solution. The positive experiences reported at airports with less technological intervention but with efficient human-centered processes suggest that a balanced approach that merges these domans is necessary. This result also supports the argument made by Rubio-Andrada et al. (2023) that the successful implementation of new technologies in airports requires careful consideration of human factors and passenger preferences.

Our study included the pre-travel stage (leaving home) and post-arrival transportation and revealed significant variations in passenger experiences across different airports and regions. This insight highlights the importance of considering the entire passenger journey, including touchpoints that extend beyond the physical boundaries of the airport. The concept of "seamless touchpoints" proposed by Wang et al. (2021) could be extended to include these pre- and post-airport phases of the journey. For example, the contrasting experiences of passengers using public transportation at the Kastrup Airport vs. those relying on taxis in other cities signify the impact of local infrastructure on the overall travel experience. Thereby, airports should collaborate with local transportation authorities and service providers to ensure a seamless journey for passengers.

The findings of this study have several important implications for airport management and service design practices. First, airports should adopt a more holistic approach to service design that considers the entire passenger journey, from pre-travel planning to post-arrival. This implementation of this policy is not without difficulties, as it requires closer collaboration with external stakeholders and local authorities. Second, in parallel to application of standardized best practices that address common pain points, airports should also maintain flexibility to accommodate local contexts and diverse passenger needs. The persona-based analysis in our study provides a framework for balancing standardization with personalization. Third, implementation of self-service technologies and digital information systems can significantly improve both efficiency and passenger satisfaction. However, technology should be thoughtfully integrated into service processes, with consideration for human factors and the role that staff-passenger interactions play. Finally, a multi-faceted communication strategy that combines clear signage, digital information systems, mobile apps and proactive staff assistance is crucial for enhancing passenger experience and reducing the stress associated with air travel.

6. CONCLUSION

Optimal airport service design is an ongoing journey that requires continuous innovation, adaptation and unwavering commitment to passenger satisfaction. This study demonstrates the significant potential of service design methodologies in enhancing customer experience within airports. By analyzing key touchpoints and pain points throughout the passenger journey across multiple international airports, various areas that call for improvement were identified and potential solutions based on different service design methodologies were discussed (thereby emphasizing the value of service design methods used by Sun et al., 2022 and Polat et al., 2024). The findings highlight the importance of efficient processes, clear communication and positive staff interactions in shaping passenger satisfaction. Key insights that can enhance managerial performance, service and operations in airports, as well as passenger satisfaction, include the importance of mapping the entire passenger journey, developing nuanced personas to understand diverse passenger needs, employing human-centered design principles to address complex challenges, and identifying and addressing pain points across various touchpoints (in this respect, this study differs from Bakır et al., 2022 that conducted statistical analysis of customer satisfaction factors among passengers, while expanding methodologies applied by Almarri et al., 2021 for a smaller sample of passengers). These aspects do not only present challenges but also offer opportunities for airports to learn from others and to implement proven strategies that served them to enhance the overall passenger experience. Additionally, optimizing boarding processes, improving signage and enhancing staff training can lead to substantial improvements in both passenger satisfaction and operational efficiency. The integration of technological innovations also offers promising avenues for further improving the airport experience, although the results of this study caution against over-reliance on technology at the expense of human-centered services.

Referring to the different elements and methods of service design in more detail, the importance of mapping the entire passenger journey cannot be overstated. This approach enables airport managers to identify and address pain points at every stage, from pre-arrival to post-departure. By visualizing the complete passenger journey and the experiences associated with its various stages, airports can develop targeted processes, devices and digital tools that create a seamless and enjoyable journey for passengers. Understanding the diverse needs of different passenger segments is especially important for designing effective services, tailored for different personas popular among the passengers that frequent the airport. By developing detailed personas that capture the unique characteristics, preferences, and pain points of various traveler types, airports can create more personalized and satisfying experiences for passengers. Hence, this study broadens the framework that was applied by Krishnan et al., 2024 to analyze the narrow context of airport lobby experiences into a full-blown analysis of multiple airport experiences, processes and domains.

Employing human-centered design principles proves invaluable in addressing the complex challenges that exist in airport operations. This approach places the needs, preferences, and behaviors of passengers at the forefront of service design, ensuring that solutions are not only efficient but also intuitive and user-friendly. Identifying and addressing pain points across various touchpoints throughout the airport journey is critical, including streamlining check-in processes, enhancing security screening procedures, improving wayfinding systems, and optimizing boarding processes. By focusing on these key areas, airports can significantly reduce stress and frustration among passengers. While our study highlights the promising potential of technological innovation in enhancing the airport experience, it also cautions against over-reliance on technology at the expense of human-centered services. While the integration of cutting-edge technologies offers exciting possibilities for streamlining processes and personalizing the passenger journey, it is crucial to strike a balance between technological advancement and the human touch. The findings suggest that while passengers appreciate the efficiency and convenience offered by technological solutions, they also value personal interactions and the reassurance provided by human staff, particularly in stressful or unfamiliar situations.

As the aviation industry continues to evolve, adopting a passenger-centric approach through service design will become a common approach used by airports to meet passenger expectations and maintain satisfaction from their services. This study provides a foundation for future research and practical applications in airport service design, contributing to the ongoing improvement of the air travel experience. However, several domains of this study can be developed by further research. Longitudinal studies can evaluate long-term impact of the planning and implementation of service design methodologies on passenger satisfaction, operational efficiency, and airport profitability. Given the global nature of air travel, research can reveal how different passenger personas affect their expectations, experiences and satisfaction. Other directions can explore various ways of integrating sustainability principles into airport service design, balancing environmental considerations with passenger comfort and operational efficiency. Other means for fulfilling these objectives involve continued research into the potential applications and implications of emerging technologies such as AI, IoT, and VR in airports.

REFERENCES

Almarri, G. A., Kamalrudin, M. A. S. S. I. L. A., Hakimi, H. A. L. I. M. A. T. O. N., & Sidek, S. A. F. I. A. H. (2021). Factors Influencing Human Centric Model For Better Passenger Experiences Using Smart Security System In Dubai Airport. *Journal of Theoretical and Applied Information Technology*, 99(4), 976–986.

Alwidian, S. (2022). Towards integrating human-centric characteristics into the goal-oriented requirements language. In *2022 IEEE 30th International Requirements Engineering Conference Workshops (REW)* (pp. 200-204). IEEE. DOI: 10.1109/REW56159.2022.00045

Bakır, M., Akan, Ş., Özdemir, E., Nguyen, P. H., Tsai, J. F., & Pham, H. A. (2022). How to achieve passenger satisfaction in the airport? Findings from regression analysis and necessary condition analysis approaches through online airport reviews. *Sustainability (Basel)*, 14(4), 2151. DOI: 10.3390/su14042151

Becker, L., & Jaakkola, E. (2020). Customer experience: Fundamental premises and implications for research. *Journal of the Academy of Marketing Science*, 48(4), 630–648. DOI: 10.1007/s11747-019-00718-x

Bellos, I., & Kavadias, S. (2021). Service design for a holistic customer experience: A process framework. *Management Science*, 67(3), 1718–1736. DOI: 10.1287/mnsc.2020.3609

Birrell, S., Payre, W., Zdanowicz, K., & Herriotts, P. (2022). Urban air mobility infrastructure design: Using virtual reality to capture user experience within the world's first urban airport. *Applied Ergonomics*, 105, 103843. DOI: 10.1016/j.apergo.2022.103843 PMID: 35810501

Bradley, C., Oliveira, L., Birrell, S., & Cain, R. (2021). A new perspective on personas and customer journey maps: Proposing systemic UX. *International Journal of Human-Computer Studies*, 148, 102583. DOI: 10.1016/j.ijhcs.2021.102583

Cankurtaran, P., & Beverland, M. B. (2020). Using design thinking to respond to crises: B2B lessons from the 2020 COVID-19 pandemic. *Industrial Marketing Management*, 88, 255–260. DOI: 10.1016/j.indmarman.2020.05.030

Chonsalasin, D., Jomnonkwao, S., & Ratanavaraha, V. (2021). Measurement model of passengers' expectations of airport service quality. *International Journal of Transportation Science and Technology*, 10(4), 342–352. DOI: 10.1016/j.ijtst.2020.11.001

Chutiphongdech, T., & Vongsaroj, R. (2022). Technical efficiency and productivity change analysis: A case study of the regional and local airports in Thailand. *Case Studies on Transport Policy*, 10(2), 870–890. DOI: 10.1016/j.cstp.2022.03.001

Da Rocha, P. M., Costa, H. G., & da Silva, G. B. (2022). Gaps, trends and challenges in assessing quality of service at airport terminals: A systematic review and bibliometric analysis. *Sustainability (Basel)*, 14(7), 3796. DOI: 10.3390/su14073796

Dalkıran, A. (2023). Airport Common-Use Check-in Operations: A Novel and Efficient Model. *Journal of the Institute of Science and Technology*, 13(2), 1026–1042.

Dam, R. F., & Siang, T. Y. (2021). *What is design thinking and why is it so popular?* Interaction Design Foundation.

Florido-Benítez, L. (2022). The World airport awards as a quality distinctive and marketing tool for airports. *Journal of Airline Operations and Aviation Management*, 1(2), 54–81. DOI: 10.56801/jaoam.v1i2.4

Foster, M. K. (2021). Design thinking: A creative approach to problem solving. *Management Teaching Review*, 6(2), 123–140. DOI: 10.1177/2379298119871468

Gao, L., Melero-Polo, I., Ruz-Mendoza, M. Á., & Trifu, A. (2022). Exploring the role of service touchpoints on the path to financial, behavioral and relational customer outcomes: Insights from a B2B service context. *Journal of Business and Industrial Marketing*, 37(11), 2260–2277. DOI: 10.1108/JBIM-02-2021-0100

Geske, A. M., Herold, D. M., & Kummer, S. (2024). Artificial Intelligence as a driver of efficiency in air passenger transport: A systematic literature review and future research avenues. *Journal of the Air Transport Research Society*, 100030.

Halpern, N., & Mwesiumo, D. (2021). Airport service quality and passenger satisfaction: The impact of service failure on the likelihood of promoting an airport online. *Research in Transportation Business & Management*, 41, 100667. DOI: 10.1016/j.rtbm.2021.100667

Heggde, G., Khanka, S., & Damodaran, A. (2022). DigiYatra–a digital experience for air travelers. *Emerald Emerging Markets Case Studies*, 12(2), 1–20. DOI: 10.1108/EEMCS-10-2021-0349

Holz, H. F., Becker, M., Blut, M., & Paluch, S. (2024). Eliminating customer experience pain points in complex customer journeys through smart service solutions. *Psychology and Marketing*, 41(3), 592–609. DOI: 10.1002/mar.21938

Jansen, B. J., Jung, S. G., Nielsen, L., Guan, K. W., & Salminen, J. (2022). How to create personas: Three persona creation methodologies with implications for practical employment. *Pacific Asia Journal of the Association for Information Systems*, 14(3), 1–28. DOI: 10.17705/1pais.14301

Kachniewska, M. (2021). Smart tourism: towards the concept of a data-based travel experience. *Handbook of Sustainable Development and Leisure Services*, 289-302.

Krishnan, R. S., Kirubha, D., Christopher, V. B., Raj, J. R. F., Gopikumar, S., & Rani, P. A. (2024). Predicting Passenger Preferences: An AI-Driven Framework for Personalized Airport Lobby Experiences. In *2024 2nd International Conference on Sustainable Computing and Smart Systems (ICSCSS)* (pp. 1616-1622). IEEE.

Kuo, P., Lord, D., Wheat, P., Yin, Y., & Jia, T. (2022). The impact of the COVID-19 pandemic on O-D flow and airport networks in the origin country and in Northeast Asia. *Journal of Air Transport Management*, 100, 102178. DOI: 10.1016/j.jairtraman.2022.102192 PMID: 35194345

Lee-Anant, C., & Monpanthong, P. (2021). Factors Influencing Airport Technology Selections in Each Service Touchpoint of Suvarnabhumi Airport, Thailand. *Turkish Journal of Computer and Mathematics Education*, 12(13), 3804–3816.

Lipkin, M., & Heinonen, K. (2022). Customer ecosystems: Exploring how ecosystem actors shape customer experience. *Journal of Services Marketing*, 36(9), 1–17. DOI: 10.1108/JSM-03-2021-0080

Liu, X., Zhang, H., Wu, S., Wang, Q., He, Z., Zhang, J., Li, R., Liu, S., & Zhang, X. (2022). Effects of buildings on wind shear at the airport: Field measurement by coherent Doppler lidar. *Journal of Wind Engineering and Industrial Aerodynamics*, 230, 105194. DOI: 10.1016/j.jweia.2022.105194

Lundin, L., & Kindström, D. (2024). Managing digitalized touchpoints in B2B customer journeys. *Industrial Marketing Management*, 121, 88–99. DOI: 10.1016/j.indmarman.2024.07.004

Ma, G., Ding, Y., & Ma, J. (2022). The impact of airport physical environment on perceived safety and domestic travel intention of Chinese passengers during the COVID-19 pandemic: The mediating role of passenger satisfaction. *Sustainability (Basel)*, 14(9), 5628. DOI: 10.3390/su14095628

Mainardes, E. W., de Melo, R. F. S., & Moreira, N. C. (2021). Effects of airport service quality on the corporate image of airports. *Research in Transportation Business & Management*, 41, 100668. DOI: 10.1016/j.rtbm.2021.100668

Monge-Zamorano, M. (2023). Cultural heritage experiences at the smart airport. *Journal of Tourism and Heritage Research*, 6(1), 281–299.

Mota, M. M., Scala, P., Schultz, M., Lubig, D., Luo, M., & Perez, E. J. (2021). The rise of the smart passenger i: analysis of impact on departing passenger flow in airports. *SESAR Innovation Days, 2021.*

Nordqvist, M. (2022). *The Future Digital Airport Experience*. Uppsala University Press.

Polat, İ., Ozturk, A., Koc, R., & Inel, Y. (2024). Designing experiences: Unveiling passenger perspectives through service design at an airport. *Journal of Global Hospitality and Tourism*, 3(1), 254–268.

Puls, R., & Wittmer, A. (2021). Managing Airports. *Aviation Systems: Management of the Integrated Aviation Value Chain*, 225-249.

Qing, Z., Sun, C., & Reneker, J. (2021). Evaluation of airport wayfinding accessibility with the use of a wheelchair simulator. *Transportation Research Record: Journal of the Transportation Research Board*, 2675(4), 52–60. DOI: 10.1177/0361198120980445

Rijaluddin, K., & Safitri, E. M. (2023). Analysis and Design of the Mentor On-Demand Service Interface Using the Design Thinking Method. *IJCONSIST JOURNALS*, 4(2), 11–16.

Rubio-Andrada, L., Celemín-Pedroche, M. S., Escat-Cortés, M. D., & Jiménez-Crisóstomo, A. (2023). Passengers satisfaction with the technologies used in smart airports: An empirical study from a gender perspective. *Journal of Air Transport Management*, 107, 102347. DOI: 10.1016/j.jairtraman.2022.102347

Schepers, J., & Streukens, S. (2022). To serve and protect: A typology of service robots and their role in physically safe services. *Journal of Service Management*, 33(2), 197–209. DOI: 10.1108/JOSM-11-2021-0409

Sia, P. Y. H., Saidin, S. S., & Iskandar, Y. H. P. (2023). Systematic review of mobile travel apps and their smart features and challenges. *Journal of Hospitality and Tourism Insights*, 6(5), 2115–2138. DOI: 10.1108/JHTI-02-2022-0087

Singh, H., Khalajzadeh, H., Paktinat, S., Graetsch, U. M., & Grundy, J. (2022). Modelling human-centric aspects of end-users with iStar. *Journal of Computer Languages*, 68, 101091. DOI: 10.1016/j.cola.2022.101091

Story, V., Zolkiewski, J., Verleye, K., Nazifi, A., Hannibal, C., Grimes, A., & Abboud, L. (2020). Stepping out of the shadows: Supporting actors' strategies for managing end-user experiences in service ecosystems. *Journal of Business Research*, 116, 401–411. DOI: 10.1016/j.jbusres.2020.04.029

Strokosch, K., & Osborne, S. P. (2023). Design of services or designing for service? The application of design methodology in public service settings. *Policy and Politics*, 51(2), 231–249. DOI: 10.1332/030557321X16750746455167

Sun, C., Li, S., Lin, Y., & Hu, W. (2022). From Visual Behavior to Signage Design: A Wayfinding Experiment with Eye-Tracking in Satellite Terminal of PVG Airport. In *Proceedings of the 2021 DigitalFUTURES: The 3rd International Conference on Computational Design and Robotic Fabrication (CDRF 2021) 3* (pp. 252-262). Springer Singapore. DOI: 10.1007/978-981-16-5983-6_24

Sun, Q., Phoebe, J., & Ziwei, L. (2022). Service design practice and its future relevance. In *The 23rd DMI: Academic design management conference proceedings* (pp. 359-368). DMI.

Tueanrat, Y., Papagiannidis, S., & Alamanos, E. (2021). Going on a journey: A review of the customer journey literature. *Journal of Business Research*, 125, 336–353. DOI: 10.1016/j.jbusres.2020.12.028

Usman, A., Azis, Y., Harsanto, B., & Azis, A. M. (2022). Airport service quality dimension and measurement: A systematic literature review and future research agenda. *International Journal of Quality & Reliability Management*, 39(10), 2302–2322. DOI: 10.1108/IJQRM-07-2021-0198

Vannucci, V., & Pantano, E. (2020). Digital or human touchpoints? Insights from consumer-facing in-store services. *Information Technology & People*, 33(1), 296–310. DOI: 10.1108/ITP-02-2018-0113

Vink, J., Koskela-Huotari, K., Tronvoll, B., Edvardsson, B., & Wetter-Edman, K. (2021). Service ecosystem design: Propositions, process model, and future research agenda. *Journal of Service Research*, 24(2), 168–186. DOI: 10.1177/1094670520952537

Wang, R., Xie, X., & Ma, H. (2021). Evolution of omni-channel business models: A new community-based omni-channel and data-enabled ecosystem. *Journal of Contemporary Marketing Science*, 4(3), 385–396. DOI: 10.1108/JCMARS-11-2020-0045

Weidig, J., Weippert, M., & Kuehnl, C. (2024). Personalized touchpoints and customer experience: A conceptual synthesis. *Journal of Business Research*, 177, 114641. DOI: 10.1016/j.jbusres.2024.114641

West, S., Gaiardelli, P., & Saccani, N. (2022). *Modern industrial services: a cookbook for design, delivery, and management.* Springer Nature. DOI: 10.1007/978-3-030-80511-1

West, S., Stoll, O., Østerlund, M., Müller-Csernetzky, P., Keiderling, F., & Kowalkowski, C. (2020). Adjusting customer journey mapping for application in industrial product-service systems. *International Journal of Business Environment*, 11(3), 275–297. DOI: 10.1504/IJBE.2020.110911

Wongyai, P. H., Ngo, T., Wu, H., Tsui, K., & Nguyen, T. H. (2024). Self-Service Technology in Aviation: A Systematic Literature Review. *Journal of the Air Transport Research Society*, 100016.

Zdanowicz, K., Herriotts, P., Payre, W., Mangurenje, D., & Birrell, S. (2022). The World's First 'Pop-Up'Urban Airport: A User-Centred Design Approach to Understand the Customer Journey. In *Human Interaction, Emerging Technologies and Future Systems V: Proceedings of the 5th International Virtual Conference on Human Interaction and Emerging Technologies, IHIET 2021, August 27-29, 2021 and the 6th IHIET: Future Systems (IHIET-FS 2021), October 28-30, 2021, France* (pp. 483-490). Springer International Publishing.

Appendix A: List of interviewed passengers

Table 3

No.	Gender	Age	Airport	Type
1	Female	41	Ben-Gurion Airport (Israel)	Business Commuter
2	Female	31	Ben-Gurion Airport (Israel)	Business Commuter
3	Female	47	Ben-Gurion Airport (Israel)	Business Commuter
4	Female	39	Ben-Gurion Airport (Israel)	Family Traveler
5	Female	31	Ben-Gurion Airport (Israel)	Leisure Traveler
6	Female	31	Ben-Gurion Airport (Israel)	Leisure Traveler
7	Female	69	Ben-Gurion Airport (Israel)	Senior Traveler
8	Female	66	Ben-Gurion Airport (Israel)	Senior Traveler
9	Male	73	Ben-Gurion Airport (Israel)	Senior Traveler
10	Female	37	Charles de Gaulle Airport (France)	Family Traveler
11	Female	40	Charles de Gaulle Airport (France)	Family Traveler
12	Female	37	Charles de Gaulle Airport (France)	Family Traveler
13	Male	24	Charles de Gaulle Airport (France)	Leisure Traveler
14	Male	27	Charles de Gaulle Airport (France)	Leisure Traveler
15	Female	32	Charles de Gaulle Airport (France)	Leisure Traveler
16	Male	30	Charles de Gaulle Airport (France)	Leisure Traveler
17	Male	72	Charles de Gaulle Airport (France)	Senior Traveler
18	Male	70	Charles de Gaulle Airport (France)	Senior Traveler
19	Female	67	Charles de Gaulle Airport (France)	Senior Traveler
20	Male	35	JFK Airport (U.S.)	Business Commuter
21	Male	42	JFK Airport (U.S.)	Business Commuter
22	Female	48	JFK Airport (U.S.)	Business Commuter
23	Male	46	JFK Airport (U.S.)	Business Commuter
24	Male	45	JFK Airport (U.S.)	Business Commuter
25	Female	25	JFK Airport (U.S.)	Leisure Traveler
26	Male	28	JFK Airport (U.S.)	Leisure Traveler
27	Male	74	JFK Airport (U.S.)	Senior Traveler
28	Male	72	JFK Airport (U.S.)	Senior Traveler
29	Male	47	Kastrup (Denmark)	Business Commuter
30	Male	49	Kastrup (Denmark)	Business Commuter
31	Male	36	Kastrup (Denmark)	Family Traveler

continued on following page

Table 3

No.	Gender	Age	Airport	Type
32	Female	35	Kastrup (Denmark)	Family Traveler
33	Female	34	Kastrup (Denmark)	Family Traveler
34	Female	38	Kastrup (Denmark)	Family Traveler
35	Male	29	Kastrup (Denmark)	Leisure Traveler
36	Female	35	Kastrup (Denmark)	Leisure Traveler
37	Female	39	Naples International Airport (Italy)	Family Traveler
38	Female	36	Naples International Airport (Italy)	Family Traveler
39	Female	36	Naples International Airport (Italy)	Family Traveler
40	Female	28	Naples International Airport (Italy)	Leisure Traveler
41	Female	26	Naples International Airport (Italy)	Leisure Traveler
42	Female	30	Naples International Airport (Italy)	Leisure Traveler
43	Female	27	Naples International Airport (Italy)	Leisure Traveler
44	Female	26	Naples International Airport (Italy)	Leisure Traveler
45	Male	68	Naples International Airport (Italy)	Senior Traveler
46	Male	73	Naples International Airport (Italy)	Senior Traveler
47	Female	45	Newark Airport (U.S.)	Business Commuter
48	Male	38	Newark Airport (U.S.)	Business Commuter
49	Male	50	Newark Airport (U.S.)	Business Commuter
50	Male	44	Newark Airport (U.S.)	Business Commuter
51	Male	43	Newark Airport (U.S.)	Business Commuter
52	Male	38	Newark Airport (U.S.)	Family Traveler
53	Male	42	Newark Airport (U.S.)	Family Traveler
54	Male	39	Newark Airport (U.S.)	Family Traveler
55	Male	71	Newark Airport (U.S.)	Senior Traveler
56	Male	51	Orlando International Airport (U.S.)	Business Commuter
57	Female	40	Orlando International Airport (U.S.)	Family Traveler
58	Male	41	Orlando International Airport (U.S.)	Family Traveler
59	Female	28	Orlando International Airport (U.S.)	Leisure Traveler
60	Male	75	Orlando International Airport (U.S.)	Senior Traveler
61	Female	67	Orlando International Airport (U.S.)	Senior Traveler
62	Male	70	Orlando International Airport (U.S.)	Senior Traveler
63	Male	75	Orlando International Airport (U.S.)	Senior Traveler
64	Female	68	Orlando International Airport (U.S.)	Senior Traveler

Chapter 11
Understanding Aviation English Course Design and Delivery Through Qualitative Research

Nicole Falzon
https://orcid.org/0009-0006-9164-0880
University of Malta, Malta

Daniel Xerri
https://orcid.org/0000-0002-5397-0647
University of Malta, Malta

ABSTRACT

The central role that English plays in civil aviation means that airline pilots, air traffic controllers and other key personnel require substantial training in its use. Aviation English (AE), which is the language used in the global aviation sector, forms part of the broader context of English for Specific Purposes (ESP). AE is designed to equip learners with the language proficiency required for effective communication in the aviation industry. With reference to one specific study, this chapter illustrates the value of qualitative research in developing an understanding of AE course design and delivery. The study involved the use of participant narratives, semi-structured interviews, and text analysis. This chapter demonstrates that qualitative research plays a critical role in informing and enhancing AE education, underscoring its value in addressing the evolving needs of the aviation sector and ensuring the highest standards of communication and safety in aviation operations.

DOI: 10.4018/979-8-3693-7403-0.ch011

1. INTRODUCTION

English is the critical language of communication in civil aviation, facilitating communication between airline pilots, air traffic controllers, and other aviation personnel (ICAO, 2023a). The specialised field of Aviation English (AE) falls under the broader category of English for Specific Purposes (ESP), which has gained prominence due to the increasing demand for tailored language skills in specific professional domains (Belcher et al., 2011; Hutchinson & Waters, 1987). The aviation industry, with its unique terminologies, protocols, and safety-critical communication needs, requires specialised language training to ensure clarity, precision, and safety in operations (ICAO, 2023a). This specialised training is essential because the nature of communication in aviation is not only technical but also time-sensitive and highly standardised, demanding a level of proficiency that goes beyond general English language skills (Cutting, 2012; Estival et al., 2016; Moder, 2013).

Despite the critical importance of AE, there is a notable gap in the literature regarding the specific instructional practices and course design strategies that most effectively meet the demands of the aviation sector. Much of the existing research focuses on the general principles of ESP and AE, but it lacks a detailed examination of how AE courses are designed and delivered across different international contexts (Kim & Elder, 2009; Tarnavska et al., 2021). This gap is significant because the effectiveness of AE instruction directly impacts aviation safety (Vaz Junior, 2018). Inadequate communication can lead to misunderstandings and errors that have serious consequences, making it imperative to explore and refine the ways in which AE is taught (Bongo et al., 2018; Kokarieva et al., 2020).

With reference to one particular study, this chapter illustrates how qualitative research was used to address the existing gap, thereby yielding insights into the instructional techniques used by AE teachers worldwide. By examining such elements as needs analysis, material selection, instructional strategies, assessment techniques, and course adaptations, this chapter discusses the qualitative approaches utilised to gain insights into AE course design and delivery. The chapter's emphasis is on how these qualitative methods facilitate a comprehensive understanding of the development and delivery of AE courses, rather than on the findings themselves.

2. THE SIGNIFICANCE OF QUALITATIVE RESEARCH

2.1 The Research Question

Just like any other research approach, qualitative research is founded on one or more well-crafted research questions. The study referred to in this chapter addressed the following research question: *How do Aviation English teachers design and deliver their courses?* This question played a central role in understanding the instructional practices within the international English Language Teaching (ELT) framework as applied to AE. The primary objective was to explore these practices by drawing on the experiences of AE teachers and their approaches to course design and delivery (Belcher, 2011; Hussin, 2002). The study sought to uncover the nuances of how AE teachers navigate the complexities of teaching a highly specialised form of English, balancing the need for technical accuracy with the pedagogical requirements of language education (Anthony, 2018; Basturkmen, 2010; Hyland, 2006).

2.2 The Role of Qualitative Research

Qualitative research methods were particularly well-suited to the study on AE course design and delivery because they allowed an in-depth exploration of complex, context-dependent phenomena (Creswell & Creswell, 2022; Flick, 2022). While quantitative research offers valuable insights through statistical analysis and measurable outcomes, it often falls short of capturing the intricacies of human communication, especially in specialised fields like AE (Denzin & Lincoln, 2017). Qualitative research, on the other hand, enables a deeper understanding of these intricacies by collecting, analysing, and interpreting non-numerical data, making it particularly appropriate for an examination of the nuanced challenges of AE instruction (Creswell & Poth, 2017; DeMarrais et al., 2024).

In the context of AE, qualitative methods allow researchers to delve into the specific challenges that teachers face when designing and delivering courses. These challenges include understanding the diverse needs of learners, selecting appropriate materials, and developing assessments that accurately measure proficiency in a language that is both technical and highly standardised (Basturkmen, 2010; Hutchinson & Waters, 1987). By focusing on the lived experiences of AE teachers, qualitative research provides insights that are grounded in the realities of teaching and learning in this specialised field (Van Manen, 2016; Woodrow, 2022).

2.3 The Process of Course Design in AE

Qualitative research can contribute to enhancing the design of AE courses because its findings offer a deeper understanding of learners' specific needs. Designing effective AE courses involves tailoring content and materials to meet learners' specific language needs and objectives (Basturkmen, 2010; Hutchinson & Waters, 1987). A thorough needs analysis is crucial to ensuring the relevance and applicability of the course content (Anthony, 2018; Dudley-Evans & John, 1998). Needs analysis in AE typically involves understanding the specific communication tasks that learners will need to perform in their professional roles, such as reading technical manuals, listening to air traffic control instructions, or writing reports. This process is critical in developing a curriculum that is aligned with the actual language demands of the aviation industry.

Teachers must select current and contextually relevant materials to prepare learners for real-world aviation scenarios (Tomlinson, 2013). This involves not only choosing textbooks and other resources that cover the necessary vocabulary and structures but also incorporating authentic materials such as flight operation manuals, safety regulations, and real-time communication recordings. A valuable approach to understanding students' needs is through the collection of qualitative data, which allows for in-depth analysis of their views and experiences and enables educators to take informed decisions (Patton, 2015).

Furthermore, assessment in AE, such as the kind of testing provided by the International Civil Aviation Organization (ICAO) and the English Language Proficiency for Aeronautical Communication (ELPAC), primarily evaluates listening and speaking skills, which are critical for effective communication in aviation (Downey et al., 2010). However, these assessments do not fully encompass the broader spectrum of skills and systems essential for aviation professionals (Wegler, 2015). For instance, the ability to comprehend and respond to complex instructions in high-stress situations, or to communicate effectively in non-standard situations, is crucial but difficult to measure through traditional forms of assessment. Therefore, ongoing assessment that gauges learners proficiency in a variety of communication contexts is essential for their language development (Hughes, 2003).

Flexibility in teaching approaches is also vital to adapt to learners' evolving needs and enhance their overall language learning experience (Jordan, 1997). In AE, this might involve varying instructional methods based on the learners' progress, integrating new technologies such as flight simulation software, or adjusting the course content to address emerging trends in the aviation industry. This flexibility ensures that AE instruction remains relevant and effective in meeting the dynamic needs of the aviation sector.

3. PARTICIPANTS

3.1 Importance of Participant Selection in Qualitative Research

Participant selection is a crucial aspect of qualitative research, particularly in specialised fields such as aviation, because it directly influences the depth, relevance, and credibility of the findings (Saunders, 2012). In the context of the study referred to in this chapter, focusing on international AE teachers as participants was particularly important for several reasons.

Firstly, the selection of participants who are exclusively AE teachers—whether employed by educational institutions or working independently as freelancers—ensured that the insights gathered were specifically relevant to the nuances of AE instruction. Given ICAO guidelines, which mandate specialised training for AE teachers, this group was uniquely positioned to provide informed perspectives on the complexities of teaching English in an aviation context (Estival et al., 2016; ICAO, 2023b).

Secondly, the diversity among the AE teachers who participated in the study—spanning nationality, age, employment status, and years of experience—provided a rich array of perspectives, which was critical for uncovering varied teaching practices across different cultural and operational contexts. This heterogeneity allowed the study to capture a broader range of instructional approaches, reflecting the global nature of aviation and the need for adaptable AE courses that meet diverse linguistic and cultural demands (Flick, 2022; Walby, 2015).

3.2 Recruitment and Participant Profiles

Recruitment is a critical aspect of qualitative research methodology. Although enrolling participants in a study can be difficult (Spratling, 2013), it plays a significant role in determining the results of the research (Newington & Metcalfe, 2014). The recruitment process of the study on AE course design and delivery leveraged professional networks and referrals; it was strategic in ensuring that the research included participants who were not only suitably qualified but also actively engaged in AE teaching. This approach likely enhanced the relevance of the data collected, as these participants were immersed in the day-to-day challenges and developments of AE instruction (Cypress, 2018; Flick, 2014).

Understanding the target population is a fundamental aspect of planning recruitment in qualitative research, helping to determine the most effective strategies (Coyne et al., 2016). Participants were drawn from various international contexts, including Europe, Asia, and South America, ensuring a wide geographical representation. This diversity was crucial in capturing the different ways AE is taught in various parts of the world, taking into account the local regulatory requirements, cultural norms, and educational traditions. Fostering participant trust and engagement was critical to the success of the recruitment process, ensuring that participants felt confident in the credibility and purpose of the research (Bonisteel et al., 2021). Each participant provided a detailed profile, including their educational background, years of experience in teaching AE, and the specific contexts in which they worked.

4. RESEARCH INSTRUMENTS

4.1 The Role of Qualitative Research in Aviation

In the field of aviation, qualitative research methods, such as interviews, focus groups, observation, and text analysis, are essential tools for exploring and understanding the complex dynamics within the industry. These methods focus on uncovering processes and patterns that emerge, rather than simply identifying final outcomes, making them particularly suitable for answering 'how' questions that describe processes or behaviors (Creswell, 2013; Yin, 2017). This is crucial in an industry where understanding the underlying factors of communication and decision-making can directly impact safety and operational efficiency (Patton, 2015; Salas et al., 2010; Stolzer, 2023).

Qualitative instruments are designed to capture the richness and depth of aviation professionals' thoughts, behaviors, and social contexts, offering insights into industry-specific challenges and practices that are often inaccessible through quantitative methods (Creswell & Poth, 2017; Denzin & Lincoln, 2017; Silverman & Patterson, 2021). The flexibility of qualitative research allows researchers to adapt their approach based on the evolving nature of aviation management studies, making these methods particularly suitable for investigating nuanced phenomena such as course design, examination preparation, safety practices, and leadership in real-world aviation settings (Baxter & Jack, 2008; Reason & Hobbs, 2017). For example, structured or semi-structured interviews provide opportunities for an in-depth exploration of the individual perspectives of different kinds of aviation professionals, from pilots to air traffic controllers (Molesworth & Estival, 2015; Xerri, 2018a). In contrast, focus groups facilitate the examination of collective views on issues like safety protocols and regulatory compliance. Observation as a research tool enables researchers to capture behaviours and interactions in operational environments like control towers or maintenance facilities, adding context and depth to a study (Guest et al., 2013; Tracy, 2024). The use of triangulation enhances the credibility and validity of qualitative research in aviation by allowing researchers to cross-verify findings and develop a comprehensive understanding of industry-specific problems (Bans-Akutey & Tiimub, 2021; Carter et al., 2014; Denzin & Lincoln, 2017; Xerri, 2018b).

4.2 Why Quantitative Methods Might Be Less Effective

While quantitative methods offer valuable insights into measurable aspects of aviation training, such as test scores, proficiency levels, and statistical trends, they are limited in their ability to explore the complex, context-dependent nature of communication in aviation (Dekker, 2017; Douglas, 2000). Quantitative approaches typically rely on standardised testing and structured data collection, which can oversimplify the realities of how aviation professionals use language in high-pressure, real-world scenarios (Kim & Elder, 2009; Molesworth & Estival, 2015). For instance, standardised tests may effectively measure a pilot's ability to recall specific terminology or respond to scripted prompts, but they cannot capture the adaptive, context-sensitive communication skills required to manage unexpected situations or navigate intercultural interactions (Bachman & Adrian, 2022; Kim & Elder, 2009).

In aviation training, particularly in AE, the critical issues often lie in the subtleties of language use—how pilots and air traffic controllers interpret and respond to ambiguous instructions, manage non-verbal cues, or negotiate meaning with colleagues from diverse linguistic backgrounds (Bullock, 2015; Barshi & Farris, 2016; Moder, 2013). These aspects of communication are deeply influenced by the operational context, cultural nuances, and individual experiences, all of which are difficult to quantify (Barshi & Farris, 2016; Kim & Elder, 2015; Tiewtrakul & Fletcher, 2010). Therefore, a purely quantitative approach might fail to address the full scope of skills necessary for effective and safe communication in aviation.

4.3 Application of Qualitative Research in AE Course Design

Given the global distribution of participants in the study referred to in this chapter, qualitative tools such as observation and focus groups were not feasible. The logistical challenges of coordinating synchronous activities across different time zones, coupled with the need for non-intrusive data collection methods that respect the confidentiality of sensitive information in a regulated industry like aviation, necessitated the use of alternative approaches. In the high stakes environment of aviation, where communication is critical to safety and efficiency, these challenges are further magnified (Monteiro & Bullock, 2020). Therefore, the study strategically employed a combination of narrative inquiry, semi-structured interviews, and text analysis to capture the nuanced and context-dependent nature of AE instruction.

4.3.1 Narrative Inquiry

Narrative inquiry was particularly valuable for understanding how AE teachers design and deliver courses that must meet rigorous international standards, such as those set by ICAO (2023b). The aviation industry's reliance on clear, precise communication makes these narratives crucial for exploring how teachers adapt their teaching to ensure that pilots and air traffic controllers achieve the required proficiency levels (Estival et al., 2016). Through the narratives, participants were able to share their experiences in navigating these challenges, providing an understanding of their teaching practices (Barkhuizen, 2014; Clandinin & Connelly, 2000; Golombek & Johnson, 2004). This method allowed the study to explore individual perspectives on how AE education is delivered in diverse and often high-pressure aviation settings.

However, a significant limitation of narrative inquiry is its inherent subjectivity. Since narratives are shaped by personal perspectives, they may not always reflect broader trends or objective realities, potentially introducing bias into the research (Bergen & Labonté, 2020; Galdas, 2017). Additionally, while narratives provide deep, qualitative insights, they often lack the breadth needed to generalise findings across a larger population of AE teachers (Smith, 2018). Despite these being considered limitations from a positivist paradigm, the interpretivist paradigm rejects the aim of generalisation, instead prioritising the depth of understanding gained from individual experiences and social contexts (Alharahsheh & Pius, 2020; Scotland, 2012). Interpretivism views reality as socially constructed and context-dependent, which allows researchers to explore the unique realities of AE teachers without seeking universal laws (Alharahsheh & Pius, 2020).

Participants were invited to submit written narratives via Google Forms, structured around prompts that guided their responses while allowing flexibility in how they expressed their experiences (Barkhuizen & Wette, 2008). This initial step was crucial for identifying prominent themes and critical issues, which then informed the development of subsequent semi-structured interviews (Kiger & Varpio, 2020; Saldaña, 2013). To safeguard the anonymity of the participants concerning the narrative prompt, the participants were provided with a pseudonym prior to filling in the Google Form, e.g., Teacher A (Nachmias et al., 2015). While filling in the Google Form, each participant selected the pseudonym that was allocated to them in order to make it easier to match their narrative to their identity as an interview participant.

4.3.2 Semi-Structured Interviews

Semi-structured interviews are crucial in qualitative research as they provide flexibility and allow for in-depth exploration of participants' insights while maintaining focus on a study's key research issues (Ruslin et al., 2022). In the study on AE course design and delivery, semi-structured interviews were conducted with 14 AE teachers from diverse international school settings and freelancing roles, focusing on critical aspects of aviation communication. This method was chosen for its ability to delve deeper into how teachers develop curricula that must address both standard aviation phraseology and the more complex plain language communication required in emergency situations. The semi-structured format allowed participants to elaborate on their experiences while ensuring that the interviews remained focused on relevant aviation-related topics, such as curriculum development, material selection, and instructional strategies specific to AE (Adeoye-Olatunde & Olenik, 2021; Cypress, 2018; Kallio et al., 2016).

However, the flexibility of semi-structured interviews can also lead to variability in the depth and focus of responses, making it difficult to compare data across participants (Kvale, 2007, 2009; Ruslin et al., 2022). Moreover, conducting these interviews remotely via Zoom introduced potential challenges related to technology, such as poor internet connections or difficulties in establishing rapport, which could affect the quality of the data collected (Baltar & Brunet, 2012; Deakin & Wakefield, 2014; Ramsey et al., 2016). These technological issues may further compound the inherent challenges in transcription, a critical process for thematic analysis, as errors in capturing the nuances of speech could impact the accuracy of the analysis (Kowal & O'Connell, 2014).

Online interviewing also poses significant data security risks, as certain platforms like chat rooms, and email services often archive data, making it accessible to service providers, system administrators, and search engines (Finlay, 2012). Therefore, interviewers must carefully choose online tools that best protect participants' data to ensure their privacy is maintained. Additionally, while tools like chat rooms and discussion forums allow for real-time transcription, they often suffer from such limitations as a lack of synchronicity, which can hinder the transcription process after the interview (Salmons, 2017).

To mitigate these challenges, the study on AE course design and delivery employed rigorous transcription protocols. Member checking, in particular, was a means to enhance credibility by seeking validation from the participants themselves. In this process, the participants were provided with transcripts of their responses, allowing them to review, edit, or suggest changes to the information presented (Polit & Tatano Beck, 2004). This active involvement of the participants created a transparent approach to data interpretation and ensured that their perspectives were accurately represented in the study's findings (Xerri, 2018b). Multiple rounds of coding were included to ensure consistency and reliability in the analysis process. Furthermore, by leveraging the strengths of online interviewing—such as the ability to reach geographically dispersed participants and reduce travel costs (Walker, 2013; Madge & O'Connor, 2002)—while addressing its inherent challenges, the research aimed to produce robust and reliable findings that accurately reflect the experiences and insights of AE teachers.

A relevant example of the effectiveness of qualitative research in AE course design is seen in a study conducted at the Escuela Superior Militar de Aviación 'Cosme Runnells Barbatto' (ESMA) in Salinas, Ecuador. This study used semi-structured interviews and student surveys to identify specific instructional challenges, such as misalignment with ICAO standards. The qualitative insights gained from both teachers and students were critical in recommending adjustments to the course design. This case demonstrates how qualitative research can uncover crucial insights that lead to significant improvements in AE instruction, ensuring that courses are better tailored to meet the specific communication needs of aviation professionals, which are essential for operational safety and effectiveness (Gutierrez Santos et al., 2018).

4.3.3 Text Analysis

Text analysis was the final qualitative instrument used in the study referred to in this chapter (Bernard & Ryan, 2014). Participants submitted sample lessons they had conducted, which were then analyzed to understand the linguistic features, discourse patterns, and instructional strategies employed within AE lessons (Fairclough, 2003; Krippendorff, 2009). In the context of aviation, where precise and unambiguous communication is critical, this method was particularly valuable for its ability to reveal how theoretical principles of AE education are applied in practice, ensuring that instructional materials are aligned with both ICAO standards and real-world operational demands. By examining actual teaching materials, the study could connect the theoretical insights gained from narratives and interviews with the practical realities of AE instruction. However, the analysis was limited by the small sample size; due to confidentiality considerations, only materials supplied by five AE teachers were included in the final analysis. This limitation further underscores the challenges of conducting qualitative studies in such sensitive and regulated environments, where obtaining consent for access to classroom settings or instructional materials often involves complex processes (Sullivan & Girginer, 2002).

4.4 Triangulation and Methodological Integration in Qualitative Research

The integration of narrative inquiry, semi-structured interviews, and text analysis in the study on AE course design and delivery aimed to capitalise on the strengths of each method, thereby enriching the depth and breadth of the research findings (Barkhuizen, 2014; Denzin & Lincoln, 2017). The sequential use of these methods—beginning with narrative collection, followed by interviews and text analysis—facilitated a comprehensive understanding of AE instructional practices. This approach also helped overcome challenges related to depth and detail in the narratives by using the interviews to explore emergent themes more thoroughly. The triangulation of multiple qualitative methods enhanced the validity and reliability of the findings by allowing for multiple perspectives and cross-validation of results (Creswell & Poth, 2017).

Triangulation is a widely recognised strategy in qualitative research, used to validate findings through the convergence of information from different sources (Flick, 2018). In the study, triangulation was employed not only to cross-check data but also to integrate diverse perspectives, thereby providing a more comprehensive understanding of AE course design and delivery (Oliver-Hoyo & Allen, 2006). By comparing the insights gained from different methods, the study was able to identify consistent themes and discrepancies, which were then explored in greater depth. This process of cross-validation is essential in qualitative research, as it strengthens the credibility of the findings and helps to mitigate potential biases inherent in any single method (Moss & Edmonds, 2005). For example, the narrative inquiry provided personal insights into how teachers perceive and navigate the challenges of teaching AE. These insights were then cross-referenced with the data obtained from semi-structured interviews, which offered a more structured exploration of these challenges. Text analysis further supported these findings by revealing how these challenges manifest themselves in actual teaching materials (Kuckartz, 2019). The convergence of these different data sources provided a robust and well-rounded understanding of AE course design, highlighting both the commonalities and the unique aspects of instructional practices across different contexts.

4.5 The Importance of Piloting a Qualitative Research Study

In aviation, the importance of piloting research instruments is particularly critical due to the industry's high stakes nature and complex operational environment. Ensuring the validity, reliability and trustworthiness of data collection methods, including the design and administration of interview questions, narratives, and other materials, is crucial as the accuracy of data directly impacts safety, regulatory compliance, and operational efficiency (Cohen et al., 2017; Morse, 2015).

Piloting these materials not only refines the instruments to capture precise and relevant information but also provides valuable insights to stakeholders, curriculum developers, and other key players. For example, during the pilot phase of the study on AE course design and delivery, feedback from participants led to the refinement of interview questions, ensuring they were better aligned with the specific challenges faced by AE teachers, thereby enhancing the relevance and accuracy of the data collected (Flick, 2022; Kezar, 2000). Furthermore, in an industry where ethical considerations are paramount, especially concerning participant safety and data security, piloting helps identify and address potential ethical issues, thereby upholding the highest standards in data collection and analysis (Denscombe, 2014; Creswell & Creswell, 2022). Ultimately, piloting research instruments and materials in aviation is essential in providing stakeholders with meaningful insights, mitigating risks, and ensuring that data-driven decisions are accurate, safe, and contribute to the overall effectiveness of aviation operations (Creswell & Creswell, 2022).

Piloting was an integral part of the study because it provided some guarantee that the research instruments were robust and capable of capturing the necessary data. The pilot phase involved testing the interview questions and narrative prompts with a small group of AE teachers who were not part of the main study. This process revealed several areas for improvement, such as the need to clarify certain questions and the importance of allowing more flexibility in the narrative prompts. As a result, the research instruments were refined to be better aligned with the specific challenges and needs of the participants, thereby enhancing the quality and relevance of the data collected.

4.6 Analysing Qualitative Data

Qualitative research excels at exploring complex human experiences, making it particularly useful for understanding nuanced topics like AE course design and delivery. Central to qualitative research is the process of data analysis, where collected data is systematically examined to identify meaningful patterns and insights. Among the various methods available, thematic analysis is widely recognised for its flexibility and applicability across different types of qualitative data, such as interview and observation data (Braun & Clarke, 2021). Thematic analysis is a method used to identify, analyse, and report patterns or themes within data, capturing key aspects relevant to the research question (Braun & Clarke, 2021). The process begins with organising and preparing the data, which involves transcribing interviews, cataloguing visual materials, and sorting data by source types, thus establishing a foundation for analysis (Guest et al., 2013; Nowell et al., 2017). In the study on

AE course design and delivery, thematic analysis was employed to systematically explore and interpret qualitative data from interviews and narratives.

The thematic analysis process in the study began with the meticulous organisation of the data, which included transcribing the interviews and documenting the narratives provided by participants. This initial phase was crucial in ensuring that all relevant data were systematically catalogued and ready for detailed analysis (Creswell & Creswell, 2022). Following this, a comprehensive reading phase facilitated deep familiarisation with the content, and the identification of preliminary patterns and recurring themes that would inform the subsequent stages of analysis (Guest et al., 2013).

Manual coding was then employed to identify meaningful data segments, which were labelled with relevant codes (Creswell & Creswell, 2022). This process involved a careful and iterative review of the data, with the codes being refined and adjusted as new patterns emerged. The manual coding approach was chosen for its flexibility and for facilitating a close connection with the data; it helped capture subtle nuances that might have been overlooked by automated coding tools (Rossman & Rallis, 2017). The coding process was closely aligned with the study's research question, ensuring that the identified themes directly addressed key aspects of AE course design and delivery. As the analysis progressed, themes began to emerge from the coded data, reflecting patterned responses and meaningful interpretations related to the research topic. These themes were then interpreted in the context of the broader research question, providing insights into the complexities of AE course design and delivery.

5. RESEARCH ETHICS IN QUALITATIVE RESEARCH

Conducting qualitative research on AE course design and delivery requires a strong commitment to ethical considerations across all phases of the process. Adhering to established ethical frameworks, such as those outlined by the British Association for Applied Linguistics (BAAL, 2021), is essential to protect participants' rights, minimise harm, and safeguard privacy (DeCosta, 2016; Hammersley & Traianou, 2012). In the aviation context, these ethical challenges are particularly acute due to the industry's tightly regulated nature and the sensitivity of information related to safety protocols (Estival et al., 2016). Researchers must take rigorous steps to ensure confidentiality, which involves anonymising data, securely storing it, and planning for its deletion after a study's completion (Denscombe, 2014). Moreover, fostering trust through transparency and informed consent is crucial; participants should be fully informed about the purpose of the research, their rights, and the potential risks involved, with the assurance that they can withdraw at any time without suffering any consequences (DeMarrais et al., 2024). The accuracy and integrity of the data collected directly impact not only the findings of the research but also its potential implications for aviation safety and communication practices. Therefore, it is imperative that researchers in this field adhere to the highest ethical standards to ensure that their work contributes positively to the industry and does not inadvertently cause harm (DeMarrais et al., 2024; Patton, 2015).

In addition to these foundational ethical practices, AE research presents specific dilemmas related to the analysis of course materials and the conduct of interviews with professionals who might fear that sharing their experiences could expose sensitive industry practices or negatively impact their reputation or that of their institution. These concerns can lead to reluctance in disclosing information, potentially limiting the depth and accuracy of the data collected (Howe & Moses, 1999). To address these concerns, the study on AE course design and delivery implemented several measures to ensure that participants felt comfortable and confident about sharing their experiences. These measures included providing detailed information about how the data would be used, ensuring that all data would be anonymised, and giving participants the option to withdraw from the study at any time (Cohen et al., 2017; Plummer, 1990). This was done in accordance with the principle that qualitative researchers focusing on AE must carefully balance the pursuit of knowledge with the responsibility to protect participants, ensuring that findings are presented in a manner that is fair, constructive, and contributes positively to the field. Upholding these rigorous ethical standards not only enhances the integrity and credibility of the research but also fosters trust and supports the broader societal good, particularly in a field as critical as aviation (Lincoln & Guba, 1985).

6. CONCLUSION

Qualitative research is indispensable for advancing knowledge about practices in the field of aviation. By capturing the complex, context-dependent nature of aviation operations, qualitative methods provide critical insights that inform best practices, enhance safety, and ensure the effective training of aviation professionals. As illustrated by the study discussed in this chapter, through methods like narrative inquiry, semi-structured interviews, and text analysis, qualitative research uncovers the complex processes involved in AE education, offering deep insights into the experiences of teachers and the specific needs of aviation professionals.

By employing a triangulated approach that integrated multiple qualitative methods, the study was able to provide a comprehensive understanding of AE course design and delivery. This understanding is essential for developing effective AE courses that not only meet current industry standards but also anticipate future challenges.

As the aviation industry continues to evolve, the need for robust qualitative research will only grow, making it a cornerstone of future advancements in aviation management. The insights gained from the discussed study provide a strong foundation for future research and practice in AE education, offering valuable guidance for educators, curriculum developers, and policymakers. By continuing to explore and refine AE instructional practices through qualitative research, the industry can ensure that aviation professionals are well-prepared to meet the linguistic demands of their roles, ultimately contributing to the safety and effectiveness of global aviation operations.

REFERENCES

Adeoye-Olatunde, O. A., & Olenik, N. L. (2021). Research and scholarly methods: Semi-structured interviews. *Journal of the American College of Clinical Pharmacy: JAACP*, 4(10), 1358–1367. DOI: 10.1002/jac5.1441

Alharahsheh, H. H., & Pius, A. (2020). A review of key paradigms: Positivism vs interpretivism. *Global Academic Journal of Humanities and Social Sciences*, 2(3), 39–43. DOI: 10.4018/IJSEM.2020010105

Anthony, L. (2018). *Introducing English for specific purposes*. Routledge. DOI: 10.4324/9781351031189

BAAL. (2021). Recommendations on good practice in applied linguistics. https://www.baal.org.uk/wp-content/uploads/2021/03/BAAL-Good-Practice-Guidelines-2021.pdf

Bachman, L., & Adrian, P. (2022). *Language assessment in practice: Developing language assessments and justifying their use in the real world*. Oxford University Press.

Baltar, F., & Brunet, I. (2012). Social research 2.0: Virtual snowball sampling method using Facebook. *Internet Research*, 22(1), 57–74. DOI: 10.1108/10662241211199960

Bans-Akutey, A., & Tiimub, B. M. (2021). Triangulation in research. *Academia Letters*, 2, 1–6. DOI: 10.20935/AL3392

Barkhuizen, G. (2014). Narrative research in language teaching and learning. *Language Teaching*, 47(4), 450–466. DOI: 10.1017/S0261444814000172

Barkhuizen, G., & Wette, R. (2008). Narrative frames for investigating the experiences of language teachers. *System*, 36(3), 372–387. DOI: 10.1016/j.system.2008.02.002

Barshi, I., & Farris, C. (2016). *Misunderstandings in ATC communication*. Routledge. DOI: 10.4324/9781315595641

Basturkmen, H. (2010). *Developing courses in English for specific purposes*. Palgrave Macmillan. DOI: 10.1057/9780230290518

Baxter, P., & Jack, S. (2008). Qualitative case study methodology: Study design and implementation for novice researchers. *The Qualitative Report*, 13(4), 544–559. DOI: 10.46743/2160-3715/2008.1573

Belcher, D. D., Johns, A. M., & Paltridge, B. (Eds.). (2011). *New directions in English for specific purposes research*. University of Michigan Press. DOI: 10.3998/mpub.371075

Bergen, N., & Labonté, R. (2020). 'Everything is perfect, and we have no problems': Detecting and limiting social desirability bias in qualitative research. *Qualitative Health Research*, 30(5), 783–792. DOI: 10.1177/1049732319889354 PMID: 31830860

Bernard, H. R., & Ryan, G. W. (2014). Text analysis: Qualitative and quantitative methods. In Bernard, H. R., & Gravlee, C. C. (Eds.), *Handbook of methods in cultural anthropology* (2nd ed., pp. 533–559). Rowman & Littlefield.

Bongo, M. F., Alimpangog, K. M. S., Loar, J. F., Montefalcon, J. A., & Ocampo, L. A. (2018). An application of DEMATEL-ANP and PROMETHEE II approach for air traffic controllers' workload stress problem: A case of Mactan Civil Aviation Authority of the Philippines. *Journal of Air Transport Management*, 68, 198–213. DOI: 10.1016/j.jairtraman.2017.10.001

Bonisteel, I., Shulman, R., Newhook, L. A., Guttmann, A., Smith, S., & Chafe, R. (2021). Reconceptualizing recruitment in qualitative research. *International Journal of Qualitative Methods*, 20, 16094069211042493. Advance online publication. DOI: 10.1177/16094069211042493

Braun, V., & Clarke, V. (2021). Thematic analysis: A practical guide. *Sage*.

Bullock, N. (2015). Defining meaningful material for the teaching of English for aeronautical communications. In Borowska, A., & Enright, A. (Eds.), *Changing perspectives on aviation English training* (pp. 35–45). Uniwersytet Warszawski.

Carter, N., Bryant-Lukosius, D., DiCenso, A., Blythe, J., & Neville, A. (2014). The use of triangulation in qualitative research. *Oncology Nursing Forum*, 41(5), 545–547. DOI: 10.1188/14.ONF.545-547 PMID: 25158659

Clandinin, D. J., & Connelly, F. M. (2000). *Narrative inquiry: Experience and story in qualitative research*. Jossey-Bass.

Cohen, L., Manion, L., & Morrison, K. (2017). *Research methods in education* (8th ed.). Routledge. DOI: 10.4324/9781315456539

Coyne, E., Grafton, E., & Reid, A. (2016). Strategies to successfully recruit and engage clinical nurses as participants in qualitative clinical research. *Contemporary Nurse*, 52(6), 669–676. DOI: 10.1080/10376178.2016.1181979 PMID: 27108654

Creswell, J. W. (2013). *Qualitative inquiry & research design: Choosing among five approaches* (3rd ed.). Sage.

Creswell, J. W., & Creswell, J. D. (2022). *Research design* (6th ed.). Sage.

Creswell, J. W., & Poth, C. (2017). Qualitative inquiry and research design: Choosing among five traditions. *Sage.*

Cutting, J. (2012). English for airport ground staff. *English for Specific Purposes*, 31(1), 3–13. DOI: 10.1016/j.esp.2011.06.002

Cypress, B. (2018). Qualitative research methods. *Dimensions of Critical Care Nursing*, 37(6), 302–309. DOI: 10.1097/DCC.0000000000000322 PMID: 30273215

Deakin, H., & Wakefield, K. (2014). Skype interviewing: Reflections of two PhD researchers. *Qualitative Research*, 14(5), 603–616. DOI: 10.1177/1468794113488126

Decosta, P. (2016). *Ethics in applied linguistics research: Language researcher narratives.* Routledge.

Dekker, S. (2017). *The safety anarchist: Relying on human expertise and innovation, reducing bureaucracy and compliance.* Routledge. DOI: 10.4324/9780203733455

DeMarrais, K., Roulston, K., & Copple, J. (2024). *Qualitative research design and methods.* Stylus Publishing.

Denscombe, M. (2014). *The good research guide.* Open University Press.

Denzin, N. K., & Lincoln, Y. S. (2017). *The SAGE handbook of qualitative research.* Sage.

Douglas, D. (2000). *Assessing language for specific purposes.* Cambridge University Press.

Downey, R., Suzuki, M., & Van Moere, A. (2010). High-stakes English-language assessments for aviation professionals: Supporting the use of a fully automated test of spoken-language proficiency. *IEEE Transactions on Professional Communication*, 53(1), 18–32. DOI: 10.1109/TPC.2009.2038734

Dudley-Evans, T., & John, M. (1998). *English for specific purposes.* Cambridge University Press.

Estival, D., Farris, C., & Molesworth, B. (2016). *Aviation English: A lingua franca for pilots and air traffic controllers.* Routledge. DOI: 10.4324/9781315661179

Fairclough, N. (2003). *Analysing discourse: Textual analysis for social research.* Routledge. DOI: 10.4324/9780203697078

Finlay, L. (2012). Five lenses for the reflexive interviewer. In Gubrium, J. F., Holstein, J. A., Marvasti, A. B., & McKinne, K. D. (Eds.), *The SAGE handbook of interview research: The complexity of the craft* (2nd ed., pp. 317–332). Sage., DOI: 10.4135/9781452218403

Flick, U. (2014). An introduction to qualitative research. *Sage*.

Flick, U. (2018). Doing triangulation and mixed methods. *Sage*.

Flick, U. (2022). *The Sage handbook of qualitative research design*. Sage. DOI: 10.4135/9781529770278

Galdas, P. (2017). Revisiting bias in qualitative research: Reflections on its relationship with funding and impact. *International Journal of Qualitative Methods*, 16(1), 1609406917748992. Advance online publication. DOI: 10.1177/1609406917748992

Golombek, P. R., & Johnson, K. E. (2004). Narrative inquiry as a mediational space: Examining emotional and cognitive dissonance in second-language teachers' development. *Teachers and Teaching*, 10(3), 307–327. DOI: 10.1080/1354060042000204388

Guest, G., Macqueen, K. M., & Namey, E. E. (2012). Applied thematic analysis. *Sage*.

Guest, G., Namey, E. E., & Mitchell, M. L. (2013). Collecting qualitative data: A field manual for applied research. *Sage*. Advance online publication. DOI: 10.4135/9781506374680

Gutierrez Santos, M., Almeida Pacheco, V., Rodriguez Reyes, R., & Veintimilla Vargas, G. (2018). Analysis of the methodology applied in aviation English course at a military school in Salinas. *Journal of Education and Human Development*, 7(3), 39–44. DOI: 10.15640/jehd.v7n3a5

Hammersley, M., & Traianou, A. (2012). Ethics in qualitative research: Controversies and contexts. *Sage*. DOI: 10.4135/9781473957619

Howe, K. R., & Moses, M. S. (1999). Chapter 2: Ethics in educational research. *Review of Research in Education*, 24(1), 21–59. DOI: 10.3102/0091732X024001021

Hughes, A. (2003). *Testing for language teachers*. Cambridge University Press.

Hussin, V. (2002). An ESP program for students of nursing. In Orr, T. (Ed.), *English for specific purposes* (pp. 25–39). TESOL Publications.

Hutchinson, T., & Waters, A. (1987). *English for specific purposes: A learning-centred approach*. Cambridge University Press. DOI: 10.1017/CBO9780511733031

Hyland, K. (2006). *English for academic purposes: An advanced resource book*. Routledge. DOI: 10.4324/9780203006603

ICAO. (2023a). *About ICAO*. https://www.icao.int/about-icao/Pages/default.aspx

ICAO. (2023b). *Future of aviation*. https://www.icao.int/Meetings/FutureOfAviation/Pages/default.aspx

Jordan, R. R. (1997). *English for academic purposes: A guide and resource book for teachers*. Cambridge University Press. DOI: 10.1017/CBO9780511733062

Kallio, H., Pietilä, A. M., Johnson, M., & Kangasniemi, M. (2016). Systematic methodological review: Developing a framework for a qualitative semi-structured interview guide. *Journal of Advanced Nursing*, 72(12), 2954–2965. DOI: 10.1111/jan.13031 PMID: 27221824

Kezar, A. (2000). The importance of pilot studies: Beginning the hermeneutic circle. *Research in Higher Education*, 41(3), 385–400. DOI: 10.1023/A:1007047028758

Kiger, M. E., & Varpio, L. (2020). Thematic analysis of qualitative data: AMEE Guide No. 131. *Medical Teacher*, 42(8), 846–854. DOI: 10.1080/0142159X.2020.1755030 PMID: 32356468

Kim, H., & Elder, C. (2009). Understanding aviation English as a lingua franca. *Australian Review of Applied Linguistics, 32*(3), 23.1–23.17. DOI: 10.1075/aral.32.3.03kim

Kim, H., & Elder, C. (2015). Interrogating the construct of aviation English: Feedback from test takers in Korea. *Language Testing*, 32(2), 129–149. DOI: 10.1177/0265532214544394

Kokarieva, A., Khomenko-Semenova, L., Glushanytsia, N., Ievtushenko, I., & Odarchenko, R. (2020). Information and communication technologies in the professional training of engineers. In *CMiGIN* (pp. 496–506). https://ceur-ws.org/Vol-2588/paper41.pdf

Kowal, S., & O'Connell, D. C. (2014). Transcription as a crucial step of data analysis. In Flick, U. (Ed.), *The SAGE handbook of qualitative data analysis* (pp. 64–79). Sage, DOI: 10.4135/9781446282243.n5

Krippendorff, K. (2009). Content analysis: An introduction to its methodology. *Sage*.

Kuckartz, U. (2019). Qualitative text analysis: A systematic approach. In Kaiser, G., & Presmeg, N. (Eds.), *Compendium for early career researchers in mathematics education* (pp. 181–197). Springer., DOI: 10.1007/978-3-030-15636-7_8

Kvale, S. (2007). Qualitative research kit: Doing interviews. *Sage*.

Kvale, S. (2009). Interviews: Learning the craft of qualitative research interviewing. *Sage*.

Lincoln, Y., & Guba, E. (1985). Naturalistic inquiry. *Sage*.

Madge, C., & O'Connor, H. (2002). On-line with e-mums: Exploring the internet as a medium for research. *Area*, 34(1), 92–102. https://www.jstor.org/stable/20004209. DOI: 10.1111/1475-4762.00060

Moder, C. L. (2013). Aviation English. In Paltridge, B., & Starfield, S. (Eds.), *The handbook of English for specific purposes* (pp. 227–242). John Wiley & Sons.

Molesworth, B. R., & Estival, D. (2015). Miscommunication in general aviation: The influence of external factors on communication errors. *Safety Science*, 73, 73–79. DOI: 10.1016/j.ssci.2014.11.004

Monteiro, A. L., & Bullock, N. (2020). A broader view of communicative competence for aeronautical communications: Implications for teaching and high-stakes testing. *The Especialist*, 41(3). Advance online publication. DOI: 10.23925/2318-7115.2020v41i3a4

Morse, J. M. (2015). Critical analysis of strategies for determining rigor in qualitative inquiry. *Qualitative Health Research*, 25(9), 1212–1222. DOI: 10.1177/1049732315588501 PMID: 26184336

Moss, S., & Edmonds, B. (2005). Sociology and simulation: Statistical and qualitative cross-validation. *American Journal of Sociology*, 110(4), 1095–1131. DOI: 10.1086/427320

Nachmias, C., Nachmias, D., & Dewaard, J. (2015). *Research methods in the social sciences*. Worth Publishers.

Newington, L., & Metcalfe, A. (2014). Factors influencing recruitment to research: Qualitative study of the experiences and perceptions of research teams. *BMC Medical Research Methodology*, 14(1), 10. Advance online publication. DOI: 10.1186/1471-2288-14-10 PMID: 24456229

Nowell, L. S., Norris, J. M., White, D. E., & Moules, N. J. (2017). Thematic analysis: Striving to meet the trustworthiness criteria. *International Journal of Qualitative Methods*, 16(1), 1609406917733847. Advance online publication. DOI: 10.1177/1609406917733847

Oliver-Hoyo, M., & Allen, D. (2006). The use of triangulation methods in qualitative educational research. *Journal of College Science Teaching*, 35(4).

Patton, M. Q. (2015). *Qualitative research & evaluation methods* (4th ed.). Sage.

Plummer, K. (1990). *Documents of life: An introduction to the problems and literature of a humanistic method*. Unwin Hyman.

Polit, D. F., & Tatano Beck, C. (2004). *Nursing research: Principles and methods* (7th ed.). Lippincott Williams & Wilkins.

Ramsey, S. R., Thompson, K. L., McKenzie, M., & Rosenbaum, A. (2016). Psychological research in the internet age: The quality of web-based data. *Computers in Human Behavior*, 58, 354–360. DOI: 10.1016/j.chb.2015.12.049

Reason, J., & Hobbs, A. (2017). *Managing maintenance error: A practical guide*. CRC Press. DOI: 10.1201/9781315249926

Rossman, G. B., & Rallis, S. F. (2017). An introduction to qualitative research: Learning in the field. *Sage (Atlanta, Ga.)*. Advance online publication. DOI: 10.4135/9781071802694

Ruslin, R., Mashuri, S., Rasak, M. S. A., Alhabsyi, F., & Syam, H. (2022). Semi-structured interview: A methodological reflection on the development of a qualitative research instrument in educational studies. *IOSR Journal of Research & Method in Education*, 12(1), 22–29. DOI: 10.9790/7388-1201052229

Salas, E., Maurino, D., & Curtis, M. (2010). Human factors in aviation: An overview. In Salas, E., & Maurino, D. (Eds.), *Human factors in aviation* (2nd ed., pp. 3–19). Elsevier., DOI: 10.1016/B978-0-12-374518-7.00001-8

Saldaña, J. (2013). *The coding manual for qualitative researchers* (2nd ed.). Sage.

Salmons, J. (2017). Using social media in data collection: Designing studies with the qualitative e-research framework. In Sloan, L., & Quan-Haase, A. (Eds.), *The SAGE handbook of social media research methods* (pp. 177–196). Sage., DOI: 10.4135/9781473983847

Saunders, M. N. (2012). Choosing research participants. In Symon, G., & Cassell, C. (Eds.), *Qualitative organizational research: Core methods and current challenges* (pp. 35–52). Sage., DOI: 10.4135/9781526435620.n3

Scotland, J. (2012). Exploring the philosophical underpinnings of research: Relating ontology and epistemology to the methodology and methods of the scientific, interpretive, and critical research paradigms. *English Language Teaching*, 5(9), 9–16. DOI: 10.5539/elt.v5n9p9

Silverman, R. M., & Patterson, K. (2021). *Qualitative research methods for community development*. Routledge. DOI: 10.4324/9781003172925

Smith, B. (2018). Generalizability in qualitative research: Misunderstandings, opportunities and recommendations for the sport and exercise sciences. *Qualitative Research in Sport, Exercise and Health*, 10(1), 137–149. DOI: 10.1080/2159676X.2017.1393221

Spratling, R. (2013). Recruitment of medically fragile children and adolescents: Lessons learned from qualitative research. *Journal of Pediatric Health Care*, 27(1), 62–65. DOI: 10.1016/j.pedhc.2012.08.001 PMID: 23084469

Stolzer, A. J., Sumwalt, R. L., & Goglia, J. J. (2023). *Safety management systems in aviation*. CRC Press. DOI: 10.1201/9781003286127

Sullivan, P., & Girginer, H. (2002). The use of discourse analysis to enhance ESP teacher knowledge: An example of Aviation English. *English for Specific Purposes*, 21(4), 97–404. DOI: 10.1016/S0889-4906(01)00042-4

Tarnavska, T., Glushanytsia, N., Bilous, N., Lyashenko, L., & Palamarchuk, K. (2021). English for aviation safety: Language training methodology. *The Asian ESP Journal*, 17(1), 5.

Tiewtrakul, T., & Fletcher, S. R. (2010). The challenge of regional accents for aviation English language proficiency standards: A study of difficulties in understanding in air traffic control–pilot communications. *Ergonomics*, 53(2), 229–239. DOI: 10.1080/00140130903470033 PMID: 20099176

Tomlinson, B. (2013). *Developing materials for language teaching*. Bloomsbury.

Tracy, S. J. (2024). *Qualitative research methods: Collecting evidence, crafting analysis, communicating impact*. John Wiley & Sons.

Van Manen, M. (2016). *Researching lived experience: Human science for an action sensitive pedagogy*. Routledge.

Vaz, C. A., Junior. (2018). Remembering Tenerife: 40 years later. *SF Journal of Aviation and Aeronaut Science, 1*(2). https://scienceforecastoa.com/Articles/SJAAS-V1-E2-1012.pdf

Walby, K. (2015). Marshall, Catherine, and Gretchen B. Rossman, Designing Qualitative Research. *Canadian Journal of Sociology*, 40(3), 399–402. DOI: 10.29173/cjs25398

Walker, D. M. (2013). The internet as a medium for health service research. Part 1. *Nurse Researcher*, 20(4), 18–21. DOI: 10.7748/nr2013.03.20.4.18.e294 PMID: 23520708

Wegler, P. (Ed.). (2015). Training needs beyond the ICAO language proficiency requirements. In A. Borowska & A. Enright (Eds.), *Changing perspectives on aviation English training* (pp. 53–60). Uniwersytet Warszawski.

Woodrow, L. (2022). *Introducing researching english for specific purposes*. Routledge. DOI: 10.4324/9781003049678

Xerri, D. (2018a). The use of interviews and focus groups in teacher research. *The Clearing House: A Journal of Educational Strategies, Issues and Ideas*, 91(3), 140–146. DOI: 10.1080/00098655.2018.1436820

Xerri, D. (2018b). Two methodological challenges for teacher-researchers: Reflexivity and trustworthiness. *The Clearing House: A Journal of Educational Strategies, Issues and Ideas*, 91(1), 37–41. DOI: 10.1080/00098655.2017.1371549

Yin, R. K. (2017). *Case study research and applications: Design and methods* (6th ed.). Sage.

KEY TERMS AND DEFINITIONS

Aviation English: Aviation English focuses on clear communication for pilots and air traffic controllers, following ICAO standards.

Course Design and Delivery: Course design plans learning objectives and materials, while delivery involves teaching and engaging students with the content.

English for Specific Purposes (ESP): ESP tailors English teaching to specific fields like business or medicine, focusing on relevant language skills and terminology.

Narrative Inquiry: Narrative inquiry studies people's stories to understand their experiences and the meanings they give to them.

Qualitative Research: An approach to research that explores experiences and behaviours through non-numerical data like interviews and observations.

Semi-Structured Interviews: Semi-structured interviews use open-ended questions with flexibility, allowing for deeper exploration of topics.

Thematic Analysis: Thematic analysis identifies patterns or themes within qualitative data to explain the research findings.

Trustworthiness: Trustworthiness in research ensures the credibility and reliability of findings through transparent and rigorous methods.

Compilation of References

Aber, J., Marzolff, I., Ries, J., & Aber, S. (2019). *Small-format aerial photography and UAS imagery: Principles, techniques and geoscience applications*. Academic Press.

Abro, G., Zulkifli, S., Masood, R., Asirvadam, V., & Laouti, A. (2022). Comprehensive Review of UAV Detection, Security, and Communication Advancements to Prevent Threats. *Drones (Basel)*, 6(10), 284. DOI: 10.3390/drones6100284

Adams, W. C. (2015). Conducting semi-structured interviews. *Handbook of practical program evaluation*, 492-505. https://doi.org/DOI: 10.1002/9781119171386.ch19

Adeoye-Olatunde, O. A., & Olenik, N. L. (2021). Research and scholarly methods: Semi-structured interviews. *Journal of the American College of Clinical Pharmacy: JAACP*, 4(10), 1358–1367. DOI: 10.1002/jac5.1441

Adler, N., & Gellman, A. (2012). Strategies for managing risk in a changing aviation environment. *Journal of Air Transport Management*, 21, 24–35. DOI: 10.1016/j.jairtraman.2011.12.014 PMID: 32834688

Afonso, F., Sohst, M., Diogo, C. M., Rodrigues, S. S., Ferreira, A., Ribeiro, I., Marques, R., Rego, F., Sohouli, A., Portugal, J., Policarpo, H., Soares, B., Ferreira, B., Fernandes, E. C., Lau, F., & Suleman, A. (2023). Strategies towards a more sustainable aviation: A systematic review. *Progress in Aerospace Sciences*, 137, 100878. DOI: 10.1016/j.paerosci.2022.100878

Ağaçsapan, B., & Çabuk, A. (2019). Sustainable aviation: GIS for airport. *GSI Journals Serie B: Advancements in Business and Economics, 2*(1).

Ageeva, E., & Foroudi, P. (2019). Tourists' destination image through regional tourism: From supply and demand sides perspectives. *Journal of Business Research. Cilt*, 101, 334–348.

Ahmad, S., & Xu, B. (2021). A cognitive mapping approach to analyse stakeholders' perspectives on sustainable aviation fuels. *Transportation Research Part D, Transport and Environment*, 100, 103076. DOI: 10.1016/j.trd.2021.103076

Air Transport Action Group [ATAG]. (2020). *Aviation: Benefits beyond borders*. https://aviationbenefits.org/

Akman Dömbekci, H., & Erişen, M. A. (2022). Nitel araştırmalarda görüşme tekniği. *Anadolu Üniversitesi Sosyal Bilimler Dergisi*, 22(Özel Sayı 2), 141-160.

Alameeri, A., Ajmal, M. M., Hussain, M., & Helo, P. T. (2017). Sustainability practices in the aviation industry: A study of UAE-based airlines. *International Journal of Sustainable Society*, 9(2), 119–147. DOI: 10.1504/IJSSOC.2017.086818

Alba, S. O., & Manana, M. (2016). Energy research in airports: A review. *Energies*, 9(5), 349. DOI: 10.3390/en9050349

Albelo Ph, D, J. L., Cuevas, H. M., Aguiar, M., Piccone, C., Petitt, K., & Villagomez, R. (2022). Defining Aviation "Skills" to Ensure Effective, Safe, and Efficient Evaluations: A Qualitative Study. *International Journal of Aviation Research*, 14(1), 1.

Alharahsheh, H. H., & Pius, A. (2020). A review of key paradigms: Positivism vs interpretivism. *Global Academic Journal of Humanities and Social Sciences*, 2(3), 39–43. DOI: 10.4018/IJSEM.2020010105

Ali, B. (2017). *Automatic Dependent Surveillance Broadcast (ADS-B)*. Routledge., DOI: 10.4324/9781315566382-4

Alkadi, R., & Shoufan, A. (2021). Unmanned Aerial Vehicles Traffic Management Solution Using Crowd-Sensing and Blockchain. *IEEE Transactions on Network and Service Management*, 20(1), 201–215. DOI: 10.1109/TNSM.2022.3201817

Almarri, G. A., Kamalrudin, M. A. S. S. I. L. A., Hakimi, H. A. L. I. M. A. T. O. N., & Sidek, S. A. F. I. A. H. (2021). Factors Influencing Human Centric Model For Better Passenger Experiences Using Smart Security System In Dubai Airport. *Journal of Theoretical and Applied Information Technology*, 99(4), 976–986.

Almashhadani, M., & Almashhadani, H. A. (2023). The impact of sustainability reports to enhance company performance and its sustainability. *International Journal of Scientific and Management Research*, 149–163.

Almofeez, S. ve diğerleri. (2024). Freight forwarder satisfaction, airport customer loyalty, and competitiveness. *Transport Policy. Cilt*, 159, 314–327.

Alparslan, F. (2024). Sustainability in aviation and safety culture. In *Contemporary sustainable organisational practices: A roadmap for transformation* (pp. 149–170). Springer Nature Switzerland. DOI: 10.1007/978-3-031-62474-2_11

Alpman, E., & Göğüş, A. Y. (2017). Havacılıkta sürdürülebilir gelişme göstergeleri. *Sürdürülebilir Havacılık Araştırmaları Dergisi*, 2(1), 1–11.

Alpman, E., & Göğüş, A. Y. (2017). Sustainable development indicators in aviation. *Journal of Sustainable Aviation Research*, 2(1), 1–11.

Alruwaili, M. (2022). *Sustainable airports: Green energy solutions to achieve carbon neutrality* (Doctoral dissertation, Cardiff University).

Alshenqeeti, H. (2014). Interviewing as a data collection method: A critical review. *English linguistics research, 3*(1), 39-45.

Altensis. (2023). Kurumsal sürdürülebilirlik. https://www.altensis.com/hizmetler/kurumsal-surdurebilirlik/surdurulebilirlik-stratejisi/

Altheide, D., Coyle, M., DeVriese, K., & Schneider, C. (2008). Emergent qualitative document analysis.

Altheide, D. L. (2000). Tracking discourse and qualitative document analysis. *Poetics*, 27(4), 287–299. DOI: 10.1016/S0304-422X(00)00005-X

Alwidian, S. (2022). Towards integrating human-centric characteristics into the goal-oriented requirements language. In *2022 IEEE 30th International Requirements Engineering Conference Workshops (REW)* (pp. 200-204). IEEE. DOI: 10.1109/REW56159.2022.00045

Ansell, P. J. (2023). Review of sustainable energy carriers for aviation: Benefits, challenges, and future viability. *Progress in Aerospace Sciences*, 141, 100919. DOI: 10.1016/j.paerosci.2023.100919

Anthony, L. (2018). *Introducing English for specific purposes*. Routledge. DOI: 10.4324/9781351031189

Araghi, Y., Kroesen, M., Molin, E., & Van Wee, B. (2016). Revealing heterogeneity in air travelers' responses to passenger-oriented environmental policies: A discrete-choice latent class model. *International Journal of Sustainable Transportation*, 10(9), 765–772. DOI: 10.1080/15568318.2016.1149645

Aras, G., & Crowther, D. (2009). Making sustainable development sustainable. *Management Decision*, 47(6), 975–988. DOI: 10.1108/00251740910966686

Armstrong, C. (2021). Key methods used in qualitative document analysis. *OSF Preprints, 1*(9).

Austin, Z., & Sutton, J. (2014). Qualitative research: Getting started. *The Canadian Journal of Hospital Pharmacy*, 67(6), 436–440. DOI: 10.4212/cjhp.v67i6.1406 PMID: 25548401

Aydin, B. (2019). Public acceptance of drones: Knowledge, attitudes, and practice. *Technology in Society*, 59, 101180. DOI: 10.1016/j.techsoc.2019.101180

BAAL. (2021). Recommendations on good practice in applied linguistics. https://www.baal.org.uk/wp-content/uploads/2021/03/BAAL-Good-Practice-Guidelines-2021.pdf

Bachiochi, P. D., & Weiner, S. P. (2004). Qualitative data collection and analysis. *Handbook of research methods in industrial and organizational psychology*, 161-183.

Bachman, L., & Adrian, P. (2022). *Language assessment in practice: Developing language assessments and justifying their use in the real world.* Oxford University Press.

Bader, G. E., & Rossi, C. A. (2002). *Focus groups: a step-by-step guide* (3rd ed.). The Bader Group.

Baffour, F. D., Francis, A. P., Chong, M. D., & Harris, N. (2024). Doing qualitative research in a controlled site: Challenges, lessons, and strategies. *Crime and Delinquency*, 70(8), 1921–1942. DOI: 10.1177/00111287221106944

Bagnoli, A., & Clark, A. (2010). Focus groups with young people: A participatory approach to research planning. *Journal of Youth Studies*, 13(1), 101–119. DOI: 10.1080/13676260903173504

Bakır, M., Akan, Ş., Özdemir, E., Nguyen, P. H., Tsai, J. F., & Pham, H. A. (2022). How to achieve passenger satisfaction in the airport? Findings from regression analysis and necessary condition analysis approaches through online airport reviews. *Sustainability (Basel)*, 14(4), 2151. DOI: 10.3390/su14042151

Baltar, F., & Brunet, I. (2012). Social research 2.0: Virtual snowball sampling method using Facebook. *Internet Research*, 22(1), 57–74. DOI: 10.1108/10662241211199960

Bans-Akutey, A., & Tiimub, B. M. (2021). Triangulation in research. *Academia Letters*, 2, 1–6. DOI: 10.20935/AL3392

Barbour, R. (2007). *Doing Focus Groups* (1st ed.). SAGE Publications. DOI: 10.4135/9781849208956

Barclay, K., Voyer, M., Mazur, N., Payne, A. M., Mauli, S., Kinch, J., Fabinyi, M., & Smith, G. (2017). The importance of qualitative social research for effective fisheries management. *Fisheries Research*, 186, 426–438. DOI: 10.1016/j.fishres.2016.08.007

Barkhuizen, G. (2014). Narrative research in language teaching and learning. *Language Teaching*, 47(4), 450–466. DOI: 10.1017/S0261444814000172

Barkhuizen, G., & Wette, R. (2008). Narrative frames for investigating the experiences of language teachers. *System*, 36(3), 372–387. DOI: 10.1016/j.system.2008.02.002

Barriball, K., & While, A. (1994). Collecting data using a semi-structured interview: A discussion paper. *Journal of Advanced Nursing*, 19(2), 328–335. DOI: 10.1111/j.1365-2648.1994.tb01088.x PMID: 8188965

Barshi, I., & Farris, C. (2016). *Misunderstandings in ATC communication*. Routledge. DOI: 10.4324/9781315595641

Barutçu, S. B., & Çolakoğlu, Ü. (2024). Havayolu taşımacılığında sürdürülebilirlik analizi: Türkiye. *Journal of Travel and Tourism Research*, 24(24), 106–124.

Barutçu, S. B., & Çolakoğlu, Ü. (2024). Havayolu taşımacılığında sürdürülebilirlik analizi; Türkiye. *Journal of Travel and Tourism Research*, 24(24), 106–124.

Barutçu, S. B., & Çolakoğlu, Ü. (2024). Sustainability Analysis in the Aviation Sector; Turkiye. *Journal of Travel and Tourism Research*, 24(24), 106–124.

Basiago, A. D. (1999). Economic, social, and environmental sustainability in development theory and urban planning practice. *The Environmentalist*, 19(2), 145–161. DOI: 10.1023/A:1006697118620

Basturkmen, H. (2010). *Developing courses in English for specific purposes*. Palgrave Macmillan. DOI: 10.1057/9780230290518

Baumgartner, R. J., & Rauter, R. (2017). Strategic perspectives of corporate sustainability management to develop a sustainable organization. *Journal of Cleaner Production*, 140, 81–92. DOI: 10.1016/j.jclepro.2016.04.146

Baumgärtner, S., & Quaas, M. (2010). What is sustainability economics? *Ecological Economics*, 69(3), 445–450. DOI: 10.1016/j.ecolecon.2009.11.019

Baxter, P., & Jack, S. (2008). Qualitative case study methodology: Study design and implementation for novice researchers. *The Qualitative Report*, 13(4), 544–559. DOI: 10.46743/2160-3715/2008.1573

Becker, L., & Jaakkola, E. (2020). Customer experience: Fundamental premises and implications for research. *Journal of the Academy of Marketing Science*, 48(4), 630–648. DOI: 10.1007/s11747-019-00718-x

Beirão, G., & Sarsfield Cabral, J. A. (2007). Understanding attitudes towards public transport and private car: A qualitative study. *Transport Policy*, 14(6), 478–489. DOI: 10.1016/j.tranpol.2007.04.009

Belcher, D. D., Johns, A. M., & Paltridge, B. (Eds.). (2011). *New directions in English for specific purposes research*. University of Michigan Press. DOI: 10.3998/mpub.371075

Bellos, I., & Kavadias, S. (2021). Service design for a holistic customer experience: A process framework. *Management Science*, 67(3), 1718–1736. DOI: 10.1287/mnsc.2020.3609

Belton, O. & Dillon, S., (2021). Futures of autonomous flight: Using a collaborative storytelling game to assess anticipatory assumptions. *Futures,* Cilt 128.

Beneigh, T. (2021). Unmanned Aircraft System Operations. Barnhart, R. K., Marshall, D. M., & Shappee, E. (Eds.). (2021). *Introduction to unmanned aircraft systems*. Crc Press.

Bergen, N., & Labonté, R. (2020). 'Everything is perfect, and we have no problems': Detecting and limiting social desirability bias in qualitative research. *Qualitative Health Research*, 30(5), 783–792. DOI: 10.1177/1049732319889354 PMID: 31830860

Berger, R. (2015). Now I see it, now I don't: Researcher's position and reflexivity in qualitative research. *Qualitative Research*, 15(2), 219–234. DOI: 10.1177/1468794112468475

Bernard, H. R., & Ryan, G. W. (2014). Text analysis: Qualitative and quantitative methods. In Bernard, H. R., & Gravlee, C. C. (Eds.), *Handbook of methods in cultural anthropology* (2nd ed., pp. 533–559). Rowman & Littlefield.

Bertaux, D. (1981). *From the life-history approach to the transformation of sociological practice*. Biography and society: The life history approach in the social sciences. California: SAGE Publications.

Bexell, M., & Jönsson, K. (2017, January). Responsibility and the United Nations' sustainable development goals. [). Routledge.]. *Forum for Development Studies*, 44(1), 13–29. DOI: 10.1080/08039410.2016.1252424

Bezerra, G. C. L., & Gomes, C. F. (2016). Measuring airport service quality: A multidimensional approach. *Journal of Air Transport Management*, 53, 85–93. DOI: 10.1016/j.jairtraman.2016.02.001

Billson, J. M. (1989). Focus groups: A practical guide for applied research. *Chinese Sociological Review*, 7(1), 24.

Billups, F. D. (2019). *Qualitative data collection tools: Design, development, and applications* (Vol. 55). Sage Publications.

Birrell, S., Payre, W., Zdanowicz, K., & Herriotts, P. (2022). Urban air mobility infrastructure design: Using virtual reality to capture user experience within the world's first urban airport. *Applied Ergonomics*, 105, 103843. DOI: 10.1016/j.apergo.2022.103843 PMID: 35810501

Blalock, G., Kadiyali, V., & Simon, D. H. (2007). The impact of post-9/11 airport security measures on the demand for air travel. *The Journal of Law & Economics*, 50(4), 731–755. DOI: 10.1086/519816

Bloomer, M. J., Cross, W., Endacott, R., O'Connor, M., & Moss, C. (2012). Qualitative observation in a clinical setting: Challenges at end of life. *Nursing & Health Sciences*, 14(1), 25–31. DOI: 10.1111/j.1442-2018.2011.00653.x PMID: 22288573

Bloor, M., Frankland, J., Thomas, M., & Robson, K. (2001). *Focus Groups in Social Research*. SAGE Publications Ltd. DOI: 10.4135/9781849209175

Boeing. (2021). *Sustainability report*. https://www.boeing.com/principles/environment/

Boermans, D. D., Jagoda, A., Lemiski, D. & Wegener, J., (2024). Environmental awareness and sustainable behavior of respondents in Germany, the Netherlands and Poland: A qualitative focus group study. *Journal of Environmental Management*, Cilt 370.

Bogicevic, V., Bujisic, M., Bilgihan, A., Yang, W., & Cobanoglu, C. (2017). The impact of traveler-focused airport technology on traveler satisfaction. *Technological Forecasting and Social Change*, 123, 351–361. DOI: 10.1016/j.techfore.2017.03.038

Bolz, K., & Nowacki, G. (2023). Air transport safety in UAV operational conditions. *Journal of Civil Engineering and Transport*, 5(1), 9–23. Advance online publication. DOI: 10.24136/tren.2023.001

Bongo, M. F., Alimpangog, K. M. S., Loar, J. F., Montefalcon, J. A., & Ocampo, L. A. (2018). An application of DEMATEL-ANP and PROMETHEE II approach for air traffic controllers' workload stress problem: A case of Mactan Civil Aviation Authority of the Philippines. *Journal of Air Transport Management*, 68, 198–213. DOI: 10.1016/j.jairtraman.2017.10.001

Bonisteel, I., Shulman, R., Newhook, L. A., Guttmann, A., Smith, S., & Chafe, R. (2021). Reconceptualizing recruitment in qualitative research. *International Journal of Qualitative Methods*, 20, 16094069211042493. Advance online publication. DOI: 10.1177/16094069211042493

Bose, T., Adhvaryu, B., & Chakraborti, S. (2019). Airport noise and urban planning: A case of Ahmedabad, India. *Population*, 3, 5–577.

Bossell, H. (1999). *Indicators for Sustainable Development: Theory, Method, Applications: A Report to the Balaton Group, International Institute for Sustainable Development*. IISD.

Bowen, G. A. (2009). Document analysis as a qualitative research method. *Qualitative Research Journal*, 9(2), 27–40. DOI: 10.3316/QRJ0902027

Bozlağan, R. (2005). Historical background of the idea of sustainable development. In *Journal of Social Policy Conferences* (No. 50, pp. 1011-1028). Istanbul University.

Bradley, C., Oliveira, L., Birrell, S., & Cain, R. (2021). A new perspective on personas and customer journey maps: Proposing systemic UX. *International Journal of Human-Computer Studies*, 148, 102583. DOI: 10.1016/j.ijhcs.2021.102583

Braun, V., & Clarke, V. (2006). Using thematic analysis in psychology. *Qualitative Research in Psychology*, 3(2), 77–101. DOI: 10.1191/1478088706qp063oa

Braun, V., & Clarke, V. (2019). Reflecting on reflexive thematic analysis. *Qualitative Research in Sport, Exercise and Health*, 11(4), 589–597. DOI: 10.1080/2159676X.2019.1628806

Braun, V., & Clarke, V. (2021). Thematic analysis: A practical guide. *Sage (Atlanta, Ga.)*.

Breen, R. L. (2006). A practical guide to focus-group research. *Journal of Geography in Higher Education*, 30(3), 463–475. DOI: 10.1080/03098260600927575

Breen, R., Lindsay, R., Jenkins, A., & Smith, P. (2001). The role of information and communication technologies in a university learning environment. *Studies in Higher Education*, 26(1), 95–114. DOI: 10.1080/03075070123233

Brinkmann, S., & Kvale, S. (2018). Doing Interviews. 2nd ed. Los Angeles; London: s.n. DOI: 10.4135/9781529716665

Britten, N. (1995). Qualitative research: Qualitative interviews in medical research. *BMJ (Clinical Research Ed.)*, 311(6999), 251–253. DOI: 10.1136/bmj.311.6999.251 PMID: 7627048

Brooks, S. M. (2013). The ethical treatment of human subjects and the Institutional Review Board process. *Interview research in political science*, 45-66.

Browne, A. L. (2016). Can people talk together about their practices? Focus groups, humour and the sensitive dynamics of everyday life. *Area*, 48(2), 198–205. DOI: 10.1111/area.12250

Brundtland, G. H., & Khalid, M. (1987). Our common future. Oxford University Press, Oxford, GB.

Bryman, A. (2016). *Social research methods* (5th ed.). Oxford University Press.

Buhusayen, B., Seet, P. S., & Coetzer, A. (2021). Front-line management during radical organisational change: Social exchange and paradox interpretations. [h]. *Sustainability (Basel)*, 13(2), 893. DOI: 10.3390/su13020893

Bulak, M. E., Zengin, F. H., & Çiftçi, F. S. (2021). Global ölçekte faaliyet gösteren havalimanlarının göreceli sürdürülebilirlik analizi. *Pamukkale Üniversitesi Mühendislik Bilimleri Dergisi*, 27(4), 504–512.

Bullock, N. (2015). Defining meaningful material for the teaching of English for aeronautical communications. In Borowska, A., & Enright, A. (Eds.), *Changing perspectives on aviation English training* (pp. 35–45). Uniwersytet Warszawski.

Busetto, L., Wick, W., & Gumbinger, C. (2020). How to use and assess qualitative research methods. *Neurological Research and Practice*, 2(1), 14. DOI: 10.1186/s42466-020-00059-z PMID: 33324920

Byrne, M. (2001). Interviewing as a data collection method. *AORN Journal*, 74(2), 233–233. DOI: 10.1016/S0001-2092(06)61533-0 PMID: 11503203

Çalışkan, A. Ö. (2012). Sustainability Reporting. *Journal of Accounting and Taxation Studies*, 5(1), 41–68.

Campbell, S., Greenwood, M., Prior, S., Shearer, T., Walkem, K., Young, S., & Walker, K. (2020). Purposive sampling: Complex or simple? Research case examples. *Journal of Research in Nursing*, 25(8), 652–661. DOI: 10.1177/1744987120927206 PMID: 34394687

Can, İ. (2019). *Finansal sürdürülebilirlik göstergeleri ve değerlendirilmesi: Borsa İstanbul sürdürülebilirlik endeksi şirketleri üzerine bir araştırma (Yüksek Lisans Tezi)*. Marmara Üniversitesi.

Cankurtaran, P., & Beverland, M. B. (2020). Using design thinking to respond to crises: B2B lessons from the 2020 COVID-19 pandemic. *Industrial Marketing Management*, 88, 255–260. DOI: 10.1016/j.indmarman.2020.05.030

Carcary, M. (2009). The research audit trial: Enhancing trustworthiness in qualitative inquiry. *Electronic Journal of Business Research Methods*, 7(1), 11–24.

Cardno, C. (2018). Policy document analysis: A practical educational leadership tool and a qualitative research method. *Educational Administration: Theory and Practice*, 24(4), 623–640.

Carraminana, D., Campana, I., Bergesio, L., Bernardos, A., & Besada, J. (2021). Sensors and Communication Simulation for Unmanned Traffic Management. *Sensors (Basel)*, 21(3), 927. DOI: 10.3390/s21030927 PMID: 33573192

Carruthers, J. (1990). A Rationale for the Use of Semi-Structured Interviews. *Journal of Educational Administration*, 28(1), 63–68. DOI: 10.1108/09578239010006046

Carter, S., & Henderson, L. (2005). Approaches to qualitative data collection in social science. *Handbook of health research methods: Investigation, measurement and analysis, 1*, 215-230.

Carter, N., Bryant-Lukosius, D., DiCenso, A., Blythe, J., & Neville, A. (2014). The use of triangulation in qualitative research. *Oncology Nursing Forum*, 41(5), 545–547. DOI: 10.1188/14.ONF.545-547 PMID: 25158659

Cartwright, J. C., Hickman, S. E., Nelson, C. A., & Knafl, K. A. (2013). Investigators' successful strategies for working with institutional review boards. *Research in Nursing & Health*, 36(5), 478–486. DOI: 10.1002/nur.21553 PMID: 23813748

CASEM. (2023). Sürdürülebilirlik nedir? https://www.casem.com.tr/surdurulebilirlik-nedir

Catterall, M., & Maclaran, P. (2006). Focus Groups in Marketing Research. In Belk, R. W. (Ed.), *Handbook of qualitative research methods in marketing* (pp. 255–267). Edward Elgar. DOI: 10.4337/9781847204127.00027

Chamata, J., & Winterton, J. (2018). A conceptual framework for the acceptance of drones. *The International Technology Management Review*, 7(1), 34–46. DOI: 10.2991/itmr.7.1.4

Chan, W. T. K., & Li, W. C. (2022). Investigating professional values among pilots, cabin crew, ground staff, and managers to develop aviation safety management systems. *International Journal of Industrial Ergonomics*, 92, 103370. DOI: 10.1016/j.ergon.2022.103370

Charmaz, K. (2006). *Constructing Grounded Theory: A Practical Guide through Qualitative Analysis* (1st ed.). SAGE Publications.

Charmaz, K. (2006). *Constructing Grounded Theory: A Practical Guide Through Qualitative Analysis*. SAGE.

Charmaz, K. (2014). *Constructing grounded theory* (2nd ed.). SAGE Publications.

Chen, C.-F., & Chen, S.-C. (2014). Measuring the effects of Safety Management System practices, morality leadership and self-efficacy on pilots' safety behaviors: Safety motivation as a mediator. *Safety Science*, 62, 376–385. DOI: 10.1016/j.ssci.2013.09.013

Chen, L., Li, Y.-Q., & Liu, C.-H. (2019). How airline service quality determines the quantity of repurchase intention - Mediate and moderate effects of brand quality and perceived value. *Journal of Air Transport Management*, 75, 185–197. DOI: 10.1016/j.jairtraman.2018.11.002

Chin, C., Gopalakrishnan, K., Balakrishnan, H., Egorov, M., & Evans, A. (2022). Protocol-Based Congestion Management for Advanced Air Mobility. *Journal of Air Transportation (Reston, Va.)*, 31(1), 35–44. DOI: 10.2514/1.D0298

Chiper, F. L., Martian, A., Vladeanu, C., Marghescu, I., Craciunescu, R., & Fratu, O. (2022). Drone detection and defense systems: Survey and a software-defined radio-based solution. *Sensors (Basel)*, 22(4), 1453. DOI: 10.3390/s22041453 PMID: 35214355

Chiu, R. L. (2003). Social sustainability, sustainable development and housing development: The experience of Hong Kong. In *Housing and social change: East-west perspectives* (pp. 221–243). Routledge.

Chmiliar, l. (2010). Multiple-case designs. In A. J. Mills, G. Eurepas & E. Wiebe (Eds.), Encyclopedia of case study research (pp 582-583). USA: SAGE Publications.

Chonsalasin, D., Jomnonkwao, S., & Ratanavaraha, V. (2021). Measurement model of passengers' expectations of airport service quality. *International Journal of Transportation Science and Technology*, 10(4), 342–352. DOI: 10.1016/j.ijtst.2020.11.001

Chou, S., Chen, C.-W., & Wong, M. (2023). When social media meets low-cost airlines: Will customer engagement increase customer loyalty? *Research in Transportation Business & Management*, 47, 100945. Advance online publication. DOI: 10.1016/j.rtbm.2023.100945

Christou, E., & Kassianidis, P. (2010). Adoption of Mobile Commerce in the Air Travel Sector: A Qualitative Survey of Attitudes. *Turizam*, 14(1), 41–52. DOI: 10.5937/Turizam1001041C

Chu, R. (2001). What online Hong Kong travelers look for on airline/travel websites? *International Journal of Hospitality Management*, 20(1), 95–100. DOI: 10.1016/S0278-4319(00)00046-3

Churchill, G. A.Jr. (1979). A Paradigm for Developing Better Measures of Marketing Constructs. *JMR, Journal of Marketing Research*, 16(1), 64–73. DOI: 10.1177/002224377901600110

Chutiphongdech, T., & Vongsaroj, R. (2022). Technical efficiency and productivity change analysis: A case study of the regional and local airports in Thailand. *Case Studies on Transport Policy*, 10(2), 870–890. DOI: 10.1016/j.cstp.2022.03.001

Ciesielska, M., Boström, K. W., & Öhlander, M. (2018). Observation methods. *Qualitative methodologies in organization studies: Volume II: Methods and possibilities*, 33-52.

Çınar, E., & Tuncal, A. (2023). The Future of UAVs in Urban Air Mobility: Public Perception and Concerns. *Türkiye İnsansız Hava Araçları Dergisi*, 5(2), 50–58. DOI: 10.51534/tiha.1381175

Çınar, E., & Tuncal, A. (2024). A comprehensive analysis of competency and training perspectives among air traffic controllers. *Aviation*, 28(2), 54–63. DOI: 10.3846/aviation.2024.21501

Clandinin, D. J., & Connelly, F. M. (2000). *Narrative inquiry: Experience and story in qualitative research*. Jossey-Bass.

Clark, M., Taylor, J., & Moore, R. (2020). Pilot fatigue and safety: A qualitative study of fatigue management systems in aviation. *Journal of Aviation Safety*, 45(2), 123–134. DOI: 10.1016/j.javiationsafety.2020.03.004

Coffey, A., & Atkinson, P. (1996). *Making sense of qualitative data: Complementary research strategies*. Sage.

Cohen, L., Manion, L., & Morrison, K. (2017). *Research methods in education* (8th ed.). Routledge. DOI: 10.4324/9781315456539

Çokluk, Ö., Yılmaz, K., & Oğuz, E. (2011). Nitel bir görüşme yöntemi: Odak grup görüşmesi. *Kuramsal Eğitimbilim*, 4(1), 95–107.

Colglazier, W. (2015). Sustainable development agenda: 2030. *Science*, 349(6252), 1048–1050. DOI: 10.1126/science.aad2333 PMID: 26339011

Collis, J., & Hussey, R. (2014). *Business Research: A practical guide for undersgraduate and postgraduate students* (4th ed.). Palgrave. DOI: 10.1007/978-1-137-03748-0

Conde, J. ve diğerleri, (2022). Applying digital twins for the management of information in turnaround event operations in commercial airports. *Advanced Engineering Informatics,* Cilt 54.

Cooper, M., & Phillips, R. (2004). Exploratory analysis of the safety climate and safety behavior relationship. *Journal of Safety Research*, 35(5), 497–512. DOI: 10.1016/j.jsr.2004.08.004 PMID: 15530924

Corbin, J., & Strauss, A. (2007). *Basics of Qualitative Research: Techniques and Procedures for Developing Grounded Theory* (3th Editio). *Sage (Atlanta, Ga.).*

Corlett, B. M. (2022). *Improvement of airline check-in and boarding processes* (Doctoral dissertation, Stellenbosch University). Stellenbosch University. https://scholar.sun.ac.za

Courage, C., & Baxter, K. (2005). *Understanding your users: a practical guide to user requirements: methods, tools, and techniques*. Morgan Kaufmann Publishers.

Cox, A., Prager, F., & Rose, A. (2011). Transportation security and the role of resilience: A foundation for operational metrics. *Transport Policy. Cilt*, 18, 307–317.

Cox, K. K., Higginbotham, J. B., & Burton, J. (1976). Applications of focus group interviews in marketing. *Journal of Marketing*, 40(1), 77–80. DOI: 10.1177/002224297604000117

Coyne, E., Grafton, E., & Reid, A. (2016). Strategies to successfully recruit and engage clinical nurses as participants in qualitative clinical research. *Contemporary Nurse*, 52(6), 669–676. DOI: 10.1080/10376178.2016.1181979 PMID: 27108654

Coyne, R. ve diğerleri. (2024). Understanding drivers' perspectives on the use of driver monitoring systems during automated driving: Findings from a qualitative focus group study. *Transportation Research Part F: Traffic Psychology and Behaviour. Cilt*, 105, 321–335.

Cracknell, A. (2017). UAVs: Regulations and law enforcement. *International Journal of Remote Sensing*, 38(8-10), 3054–3067. DOI: 10.1080/01431161.2017.1302115

Creswell, J. W. (2018). *Qualitative Inquiry and Research Design: Choosing Among Five Approaches* (M. Bütün & S. B. Demir (eds.); 4.). Siyasal Publising.

Creswell, J. W., & Miller, D. L. (2000) Determining validity in qualitative inquiry. *The ory into Practice,* 39(3), 124–130.

Creswell, J. W. (2002). *Educational research: Planning, conducting, and evaluating quantitative.* Prentice Hall Upper Saddle River.

Creswell, J. W. (2007). *Qualitative Inquiry & Research Design: Choosing Among Five Approaches* (2nd editio). *Sage (Atlanta, Ga.).*

Creswell, J. W. (2013). *Research Design: Qualitative, Quantitative and Mixed Method Approaches* (4th Editio). *Sage (Atlanta, Ga.).*

Creswell, J. W. (2018). *Nitel araştırma yöntemleri: Beş yaklaşıma göre nitel araştırma ve araştırma deseni (M. Bütün & S. Demir, Çev* (Kitabevi, S., Ed.).

Creswell, J. W., & Creswell, J. D. (2022). *Research design* (6th ed.). Sage.

Creswell, J. W., Fetters, M. D., & Ivankova, N. V. (2004). Designing a mixed methods study in primary care. *Annals of Family Medicine*, 2(1), 7–12. DOI: 10.1370/afm.104 PMID: 15053277

Creswell, J. W., & Poth, C. (2017). Qualitative inquiry and research design: Choosing among five traditions. *Sage (Atlanta, Ga.).*

Creswell, J. W., & Poth, C. N. (2018). *Qualitative inquiry & research design: Choosing among five approaches* (4th ed.). Sage.

Creswell, J. W., & Poth, C. N. (2018). *Qualitative inquiry and research design: Choosing among five approaches* (4th ed.). Sage Publications.

Crowe, S., Cresswell, K., Robertson, A., Huby, G., Avery, A., & Sheikh, A. (2011). The case study approach. *BMC Medical Research Methodology*, 11(1), 100–100. DOI: 10.1186/1471-2288-11-100 PMID: 21707982

Curtis, S., Gesler, W., Smith, G., & Washburn, S. (2000). Approaches to sampling and case selection in qualitative research: Examples in the geography of health. *Social Science & Medicine*, 50(7-8), 1001–1014. DOI: 10.1016/S0277-9536(99)00350-0 PMID: 10714922

Cusick, S. K., Cortes, A. I., & Rodrigues, C. C. (2017). Commercial Aviation Safety. 6th ed. s.l.:McGraw-Hill Education.

Cutting, J. (2012). English for airport ground staff. *English for Specific Purposes*, 31(1), 3–13. DOI: 10.1016/j.esp.2011.06.002

Cypress, B. (2018). Qualitative research methods. *Dimensions of Critical Care Nursing*, 37(6), 302–309. DOI: 10.1097/DCC.0000000000000322 PMID: 30273215

D'Amato, E., Notaro, I., & Mattei, M. (2018, June). Distributed collision avoidance for unmanned aerial vehicles integration in the civil airspace. In *2018 International Conference on Unmanned Aircraft Systems (ICUAS)* (pp. 94-102). IEEE. https://doi .org/DOI: 10.1109/ICUAS.2018.8453432

Da Rocha, P. M., Costa, H. G., & da Silva, G. B. (2022). Gaps, trends and challenges in assessing quality of service at airport terminals: A systematic review and bibliometric analysis. *Sustainability (Basel)*, 14(7), 3796. DOI: 10.3390/su14073796

Dalbudak, E., & Rençber, Ö. F. (2022). Çok kriterli karar verme yöntemleri üzerine literatür incelemesi. *Gaziantep Üniversitesi İktisadi ve İdari Bilimler Fakültesi Dergisi*, 4(1), 1–17. DOI: 10.55769/gauniibf.1068692

Dalgıç Turhan, G., Özen, T., & Albayrak, R. S. (2018). Kurumsal sürdürülebilirlik kavramı, stratejik önemi ve sürdürülebilirlik performansı ölçümü. *Ege Stratejik Araştırmalar Dergisi*, 9(1), 17–37. DOI: 10.18354/esam.304155

Dalkıran, A. (2023). Airport Common-Use Check-in Operations: A Novel and Efficient Model. *Journal of the Institute of Science and Technology*, 13(2), 1026–1042.

Dam, R. F., & Siang, T. Y. (2021). *What is design thinking and why is it so popular?* Interaction Design Foundation.

Davis, A., Lee, J., & King, T. (2021). Ground handling perspectives on operational efficiency: A qualitative study. *Aviation Operations Journal*, 38(1), 56–71. DOI: 10.1109/aoj.2021.08.002

de Castro Carvalho, I., Calijuri, M. L., Assemany, P. P., Silva, M. D. F. M., Neto, R. F. M., da Fonseca Santiago, A., & de Souza, M. H. B. (2013). Sustainable airport environments: A review of water conservation practices in airports. *Resources, Conservation and Recycling*, 74, 27–36. DOI: 10.1016/j.resconrec.2013.02.016

De Grosbois, D. (2016). Corporate social responsibility reporting in the cruise tourism industry: A performance evaluation using a new institutional theory based model. *Journal of Sustainable Tourism*, 24(2), 245–269. DOI: 10.1080/09669582.2015.1076827

De Villiers, C., Farooq, M. B., & Molinari, M. (2022). Qualitative research interviews using online video technology–challenges and opportunities. *Meditari Accountancy Research*, 30(6), 1764–1782. DOI: 10.1108/MEDAR-03-2021-1252

Deakin, H., & Wakefield, K. (2014). Skype interviewing: Reflections of two PhD researchers. *Qualitative Research*, 14(5), 603–616. DOI: 10.1177/1468794113488126

Decosta, P. (2016). *Ethics in applied linguistics research: Language researcher narratives*. Routledge.

DeJonckheere, M., & Vaughn, L. (2019). Semistructured interviewing in primary care research: A balance of relationship and rigour. *Family Medicine and Community Health*, 7(2), e000057. Advance online publication. DOI: 10.1136/fmch-2018-000057 PMID: 32148704

Dekker, S. (2017). *The field guide to understanding human error* (3rd ed.). CRC Press.

Dekker, S. (2017). *The field guide to understanding 'human error* (3rd ed.). CRC Press.

Dekker, S. (2017). *The safety anarchist: Relying on human expertise and innovation, reducing bureaucracy and compliance*. Routledge. DOI: 10.4324/9780203733455

Deliens, T., Clarys, P., De Bourdeaudhuij, I., & Deforche, B. (2014). Determinants of eating behaviour in university students: A qualitative study using focus group discussions. *BMC Public Health*, 14(1), 1–12. DOI: 10.1186/1471-2458-14-53 PMID: 24438555

Deliens, T., Deforche, B., De Bourdeaudhuij, I., & Clarys, P. (2015). Determinants of physical activity and sedentary behaviour in university students: A qualitative study using focus group discussions. *BMC Public Health*, 15(1), 1–9. DOI: 10.1186/s12889-015-1553-4 PMID: 25881120

DeMarrais, K. (2004). Qualitative interview studies: Learning through experience. In DeMarrais, K., & Lapan, S. D. (Eds.), *Foundations for research* (pp. 51–68). Erlbaum.

DeMarrais, K., Roulston, K., & Copple, J. (2024). *Qualitative research design and methods*. Stylus Publishing.

Demirci, S., Doğan, T. G., & Eroğlu, E. Havalimanlarında sürdürülebilirlik: Singapur Changi Havalimanı. *Düzce Üniversitesi Süs ve Tıbbi Bitkiler Botanik Bahçesi Dergisi, 2*(1), 14-26.

Demirkaya, G., & Kuşhan, M. (2023). Green airports. *Makina, 31*(178).

Demir, Ö. ve Sezgin, E. E. (. (2014). Institutionalization and sustainability in family businesses: A research conducted in TRB1 region. *Electronic Turkish Studies*, 9(5), 707–725. DOI: 10.7827/TurkishStudies.6904

Denny, E., & Weckesser, A. (2022). How to do qualitative research? *BJOG*, 129(7), 1166–1167. DOI: 10.1111/1471-0528.17150 PMID: 35430773

Denscombe, M. (2014). *The good research guide*. Open University Press.

Denzin, N. K. (2017). *The research act: A theoretical introduction to sociological methods* (4th ed.). Routledge. DOI: 10.4324/9781315134543

Denzin, N. K., & Lincoln, Y. S. (2017). *The Sage handbook of qualitative research* (5th ed.). Sage Publications.

Denzin, N. K., & Lincoln, Y. S. (2017). *The SAGE handbook of qualitative research*. Sage.

Dey, I. (1993). *Qualitative Data Analysis: A User-friendly Guide for Social Scientists* (1st Editio). Routledge.

Dichabeng, P., Merat, N., & Markkula, G. (2021). Factors that influence the acceptance of future shared automated vehicles – A focus group study with United Kingdom drivers. *Transportation Research Part F: Traffic Psychology and Behaviour*, 82, 121–140. DOI: 10.1016/j.trf.2021.08.009

DiCicco-Bloom, B., & Crabtree, B. F. (2006). The qualitative research interview. *Medical Education*, 40(4), 314–321. DOI: 10.1111/j.1365-2929.2006.02418.x PMID: 16573666

Diedrich, A., Upham, P., Levidow, L., & van den Hove, S. (2011). Framing environmental sustainability challenges for research and innovation in European policy agendas. *Environmental Science & Policy*, 14(8), 935–939. DOI: 10.1016/j.envsci.2011.07.012

Diiulio, J., Militello, L., & Klein, D. (2020). UAS integration in congested terminal airspace: Challenges posed to pilots. *Journal of Unmanned Vehicle Systems*, 8(2), 79–88. DOI: 10.1139/juvs-2019-0015

Doganis, R. (2010). *Flying Off Course* (4th ed.). Rouledge.

Donehue, P., & Baker, D. (2012). Remote, rural, and regional airports in Australia. *Transport Policy*, 24, 232–239. DOI: 10.1016/j.tranpol.2012.08.007

Doody, O., & Noonan, M. (2013). Preparing and conducting interviews to collect data. *Nurse Researcher*, 20(5), 28–32. DOI: 10.7748/nr2013.05.20.5.28.e327 PMID: 23687846

dos Santos, A. J., Mancini, S. D., Roveda, J. A. F., Ewbank, H., & Roveda, S. R. M. M. (2020). A fuzzy assessment method to airport waste management: A case study of Congonhas Airport, Brazil. *Journal of Air Transport Management*, 87, 101838. DOI: 10.1016/j.jairtraman.2020.101838

Douglas, D. (2000). *Assessing language for specific purposes*. Cambridge University Press.

Downey, R., Suzuki, M., & Van Moere, A. (2010). High-stakes English-language assessments for aviation professionals: Supporting the use of a fully automated test of spoken-language proficiency. *IEEE Transactions on Professional Communication*, 53(1), 18–32. DOI: 10.1109/TPC.2009.2038734

Dudley-Evans, T., & John, M. (1998). *English for specific purposes*. Cambridge University Press.

Dunn, J., & Holt, N. (2004). A Qualitative Investigation of a Personal-Disclosure Mutual-Sharing Team Building Activity. *The Sport Psychologist*, 18(4), 363–380. DOI: 10.1123/tsp.18.4.363

Easton, K. L., McComish, J. F., & Greenberg, R. (2000). Avoiding common pitfalls in qualitative data collection and transcription. *Qualitative Health Research*, 10(5), 703–707. DOI: 10.1177/104973200129118651 PMID: 11066874

Eibfeldt, H., Vogelpohl, V., Stolz, M., Papenfub, A., Biella, M., Belz, J., & Kügler, D. (2020). The acceptance of civil drones in Germany. *CEAS Aeronautical Journal*, 11(3), 665–676. DOI: 10.1007/s13272-020-00447-w

Eid, A., Salah, M., Barakat, M., & Obrecht, M. (2022). Airport sustainability awareness: A theoretical framework. *Sustainability (Basel)*, 14(19), 11921. DOI: 10.3390/su141911921

Elbardan, H. (2023). Interview/Interpretive Approach in Air Travel Research: Research Process, Sampling and Trustworthiness Criteria. In A. Hassan, N. A. A. Rahman, & N. M. Nur (Eds.), *Air Travel Industry* (1st Editio, pp. 87–98). Routledge.

Elhmoud, E. R., & Kutty, A. A. (2020). Sustainability assessment in aviation industry: A mini-review on the tools, models and methods of assessment.

Elkington, J. (2006). Governance for sustainability. *Corporate Governance*, 14(6), 522–529. DOI: 10.1111/j.1467-8683.2006.00527.x

Elkington, J., & Rowlands, I. H. (1999). Cannibals with forks: The triple bottom line of 21st century business. *Alternatives Journal*, 25(4), 42–44.

Enexion. (2023). Stratejik sürdürülebilirlik. https://www.enexion.com.tr/stratejik-surdurulebilirlik-8/

Engel, E., Fischer, R., & Galetovic, A. (2018). The joy of flying: Efficient airport PPP contracts. *Transportation Research Part B: Methodological*, 114, 131–146. DOI: 10.1016/j.trb.2018.05.001

Eriksson, P., & Kovalainen, A. (2008). *Qualitative methodsin business research.* SAGE Publications Ltd. DOI: 10.4135/9780857028044

Ertan, Y. (2018). Sustainability Reporting in Turkey (2005-2017). *Journal of Accounting and Taxation Studies*, 11(3), 463–478.

Escarus. (2024). Sürdürülebilirlik stratejisi geliştirme. https://escarus.com/surdurulebilirlik-stratejisi-gelistirme/

ESG. (2012). Sürdürülebilirlik stratejisi. https://www.esgturkey.com/surdurulebilirlik-stratejisi/

Estival, D., Farris, C., & Molesworth, B. (2016). *Aviation English: A lingua franca for pilots and air traffic controllers.* Routledge. DOI: 10.4324/9781315661179

Fairclough, N. (2003). *Analysing discourse: Textual analysis for social research.* Routledge. DOI: 10.4324/9780203697078

Farhadi, F., Ghoniem, A., & Al-Salem, M. (2014). Runway capacity management - An empirical study with application to Doha International Airport. *Transportation Research Part E, Logistics and Transportation Review*, 68, 53–63. DOI: 10.1016/j.tre.2014.05.004

Farooq, M. B., & De Villiers, C. (2017). Telephonic qualitative research interviews: When to consider them and how to do them. *Meditari Accountancy Research*, 25(2), 291–316. DOI: 10.1108/MEDAR-10-2016-0083

Fasano, G., Crespillo, O., Sabatini, R., Roy, A., & Ogan, R. (2023). From the Editors of the Special Issue on Urban Air Mobility and UAS Airspace Integration: Vision, Challenges, and Enabling Avionics Technologies. *IEEE Aerospace and Electronic Systems Magazine*, 38(5), 4–5. Advance online publication. DOI: 10.1109/MAES.2023.3266992

Federal Aviation Administration [FAA] (2024). *Drones by the Numbers (as of 2/29/24).* Retrieved May 18, 2024, from https://www.faa.gov/uas

Fernández, X. L., Coto-Millán, P., & Díaz-Medina, B. (2018). The impact of tourism on airport efficiency: The Spanish case. *Utilities Policy. Cilt*, 55, 52–58.

Fern, E. F. (2001). *Advanced Focus Group Research.* Sage Publications, Inc. DOI: 10.4135/9781412990028

Ferroff, C. V., Mavin, T. J., Bates, P. R., & Murray, P. S. (2012). A case for social constructionism in aviation safety and human performance research. *Aeronautica*, 2(1), 1–12.

Finlay, L. (2012). Five lenses for the reflexive interviewer. In Gubrium, J. F., Holstein, J. A., Marvasti, A. B., & McKinne, K. D. (Eds.), *The SAGE handbook of interview research: The complexity of the craft* (2nd ed., pp. 317–332). Sage., DOI: 10.4135/9781452218403

Flick, U. (2018). Doing qualitative data collection–charting the routes. *The SAGE handbook of qualitative data collection*, 1-16.

Flick, U. (2014). An introduction to qualitative research. *Sage (Atlanta, Ga.).*

Flick, U. (2018). Doing triangulation and mixed methods. *Sage (Atlanta, Ga.).*

Flick, U. (2022). *The Sage handbook of qualitative research design*. Sage. DOI: 10.4135/9781529770278

Florido-Benítez, L. (2022). The World airport awards as a quality distinctive and marketing tool for airports. *Journal of Airline Operations and Aviation Management*, 1(2), 54–81. DOI: 10.56801/jaoam.v1i2.4

Fodness, D., & Murray, B. (2007). Passengers' expectations of airport service quality. *Journal of Services Marketing*, 21(7), 492–506. DOI: 10.1108/08876040710824852

Fogarty, G. J., & Shaw, A. (2010). Safety climate and the Theory of Planned Behavior: Towards the prediction of unsafe behavior. *Accident; Analysis and Prevention*, 42(5), 1455–1459. DOI: 10.1016/j.aap.2009.08.008 PMID: 20538101

Foot, D. K., & Ross, S. (2017). Social sustainability. In Sloan, G. (Ed.), *Teaching business sustainability* (pp. 107–125). Springer. DOI: 10.4324/9781351281805-7

Foroudi, P., Hafeez, K., & Foroudi, M. (2017). Evaluating the impact of corporate logos towards corporate reputation: A case of Persia and Mexico. *Qualitative Market Research*, 20(2), 158–180. DOI: 10.1108/QMR-05-2015-0043

Fossey, E., Harvey, C., McDermott, F., & Davidson, L. (2002). Understanding and Evaluating Qualitative Research. *The Australian and New Zealand Journal of Psychiatry*, 36(6), 717–732. DOI: 10.1046/j.1440-1614.2002.01100.x PMID: 12406114

Foster, M. K. (2021). Design thinking: A creative approach to problem solving. *Management Teaching Review*, 6(2), 123–140. DOI: 10.1177/2379298119871468

Frangapoulos, C., (2006), Exergy, energy system analysis, and optimization, in Encyklopedia of Life Supporting System.

Freathy, P., & O'Connell, F. (1999). Planning for profit: The commercialization of European airports. *Long Range Planning*, 32(6), 587–597. DOI: 10.1016/S0024-6301(99)00075-8

Fu, X. ve diğerleri, (2020). On-time performance policy in the Chinese aviation market - An innovation or disruption?. *Transport Policy,* Cilt 95.

Fuerst, F., Gross, S., & Klose, U. (2011). The sky is the limit? The determinants and constraints of European airports commercial revenues. *Journal of Air Transport Management*, 17(5), 278–283. DOI: 10.1016/j.jairtraman.2011.03.001

Galdas, P. (2017). Revisiting bias in qualitative research: Reflections on its relationship with funding and impact. *International Journal of Qualitative Methods*, 16(1), 1609406917748992. Advance online publication. DOI: 10.1177/1609406917748992

Ganti, S., & Kim, Y. (2016). Implementation of detection and tracking mechanism for small UAS. *2016 International Conference on Unmanned Aircraft Systems (ICUAS),* 1254-1260. https://doi.org/DOI: 10.1109/ICUAS.2016.7502513

Gao, L., Melero-Polo, I., Ruz-Mendoza, M. Á., & Trifu, A. (2022). Exploring the role of service touchpoints on the path to financial, behavioral and relational customer outcomes: Insights from a B2B service context. *Journal of Business and Industrial Marketing*, 37(11), 2260–2277. DOI: 10.1108/JBIM-02-2021-0100

García, J. L., & Sanz, J. M. (2018). Climate Change, Ethics and Sustainability: An innovative Approach. *Journal of Innovation & Knowledge*, 70-75.

Geertz, C. (1973). *The interpretation of cultures: Selected essays.* Basic Books.

Gemmer, E. (2019). Aviation sustainability. Honors Projects. 425. https://scholarworks.bgsu.edu/honorsprojects/425

Gerede, E. (2015). A qualitative study on the exploration of challenges to the implementation of the Safety Management System in aircraft maintenance organizations in Turkey. *Journal of Air Transport Management*, 47, 230–240. DOI: 10.1016/j.jairtraman.2015.06.006

Geske, A. M., Herold, D. M., & Kummer, S. (2024). Artificial Intelligence as a driver of efficiency in air passenger transport: A systematic literature review and future research avenues. *Journal of the Air Transport Research Society*, 100030.

Gibbs, L., Kealy, M., Willis, K., Green, J., Welch, N., & Daly, J. (2007). What have sampling and data collection got to do with good qualitative research? *Australian and New Zealand Journal of Public Health*, 31(6), 540–544. DOI: 10.1111/j.1753-6405.2007.00140.x PMID: 18081574

Gibbs, L., Slevitch, L., & Washburn, I. (2017). Competency-based training in aviation: The impact on flight attendant performance and passenger satisfaction. *Journal of Aviation/Aerospace Education Research*, 26(2), 55–80. DOI: 10.15394/jaaer.2017.1716

Gilgun, J. (1994). A Case for Case Studies in Social Work Research. *Social Work*, 39, 371–380. DOI: 10.1093/sw/39.4.371

Gillen, D. & Morrison, W. G., (2015). Aviation security: Costing, pricing, finance and performance. *Journal of Air Transport Management,* Cilt 48.

Gillen, D., Morrison, W. G., & Stewart, C. (2020). *Airline economics in the twenty-first century*. Routledge.

Gillen, M., Sinclair, R. R., & Cleary, P. D. (2020). Leadership in high-stress industries: Addressing employee morale and performance. *Safety Science*, 123, 104561. DOI: 10.1016/j.ssci.2020.104561

Gillett, K., O'Neill, B., & Bloomfield, J. G. (2016). Factors influencing the development of end-of-life communication skills: A focus group study of nursing and medical students. *Nurse Education Today*, 36, 395–400. DOI: 10.1016/j.nedt.2015.10.015 PMID: 26530442

Gill, P., Stewart, K., Treasure, E., & Chadwick, B. (2008). Methods of data collection in qualitative research: Interviews and focus groups. *British Dental Journal*, 204(6), 291–295. DOI: 10.1038/bdj.2008.192 PMID: 18356873

Giovannoni, E., & Fabietti, G. (2013). What is sustainability? A review of the concept and its applications. In *Integrated reporting: Concepts and cases that redefine corporate accountability* (pp. 21–40). Springer. DOI: 10.1007/978-3-319-02168-3_2

Girija, S., Banerji, B., Batra, N., Paruchuru, M., & Yeediballi, T. (2024). Making frugal innovations inclusive: A gendered approach. *Journal of Cleaner Production*, 434, 140182. DOI: 10.1016/j.jclepro.2023.140182

Gladwin, T. N., Kennelly, J. J., & Krause, T. (1995). Shifting paradigms for sustainable development: Implications for management theory and research. *Academy of Management Review*, 20(4), 874–907. DOI: 10.2307/258959

Glaser, B. G., & Strauss, A. L. (1967). *The discovery of grounded theory: Strategies for qualitative research*. Aldine.

Glesne, C. (2015). *Nitel Araştırmaya Giriş* (A. Ersoy & P. Yalçınoğlu (eds.); 5. Basım). Anı Yayıncılık.

Glesne, C. (2015). *Introduction to Qualitative Research* (Ersoy, A., & Yalçınoğlu, P., Eds.; 5th ed.). Anı Publishing.

Golombek, P. R., & Johnson, K. E. (2004). Narrative inquiry as a mediational space: Examining emotional and cognitive dissonance in second-language teachers' development. *Teachers and Teaching*, 10(3), 307–327. DOI: 10.1080/1354060042000204388

Goodland, R. (1995). The concept of environmental sustainability. [REMOVED HYPERLINK FIELD]. *Annual Review of Ecology and Systematics*, 26(1), 1–24. DOI: 10.1146/annurev.es.26.110195.000245

Gössling, S., Cohen, S. A., & Hares, A. (2016). Inside the black box: EU policy officers' perspectives on transport and climate change mitigation. *Journal of Transport Geography*. *Cilt*, 57, 83–93.

Grady, M. P. (1998). Qualitative and action research: A practitioner handbook. Phi Delta Kappa International. *Archives of Sexual Behavior*, 41, 1319–1320.

Graham, A. (2013). Understanding the low cost carrier and airport relationship: A critical analysis of the salient issues. *Tourism Management*, 36, 66–76. DOI: 10.1016/j.tourman.2012.11.011

Graham, A. (2023). *Managing Airports: An International Perspective* (6th ed.). Routledge.

Graham, A., & Morrell, P. (2016). *Airport Finance and Investment in the Global Economy*. Routledge. DOI: 10.4324/9781315566498

Green, J., & Thorogood, N. (2018). *Qualitative Methods for Health Research*. SAGE.

Guan, X., Lyu, R., Shi, H., & Chen, J. (2020). A survey of safety separation management and collision avoidance approaches of civil UAS operating in integration national airspace system. *Chinese Journal of Aeronautics*, 33(11), 2851–2863. DOI: 10.1016/j.cja.2020.05.009

Guba, E. G., & Lincoln, Y. S. (1982). Epistemological and methodological bases of naturalistic inquiry. *Educational Communication and Technology*, 30(4), 233–252. DOI: 10.1007/BF02765185

Guest, G., Macqueen, K. M., & Namey, E. E. (2012). Applied thematic analysis. *Sage (Atlanta, Ga.)*.

Guest, G., Namey, E. E., & Mitchell, M. L. (2013). Collecting qualitative data: A field manual for applied research. *Sage (Atlanta, Ga.)*. Advance online publication. DOI: 10.4135/9781506374680

Gui, G., Zhou, Z., Wang, J., Liu, F., & Sun, J. (2020). Machine Learning Aided Air Traffic Flow Analysis Based on Aviation Big Data. *IEEE Transactions on Vehicular Technology*, 69(5), 4817–4826. DOI: 10.1109/TVT.2020.2981959

Guimarans, D., Arias, P., Tomasella, M., & Wu, C. L. (2019). A review of sustainability in aviation: A multidimensional perspective. *Sustainable transportation and smart logistics*, 91-121.

Guo, D., Ramos, R. L. & Wang, F., (2024). Qualitative online interviews: Voices of applied linguistics researchers. *Research Methods in Applied Linguistics,* 3(3).

Gutierrez Santos, M., Almeida Pacheco, V., Rodriguez Reyes, R., & Veintimilla Vargas, G. (2018). Analysis of the methodology applied in aviation English course at a military school in Salinas. *Journal of Education and Human Development*, 7(3), 39–44. DOI: 10.15640/jehd.v7n3a5

Hahlweg, P., Härter, M., Nestoriuc, Y., & Scholl, I. (2017). How are decisions made in cancer care? A qualitative study using participant observation of current practice. *BMJ Open*, 7(9), e016360. DOI: 10.1136/bmjopen-2017-016360 PMID: 28963286

Halpern, N., & Graham, A. (2013). *Airport Marketing*. Routledge. DOI: 10.4324/9780203117903

Halpern, N., & Mwesiumo, D. (2021). Airport service quality and passenger satisfaction: The impact of service failure on the likelihood of promoting an airport online. *Research in Transportation Business & Management*, 41, 100667. DOI: 10.1016/j.rtbm.2021.100667

Hameed, I., Chatterjee, R. S., Zainab, B., Tzhe, A. X., Yee, L. S., & Khan, K. (2024). Navigating loyalty and trust in the skies: The mediating role of customer satisfaction and image for sustainable airlines. *Sustainable Futures : An Applied Journal of Technology, Environment and Society*, 8, 100299. DOI: 10.1016/j.sftr.2024.100299

Hammersley, M., & Traianou, A. (2012). Ethics in qualitative research: Controversies and contexts. *Sage (Atlanta, Ga.).* Advance online publication. DOI: 10.4135/9781473957619

Hanaysha, J. R. (2024). An examination of the linkages among website characteristics, brand love, and word of mouth in the airline industry. *Transport Economics and Management*, 2, 242–248. DOI: 10.1016/j.team.2024.09.001

Han, H. (2013). Effects of in-flight ambience and space/function on air travelers' decision to select a low-cost airline. *Tourism Management*, 37, 125–135. DOI: 10.1016/j.tourman.2013.01.008

Han, H., Yu, J., & Kim, W. (2019). Passenger experience and loyalty in the airline industry: A critical review. *Journal of Travel Research*, 58(4), 564–579. DOI: 10.1177/0047287518797156

Han, H., Yu, J., & Kim, W. (2019). Passengers' experiences and satisfaction in air travel: A qualitative analysis. *Journal of Air Transport Management*, 74, 84–93.

Hanson, D., Toru Delibasi, T., Gatti, M., & Cohen, S. (2022). How do changes in economic activity affect air passenger traffic? The use of state-dependent income elasticities to improve aviation forecasts. *Journal of Air Transport Management*, 98, 1–14. DOI: 10.1016/j.jairtraman.2021.102147

Hartley, J. (2004). Case study research. In Cassell, C., & Symon, G. (Eds.), *Essential guide to qualitative methods in organizational research* (pp. 323–333). Sage. DOI: 10.4135/9781446280119.n26

Hartmann, J. H. (2014). *Feasibility study of Air Traffic Control Towers around the globe: International research regarding the local influences providing an optimal structural design for air traffic control towers around the globe in an economical perspective*, Master thesis, TU Delft, Delft University of Technology.

Hatfield, M., Cahill, C., Webley, P., Garron, J., & Beltran, R. (2020). Integration of Unmanned Aircraft Systems into the National Airspace System-Efforts by the University of Alaska to Support the FAA/NASA UAS Traffic Management Program. *Remote Sensing (Basel)*, 12(19), 3112. DOI: 10.3390/rs12193112

Hauptvogel, D. ve diğerleri, (2024). Fairness perspectives of airport residents: A qualitative approach. *Transportation Research Interdisciplinary Perspectives,* Cilt 27.

Heggde, G., Khanka, S., & Damodaran, A. (2022). DigiYatra–a digital experience for air travelers. *Emerald Emerging Markets Case Studies*, 12(2), 1–20. DOI: 10.1108/EEMCS-10-2021-0349

Heiets, I. et al., (2024). Social media activities of airlines: What makes them successful?. *Journal of the Air Transport Research Society,* Volume 2.

Heiets, I., La, J., Zhou, W., Xu, S., Wang, X., & Xu, Y. (2022). Digital transformation of the airline industry. *Research in Transportation Economics*, 92, 101186. DOI: 10.1016/j.retrec.2022.101186

Hennink, M. (2007). *International Focus Group Research: A Handbook for the Health and Social Sciences*. Cambridge University Press. DOI: 10.1017/CBO9780511619458

Hennink, M., Hutter, I., & Bailey, A. (2020). Qualitative research methods. *Sage (Atlanta, Ga.)*.

Henriques, T. A., & O'Neill, H. (2023). Design science research with focus groups – a pragmatic meta-model. *International Journal of Managing Projects in Business*, 16(1), 119–140. DOI: 10.1108/IJMPB-01-2020-0015

Heyes, G., Hooper, P., Raje, F., & Sheppard, J. (2021). The case for a design-led, end-user focused airport noise management process. *Transportation Research Part D, Transport and Environment*, 95, 102847. DOI: 10.1016/j.trd.2021.102847

Heyes, G., Urquhart, C., Hooper, P., & Thomas, C. (2023). Comprehensive Strategic Analysis for Sustainability: An Aviation Industry Case Study. *Sustainability (Basel)*, 15(11), 8806. DOI: 10.3390/su15118806

Hilburn, B. G., & Pesmen, B. S. (2023). *Foreign Object Debris Detection System Cost-Benefit Analysis (No. DOT/FAA/TC-22/47).*

Hilditch, C. J., Gregory, K. B., Arsintescu, L., Bathurst, N. G., Nesthus, T. E., Baumgartner, H. M., Lamp, A. C. M., Barger, L. K., & Flynn-Evans, E. E. (2023). Perspectives on fatigue in short-haul flight operations from US pilots: A focus group study. *Transport Policy*, 136, 11–20. DOI: 10.1016/j.tranpol.2023.03.004

Hinkes, C., & Christoph-Schulz, I. (2019). Consumer attitudes toward palm oil: Insights from focus group discussions. *Journal of Food Products Marketing*, 25(9), 875–895. DOI: 10.1080/10454446.2019.1693468

Ho, F., Geraldes, R., Gonçalves, A., Rigault, B., Sportich, B., Kubo, D., Cavazza, M., & Prendinger, H. (2019). Decentralized Multi-Agent Path Finding for UAV Traffic Management. *IEEE Transactions on Intelligent Transportation Systems*, 23(2), 997–1008. DOI: 10.1109/TITS.2020.3019397

Hofisi, C., Hofisi, M., & Mago, S. (2014). Critiquing interviewing as a data collection method. *Mediterranean Journal of Social Sciences*, 5(16), 60–64. DOI: 10.5901/mjss.2014.v5n16p60

Hofstede, G. (2001). *Culture's consequences: Comparing values, behaviors, institutions, and organizations across nations* (2nd ed.). Sage.

Holz, H. F., Becker, M., Blut, M., & Paluch, S. (2024). Eliminating customer experience pain points in complex customer journeys through smart service solutions. *Psychology and Marketing*, 41(3), 592–609. DOI: 10.1002/mar.21938

Hotle, S., & Mumbower, S. (2021). The impact of COVID-19 on domestic U.S. air travel operations and commercial airport service. *Transportation Research Interdisciplinary Perspectives*, 9, 100277. DOI: 10.1016/j.trip.2020.100277

Hou, M., Wang, K. & Yang, H., (2021). Hub airport slot Re-allocation and subsidy policy to speed up air traffic recovery amid COVID-19 pandemic - case on the Chinese airline market. *Journal of Air Transport Management,* Cilt 93.

Howe, K. R., & Moses, M. S. (1999). Chapter 2: Ethics in educational research. *Review of Research in Education*, 24(1), 21–59. DOI: 10.3102/0091732X024001021

Hsu, C. H. C., Oh, H., & Assaf, A. G. (2011). A Customer-Based Brand Equity Model for Upscale Hotels. *Journal of Travel Research*, 51(1), 81–93. DOI: 10.1177/0047287510394195

http-1. Türk Hava Yolları (THY). Sürdürülebilirlik. https://investor.turkishairlines.com/tr/kurumsal-yonetim/surdurulebilirlik

http-2. Türk Hava Yolları (THY), CO2mission. https://www.turkishairlines.com/tr-int/co2mission/

http-3. Türk Hava Yolları (THY), Sürdürülebilir havacılık yakıtı (SAF) nedir? https://www.turkishairlines.com/tr-int/surdurulebilir-hava-yakiti/

http-4. Pegasus Yatırımcı İlişkileri. Sürdürülebilirlik politikası. https://www.pegasusyatirimciiliskileri.com/medium/image/kurumsal-surdurulebilirlik-politikasi_1023/view.aspx

Hughes, A. (2003). *Testing for language teachers*. Cambridge University Press.

Hunt, V., & Zellweger, A. (1987). The FAA's Advanced Automation System: Strategies for Future Air Traffic Control Systems. *Computer*, 20(2), 19–32. DOI: 10.1109/MC.1987.1663475

Hutchinson, T., & Waters, A. (1987). *English for specific purposes: A learning-centred approach*. Cambridge University Press. DOI: 10.1017/CBO9780511733031

Hyland, K. (2006). *English for academic purposes: An advanced resource book*. Routledge. DOI: 10.4324/9780203006603

ICAO. (2023a). *About ICAO*. https://www.icao.int/about-icao/Pages/default.aspx

ICAO. (2023b). *Future of aviation*. https://www.icao.int/Meetings/FutureOfAviation/Pages/default.aspx

ICAO. (2024). Future of aviation. https://www.icao.int/Meetings/FutureOfAviation/Pages/default.aspx

Idowu, S. O., & Louche, C. (Eds.). (2011). *Theory and practice of corporate social responsibility* (p. 284). Springer.

İGA. (2024). *Sürdürülebilirlik.* İGA Havalimanı. Erişim adresi: https://www.igairport.aero/surdurulebilirlik/

İlerisoy, M. (2023). Yorumlayıcı fenomenolojik analiz. *İstanbul Ticaret Üniversitesi Sosyal Bilimler Dergisi*, 22(46), 509-527.

International Air Transport Association. (2021). The impact of COVID-19 on aviation. Retrieved from IATA website

International Civil Aviation Organization (ICAO). (2019). *Environmental protection.* https://www.icao.int/environmental-protection/Pages/default.aspx

International Civil Aviation Organization (ICAO). (2024). *Aircraft noise: Balanced approach to aircraft noise management.* https://www.icao.int/environmental-protection/Pages/noise.aspx

International Civil Aviation Organization [ICAO]. (2022). *Annex 1, Personnel Licensing* (Thirteenth Edition). Montréal, Quebec, Canada.

International Civil Aviation Organization [ICAO]. (2024). *Increased use of unmanned aircraft systems (UAS).* Retrieved May 18, 2024, from https://www.icao.int/annual-report-2021/Pages/emerging-and-cross-cutting-aviation-issues-increased-use-of-unmanned-aircraft-systems-uas.aspx

International Civil Aviation Organization. (2011). Aviation & Sustainability. Determining the complex environmental, economic and social impacts that are defining aviation's future. *ICAO Journal*, 66(6), 1–36.

Irtak, K. (2023). THY'ye "En sürdürülebilir bayrak taşıyıcı hava yolu" ödülü. Anadolu Ajansı. https://www.aa.com.tr/tr/gundem/thyye-en-surdurulebilir-bayrak-tasiyici-hava-yolu-odulu/2936491

Işık, E., & Semerci, Ç. (2019). Nitel araştırmalarda veri üçgenlemesi olarak odak grup görüşmesi, bireysel görüşme ve gözlem. *Turkish Journal of Educational Studies*, 6(3), 53–66. DOI: 10.33907/turkjes.607997

Israel, G. D., & Galindo-Gonzalez, S. (2008). Using Focus Group Interviews for Planning or Evaluating Extension Programs: AEC 387/PD036, rev. 5/2008. *EDIS, 2008*(5).

Jaffary, Z. & Wiedemann, M., (2024). Towards net zero aviation: Exploring safe hydrogen refuelling at airports. *Journal of the Air Transport Research Society.*

Jain, N. (2021). Survey versus interviews: Comparing data collection tools for exploratory research. *The Qualitative Report*, 26(2), 541–554. DOI: 10.46743/2160-3715/2021.4492

Jansen, B. J., Jung, S. G., Nielsen, L., Guan, K. W., & Salminen, J. (2022). How to create personas: Three persona creation methodologies with implications for practical employment. *Pacific Asia Journal of the Association for Information Systems*, 14(3), 1–28. DOI: 10.17705/1pais.14301

Jenkins, E. K., Slemon, A., Haines-Saah, R. J., & Oliffe, J. (2018). A guide to multisite qualitative analysis. *Qualitative Health Research*, 28(12), 1969–1977. DOI: 10.1177/1049732318786703 PMID: 30024317

Jeremić, V., & Sachs, J. D. (2014). The United Nations in the age of sustainable development. *The Economic and social review, 45*(2, Summer), 161-188.

Jia, X., Macário, R., & Buyle, S. (2023). Expanding horizons: A review of sustainability evaluation methodologies in the airport sector and beyond. *Sustainability (Basel)*, 15(15), 11584. DOI: 10.3390/su151511584

Jittrapirom, P., Boonsiripant, S. & Phamornmongkhonchai, M., (2021). Aligning stakeholders' mental models on carsharing system using remote focus group method. *Transportation Research Part D: Transport and Environment,* Cilt 101.

Johnson, R. (1997). Examining the Validity Structure of Qualitative Research. *Education 3-13*, 118, 282.

Johnson, L., & Miller, S. (2019). Exploring customer experiences in airport services: Insights from focus groups. *Journal of Airport Management*, 27(4), 112–124. DOI: 10.1016/j.jairports.2019.07.002

Jordan, B. (2019). Ethnographic research in high-stakes environments: Lessons from the aviation industry. *Journal of Organizational Ethnography*, 8(3), 345–360. DOI: 10.1108/JOE-03-2019-0014

Jordan, R. R. (1997). *English for academic purposes: A guide and resource book for teachers*. Cambridge University Press. DOI: 10.1017/CBO9780511733062

Jung, H. (2017). Focus group interaction in evaluation research. *Applied Linguistics Review*, 9(4), 563–587. DOI: 10.1515/applirev-2017-0023

Kabashkin, I., Perekrestov, V., Tyncherov, T., Shoshin, L., & Susanin, V. (2024). Framework for Integration of Health Monitoring Systems in Life Cycle Management for Aviation Sustainability and Cost Efficiency. *Sustainability (Basel)*, 16(14), 6154. DOI: 10.3390/su16146154

Kachniewska, M. (2021). Smart tourism: towards the concept of a data-based travel experience. *Handbook of Sustainable Development and Leisure Services*, 289-302.

Kale, U., Herrera, M., & Nagy, A. (2021). Examining pragmatic failure and other language-related risks in global aviation. *Aircraft Engineering and Aerospace Technology*, 93(8), 1313–1322. DOI: 10.1108/AEAT-03-2021-0081

Kallio, H., Pietilä, A. M., Johnson, M., & Kangasniemi, M. (2016). Systematic methodological review: Developing a framework for a qualitative semi-structured interview guide. *Journal of Advanced Nursing*, 72(12), 2954–2965. DOI: 10.1111/jan.13031 PMID: 27221824

Kanki, B. G., Anca, J., & Chideste, T. R. (2019). Crew Resource Management. 3rd ed. s.l.:Elsevier Academic Press. DOI: 10.1016/B978-0-12-812995-1.00004-X

Kanmaz, U. (2024). Qualitative Impact Analysis of Sustainable Aviation Fuels on Aviation Sustainability. [JSHSR]. *International Journal of Social and Humanities Sciences Research*, 11(107), 920–934.

Kantenbacher, J., Hanna, P., Cohen, S., Miller, G., & Scarles, C. (2018). Public attitudes about climate policy options for aviation. *Environmental Science & Policy*, 81, 46–53. DOI: 10.1016/j.envsci.2017.12.012

Kaplanski, G., & Levy, H. (2010). Sentiment and stock prices: The case of aviation disasters. *Journal of Financial Economics*, 95(2), 174–201. DOI: 10.1016/j.jfineco.2009.10.002

Kaplowitz, M. D., & Hoehn, J. P. (2001). Do focus groups and individual interviews reveal the same information for natural resource valuation? *Ecological Economics*, 36(2), 237–247. DOI: 10.1016/S0921-8009(00)00226-3

Karagiannis, I., Vouros, P., Skouloudis, A., & Evangelinos, K. (2019). Sustainability reporting, materiality, and accountability assessment in the airport industry. *Business Strategy and the Environment*, 28(7), 1370–1405. DOI: 10.1002/bse.2321

Karakuş, G. (2023). A study on Turkish Airlines sustainability reports. Aerospace Research Letters (ASREL). *Journal*, 2(2), 86–113. DOI: 10.56753/ASREL.2023.2.4

Karam, H., Anwama, E., Davidson, I. E. A., Alfazari, H., Krykhtine, F., & Mora-Camino, F. (2022, September). A Methodology for Evaluating Aviation Sustainability Perspectives. In *International Seminar on Aeronautics and Energy* (pp. 277-289). Singapore: Springer Nature Singapore.

Karaman, A. S., Kilic, M., & Uyar, A. (2018). Sustainability reporting in the aviation industry: Worldwide evidence. *Sustainability Accounting. Management and Policy Journal*, 9(4), 362–391.

Kavacık, M., Zafer, S., Yıldız, A., & Karaman, D. (2012). Sustainable development in aviation industry and the case of Turkish airlines. *Journal of Animal and Veterinary Advances*, 9(3), 547–550.

Kayhan, S., Ergün, N., & Gerede, E. (2018). Research determining issues on the administrative success of security services at civil airports in Turkey. In *Security Journal* (Vol. 31, Issue 2). DOI: 10.1057/s41284-017-0111-4

Kegler, J. J., Neves, E. T., Lacerda, M. R., & Oliveira, D. C. (2023). Focus group in the development of concepts for a Nursing model: Experience report. *Revista Brasileira de Enfermagem*, 76(4), e20220689. DOI: 10.1590/0034-7167-2022-0689 PMID: 37820149

Keller, R. M. (2016). Ontologies for aviation data management. *AIAA/IEEE Digital Avionics Systems Conference - Proceedings, 2016-Decem*, 1–9. DOI: 10.1109/DASC.2016.7777971

Kesore, L. V. (2020). *Sustainability reporting in the airline industry: a comparative case study analysis of four selected European passenger airlines and their countries of registration on the basis of the airlines' annual reports and sustainability report from 2018*. University of Twente.

Kezar, A. (2000). The importance of pilot studies: Beginning the hermeneutic circle. *Research in Higher Education*, 41(3), 385–400. DOI: 10.1023/A:1007047028758

Khan, N. & Efthymiou, M., (2021). The use of biometric technology at airports: The case of customs and border protection (CBP). *International Journal of Information Management Data Insights,* Volume 1.

Khan, N., Saini, K., Khan, M. A., & Khajuria, G. (2022). Marketing Practices for Fund Raising and Sustainability of NGOs: Focus Group Discussions. *World Journal of Management and Economics, 16*(8).

Khan, T. H., & MacEachen, E. (2022). An alternative method of interviewing: Critical reflections on videoconference interviews for qualitative data collection. *International Journal of Qualitative Methods*, 21, 16094069221090063. DOI: 10.1177/16094069221090063

Kidd, C. V. (1992). The evolution of sustainability. *Journal of Agricultural & Environmental Ethics*, 5(1), 1–26. DOI: 10.1007/BF01965413

Kiger, M. E., & Varpio, L. (2020). Thematic analysis of qualitative data: AMEE Guide No. 131. *Medical Teacher*, 42(8), 846–854. DOI: 10.1080/0142159X.2020.1755030 PMID: 32356468

Kılıç, M., Uyar, A., & Karaman, A. S. (2019). What impacts sustainability reporting in the global aviation industry? An institutional perspective. *Transport Policy*, 79, 54–65. DOI: 10.1016/j.tranpol.2019.04.017

Kim, H., & Elder, C. (2009). Understanding aviation English as a lingua franca. *Australian Review of Applied Linguistics, 32*(3), 23.1–23.17. DOI: 10.1075/aral.32.3.03kim

Kimathi, T. W. (2024). Leadership, corporate social responsibility, and sustainability in airlines: Case study for Kenya Airways.

Kim, H., & Elder, C. (2015). Interrogating the construct of aviation English: Feedback from test takers in Korea. *Language Testing*, 32(2), 129–149. DOI: 10.1177/0265532214544394

King, N., & Horrocks, C. (2010). *Interviews in Qualitative Research.* SAGE Publications.

Kistamah, H., & Matsuo, M. (2024). Journal of the Air Transport Research Society Above the clouds : Unveiling insights through semi-structured interviews with airline employees from the African, Asian, and North American regions surrounding the COVID-19 pandemic era. *Journal of the Air Transport Research Society*, 3(September), 100047. DOI: 10.1016/j.jatrs.2024.100047

Kneale, P. (2002). Developing and embedding reflective portfolios in geography. *Journal of Geography in Higher Education*, 26(1), 81–94. DOI: 10.1080/03098260120110386

Kochan, A., Kruk, A., & Rutkowska, P. (2018). Prospects for unmanned aerial vehicles market development in Poland. *WUT Journal of Transportation Engineering*, 123, 41–53. DOI: 10.5604/01.3001.0013.7352

Koç, S., & Durmaz, V. (2015). Airport corporate sustainability: An analysis of indicators reported in the sustainability practices. *Procedia: Social and Behavioral Sciences*, 181, 158–170. DOI: 10.1016/j.sbspro.2015.04.877

Koech, A. K., Buyle, S., & Macário, R. (2023). Airline brand awareness and perceived quality effect on the attitudes towards frequent-flyer programs and airline brand choice - Moderating effect of frequent-flyer programs. *Journal of Air Transport Management*, 107, 102342. DOI: 10.1016/j.jairtraman.2022.102342

Kokarieva, A., Khomenko-Semenova, L., Glushanytsia, N., Ievtushenko, I., & Odarchenko, R. (2020). Information and communication technologies in the professional training of engineers. In *CMiGIN* (pp. 496–506). https://ceur-ws.org/Vol-2588/paper41.pdf

Koonce, J. M. (2002). *Human Factors in the Training of Pilots* (1st ed.). CRC Press.

Korul, V. (2001). Management of Airports' Relationships with the Environment and Analysis of Environmental Pollution Practices at Airports Open to International Traffic in Turkey (Master's thesis, Anadolu University), Turkey.

Kowal, S., & O'Connell, D. C. (2014). Transcription as a crucial step of data analysis. In Flick, U. (Ed.), *The SAGE handbook of qualitative data analysis* (pp. 64–79). Sage., DOI: 10.4135/9781446282243.n5

Krippendorff, K. (2009). Content analysis: An introduction to its methodology. *Sage (Atlanta, Ga.).*

Krippendorff, K. (2018). *Content analysis: An introduction to its methodology* (4th ed.). Sage Publications.

Krishnan, R. S., Kirubha, D., Christopher, V. B., Raj, J. R. F., Gopikumar, S., & Rani, P. A. (2024). Predicting Passenger Preferences: An AI-Driven Framework for Personalized Airport Lobby Experiences. In *2024 2nd International Conference on Sustainable Computing and Smart Systems (ICSCSS)* (pp. 1616-1622). IEEE.

Krosnick, J. A., & Presser, S. (2017). Question and questionnaire design. In Marsden, P., & Wright, J. (Eds.), *Handbook of survey research* (2nd ed., pp. 263–314). Emerald Group Publishing.

Krueger, R. A. (1998). *Analyzing & Reporting Focus Group Results.* basım yeri bilinmiyor:SAGE Publications, Inc.

Krueger, R. A. (1998). *Developing Questions for Focus Groups.* SAGE. DOI: 10.4135/9781483328126

Krueger, R. A. (2014). *Focus groups: A practical guide for applied research.* Sage publications.

Krueger, R. A., & Casey, M. A. (2015). *Focus Groups: A Practical Guide for Applied Research* (5th ed.). SAGE Publications, Inc.

Krueger, R. A., & Casey, M. A. (2015). *Focus groups: a practical guide for applied research* (5th ed.). SAGE.

Krueger, R. A., Casey, M. A., Donner, J., Kirsch, S., & Maack, J. N. (2001). Social analysis: Selected tools and techniques. *World Development*, 36, 4–23.

Kuckartz, U., & Rädiker, S. (2019). Analyzing Qualitative Data with MAXQDA. 1 dü. basım yeri bilinmiyor:Springer. DOI: 10.1007/978-3-030-15671-8

Kuckartz, U. (2019). Qualitative text analysis: A systematic approach. In Kaiser, G., & Presmeg, N. (Eds.), *Compendium for early career researchers in mathematics education* (pp. 181–197). Springer., DOI: 10.1007/978-3-030-15636-7_8

Kuhlman, T., & Farrington, J. (2010). What is sustainability? [REMOVED HYPERLINK FIELD]. *Sustainability (Basel)*, 2(11), 3436–3448. DOI: 10.3390/su2113436

Kuo, P., Lord, D., Wheat, P., Yin, Y., & Jia, T. (2022). The impact of the COVID-19 pandemic on O-D flow and airport networks in the origin country and in Northeast Asia. *Journal of Air Transport Management*, 100, 102178. DOI: 10.1016/j.jairtraman.2022.102192 PMID: 35194345

Kurt, Y. (2016). An Overview of Qualitative Research Methods in Aviation Management Research in Turkey. *Proceedings of ADVED 2016 2nd International Conference on Advances in Education and Social Sciences.*

Kurt, Y., & Gerede, E. (2018). An Assessment of Aviation Safety Management System Applications From the New Institutional Theory Perspective. *International Journal of Management Economics and Business*, 14(1), 97–122. DOI: 10.17130/ijmeb.2018137576

KuveytTürk. (2023). Sürdürülebilir finans ve çevresel etkileri. https://www.kuveytturk.com.tr/blog/surdurulebilirlik/surdurulebilir-finans-ve-cevresel-etkileri

Kvale, S. (1996). *Interviews: an introduction to qualitative research interviewing.* SAGE.

Kvale, S. (2007). Qualitative research kit: Doing interviews. *Sage (Atlanta, Ga.).*

Kvale, S. (2009). Interviews: Learning the craft of qualitative research interviewing. *Sage (Atlanta, Ga.).*

Kvale, S., & Brinkmann, S. (2018). Doing Interviews. *Sage (Atlanta, Ga.).*

Labib, N., Danoy, G., Musial, J., Brust, M., & Bouvry, P. (2019). Internet of Unmanned Aerial Vehicles—A Multilayer Low-Altitude Airspace Model for Distributed UAV Traffic Management. *Sensors (Basel)*, 19(21), 4779. DOI: 10.3390/s19214779 PMID: 31684133

Lamnatou, C., & Chemisana, D. (2015). A critical analysis of factors affecting photovoltaic-green roof performance. *Renewable & Sustainable Energy Reviews*, 43, 264–280. DOI: 10.1016/j.rser.2014.11.048

Lange, A. (2019). Does cargo matter? The impact of air cargo operations on departure on-time performance for combination carriers. *Transportation Research Part A: Policy and Practice. Cilt*, 119, 214–223.

Langford, J., & McDonagh, D. (2003). *Focus groups: supporting effective product development* (1st ed.). Taylor & Francis. DOI: 10.4324/9780203302743_chapter_12

Leal Filho, W., Ng, A. W., Sharifi, A., Janová, J., Özuyar, P. G., Hemani, C., Heyes, G., Njau, D., & Rampasso, I. (2023). Global tourism, climate change and energy sustainability: Assessing carbon reduction mitigating measures from the aviation industry. *Sustainability Science*, 18(2), 983–996. DOI: 10.1007/s11625-022-01207-x PMID: 36105893

LeCompte, M. D., & Schensul, J. J. (2010). *Designing and conducting ethnographic research: An introduction* (2nd ed.). AltaMira Press.

Lee-Anant, C., & Monpanthong, P. (2021). Factors Influencing Airport Technology Selections in Each Service Touchpoint of Suvarnabhumi Airport, Thailand. *Turkish Journal of Computer and Mathematics Education*, 12(13), 3804–3816.

Lee, D. S., Fahey, D. W., Skowron, A., Allen, M. R., Burkhardt, U., Chen, Q., Doherty, S. J., Freeman, S., Forster, P. M., Fuglestvedt, J., Gettelman, A., De León, R. R., Lim, L. L., Lund, M. T., Millar, R. J., Owen, B., Penner, J. E., Pitari, G., Prather, M. J., & Wilcox, L. J. (2021). The contribution of global aviation to anthropogenic climate forcing for 2000 to 2018. *Atmospheric Environment*, 244, 117834. DOI: 10.1016/j.atmosenv.2020.117834 PMID: 32895604

Lee, J., Yu, J., Radic, A., & Han, H. (2024). Uncovering air traveler purchase behavior: Influence of airline goods product characteristics towards repurchase intention. *Journal of Retailing and Consumer Services*, 79, 103858. DOI: 10.1016/j.jretconser.2024.103858

Levontin, P., Kulmala, S., Haapasaari, P., & Kuikka, S. (2011). Integration of biological, economic, and sociological knowledge by Bayesian belief networks: The interdisciplinary evaluation of potential management plans for Baltic salmon. *ICES Journal of Marine Science*, 68(3), 632–638. DOI: 10.1093/icesjms/fsr004

Liamputtong, P. (2011). *Focus Group Methodology: Principles and Practice* (1st ed.). SAGE Publications Ltd. DOI: 10.4135/9781473957657

Liang, V. W. (2015). *Engaging with sustainability in everyday transitions: An ethnographic study of an international airport operator*. University of Manchester.

Lincoln, Y. S., & Guba, E. G. (1985). *Naturalistic Inquiry*. Sage Publishing. https://books.google.com.tr/books?hl=tr&lr=&id=2oA9aWlNeooC&oi=fnd&pg=PA7&dq=Lincoln+ve+Guba+(1985)+&ots=0uquW8Pcxk&sig=wdl6zKzPxh_wwFtsjqMFHDZnAQ0&redir_esc=y#v=onepage&q=Lincoln ve Guba (1985)&f=false

Lincoln, Y. (2009). Ethical practices in qualitative research. In Mertens, D., & Ginsberg, P. (Eds.), *The Handbook of Social Research Ethics* (pp. 150–169). Sage Publications. DOI: 10.4135/9781483348971.n10

Lincoln, Y. S., & Guba, E. G. (1985). *Naturalistic inquiry*. SAGE Publications. DOI: 10.1016/0147-1767(85)90062-8

Lincoln, Y., & Guba, E. (1985). Naturalistic inquiry. *Sage (Atlanta, Ga.).*

Lipkin, M., & Heinonen, K. (2022). Customer ecosystems: Exploring how ecosystem actors shape customer experience. *Journal of Services Marketing*, 36(9), 1–17. DOI: 10.1108/JSM-03-2021-0080

Litman, T. (2011). Sustainability and livability: Summary of definitions, goals, objectives and performance indicators. https://www.vtpi.org/sus_liv.pdf

Littlechild, S. (2018). Economic regulation of privatised airports: Some lessons from UK experience. *Transportation Research Part A, Policy and Practice*, 114, 100–114. DOI: 10.1016/j.tra.2018.01.024

Liu, Y., Zhang, X., & Guan, X. (2018). Framework Design of Unmanned Aerial Systems (UAS) Integration in the National Airspace System (NAS). *2018 13th World Congress on Intelligent Control and Automation (WCICA)*, 1323-1327. https://doi.org/DOI: 10.1109/WCICA.2018.8630592

Liu, X., Zhang, H., Wu, S., Wang, Q., He, Z., Zhang, J., Li, R., Liu, S., & Zhang, X. (2022). Effects of buildings on wind shear at the airport: Field measurement by coherent Doppler lidar. *Journal of Wind Engineering and Industrial Aerodynamics*, 230, 105194. DOI: 10.1016/j.jweia.2022.105194

Lo Iacono, V., Symonds, P., & Brown, D. H. (2016). Skype as a tool for qualitative research interviews. *Sociological Research Online*, 21(2), 103–117. DOI: 10.5153/sro.3952

lo Storto, C. (2018). The analysis of the cost-revenue production cycle efficiency of the Italian airports: A NSBM DEA approach. *Journal of Air Transport Management*, 72, 77–85. DOI: 10.1016/j.jairtraman.2018.05.010

Longhurst, R. (2009). Interviews: In-depth, semi-structured. *International encyclopedia of human geography*, 580-584. https://doi.org/DOI: 10.1016/B978-008044910-4.00458-2

Lopez, V., & Whitehead, D. (2013). Sampling data and data collection in qualitative research. *Nursing & midwifery research: Methods and appraisal for evidence-based practice, 123*, 140.

Lundin, L., & Kindström, D. (2024). Managing digitalized touchpoints in B2B customer journeys. *Industrial Marketing Management*, 121, 88–99. DOI: 10.1016/j.indmarman.2024.07.004

Lyall-Wilson, B., Kim, N., & Hohman, E. (2019). Modeling human factors topics in aviation reports. *Proceedings of the Human Factors and Ergonomics Society Annual Meeting*. DOI: 10.1177/1071181319631095

Lynes, J. K., & Andrachuk, M. (2008). Motivations for corporate social and environmental responsibility: A case study of Scandinavian Airlines. *Journal of International Management*, 14(4), 377–390. DOI: 10.1016/j.intman.2007.09.004

Machnik-Słomka, J. (2017). Entrepreneurship of aviation industry in the perspective of sustainable development. *International Multidisciplinary Scientific GeoConference SGEM*, 17, 301–308.

Macias, M., Barrado, C., Pastor, E., & Royo, P. (2019). The future of drones and their public acceptance. In *2019 IEEE/AIAA 38th Digital Avionics Systems Conference (DASC)* (pp. 1-8). IEEE. DOI: 10.1109/DASC43569.2019.9081623

Mackieson, P., Shlonsky, A., & Connolly, M. (2019). Increasing rigor and reducing bias in qualitative research: A document analysis of parliamentary debates using applied thematic analysis. *Qualitative Social Work: Research and Practice*, 18(6), 965–980. DOI: 10.1177/1473325018786996

Mack, N., Woodsong, C., MacQueen, K. M., & Guest, G. (2005). *Qualitative research methods*. Family Health International.

MacSween, S. (2003). A public opinion survey-unmanned aerial vehicles for cargo, commercial, and passenger transportation. In *2nd AIAA" Unmanned Unlimited" Conf. and Workshop & Exhibit* (p. 6519). DOI: 10.2514/6.2003-6519

Macsween-George, S. L. (2003). Will the public accept UAVs for cargo and passenger transportation? In *2003 IEEE Aerospace Conference Proceedings (Cat. No. 03TH8652)* (Vol. 1, pp. 1-367). IEEE. DOI: 10.1109/AERO.2003.1235066

Madge, C., & O'Connor, H. (2002). On-line with e-mums: Exploring the internet as a medium for research. *Area*, 34(1), 92–102. https://www.jstor.org/stable/20004209. DOI: 10.1111/1475-4762.00060

Ma, G., Ding, Y., & Ma, J. (2022). The impact of airport physical environment on perceived safety and domestic travel intention of Chinese passengers during the COVID-19 pandemic: The mediating role of passenger satisfaction. *Sustainability (Basel)*, 14(9), 5628. DOI: 10.3390/su14095628

Mahama, A. J., & Iyer, S. S. (2023). Circular Supply Chain Case Studies in Aviation Sustainability. In *Handbook of Research on Designing Sustainable Supply Chains to Achieve a Circular Economy* (pp. 351–376). IGI Global. DOI: 10.4018/978-1-6684-7664-2.ch017

Mainardes, E. W., de Melo, R. F. S., & Moreira, N. C. (2021). Effects of airport service quality on the corporate image of airports. *Research in Transportation Business & Management*, 41, 100668. DOI: 10.1016/j.rtbm.2021.100668

Mak, B., & Chan, W. W. (2006). Environmental reporting of airlines in the Asia Pacific region. *Journal of Sustainable Tourism*, 14(6), 618–628. DOI: 10.2167/jost586.0

Man, M., & Bogeanu-Popa, M. M. (2020). Impact of non-financial information on sustainable reporting of organisations' performance: Case study on the companies listed on the bucharest stock exchange. *Sustainability (Basel)*, 12(6), 2179. DOI: 10.3390/su12062179

Markatos, D. N., Malefaki, S., & Pantelakis, S. G. (2023). Sensitivity analysis of a hybrid MCDM model for sustainability assessment—An example from the aviation industry. *Aerospace (Basel, Switzerland)*, 10(4), 385. DOI: 10.3390/aerospace10040385

Markatos, D. N., & Pantelakis, S. G. (2022). Assessment of the impact of material selection on aviation sustainability, from a circular economy perspective. *Aerospace (Basel, Switzerland)*, 9(2), 52. DOI: 10.3390/aerospace9020052

Martin, E. A., Winter, S., & Holder, T. (2017). An Exploration of Trainee Practitioners' Experiences When Using Observation. *The Sport Psychologist*, 31(2), 160–172. DOI: 10.1123/tsp.2016-0019

Matolak, D. (2015). Unmanned aerial vehicles: Communications challenges and future aerial networking. *2015 International Conference on Computing, Networking and Communications (ICNC)*, 567-572. https://doi.org/DOI: 10.1109/ICCNC.2015.7069407

Matus, P., & Materna, M. (2021). Approaches of chosen air navigation services providers to UAV integration into air traffic control systems. *Práce a štúdie - Vydanie 10*. https://doi.org/DOI: 10.26552/pas.Z.2021.2.26

Maxwell, J. (2012). *A realist approach for qualitative research*. SAGE.

Maxwell, J. A. (2012). *Qualitative Research Design: An Interactive Approach* (3th Editio). *Sage (Atlanta, Ga.)*.

May, T. (2002). *Qualitative Research in Action*. SAGE Publications. DOI: 10.4135/9781849209656

McCulloch, G. (2004). *Documentary research: In education, history and the social sciences* (1st Editio). Routledge.

McKenzie, S. (2004). Social sustainability: Towards some definitions. Hawke Research Institute, Working Paper Series No. 27, University of South Australia.

McLafferty, I. (2004). Focus group interviews as a data collecting strategy. *Journal of Advanced Nursing*, 48(2), 187–194. DOI: 10.1111/j.1365-2648.2004.03186.x PMID: 15369499

McManners, P. (2016). Developing policy integrating sustainability: A case study into aviation. *Environmental Science & Policy. Cilt*, 57, 86–92.

McManners, P. (2016). The action research case study approach: A methodology for complex challenges such as sustainability in aviation. *Action Research*, 14(2), 201–216. DOI: 10.1177/1476750315597979

McManners, P. J. (2016). Developing policy integrating sustainability: A case study into aviation. *Environmental Science & Policy*, 57, 86–92. DOI: 10.1016/j.envsci.2015.11.016

Merriam, S. B., & Tisdell, E. J. (2016). *Qualitative Research: A Guide to Design and Implementation* (4th Editio). Jossey-Bass.

Merriam, S. (2013). *Qualitative Research: A Guide to Design and Implementation* (Turan, S., Ed.). Nobel Publishing.

Mertens, D. M. (2018). Ethics of qualitative data collection. *The SAGE handbook of qualitative data collection*, 33-48.

Mertens, D. M., & Hesse-Biber, S. (2013). Mixed methods and credibility of evidence in evaluation. *New Directions for Evaluation*, 2013(138), 5–13. DOI: 10.1002/ev.20053

Merton, R. K. & Kendall, P. L., (1946). The Focused Interview. *American Journal of Sociology,* Cilt 51.

Merton, R. K., Fiske, M., & Kendall, P. L. (1956). The focused interview: A manual of problems and procedures. Free Pree. In *Krueger, R. A., & Casey, M. A. (2015). Focus groups: a practical guide for applied research* (5th ed.). SAGE.

Methods Guides How to run a focus group. (n.d.). https://sites.reading.ac.uk/wp-content/uploads/sites/42/2020/06/Focus-Groups-Guide-Final.pdf

Meurer, W. J., Frederiksen, S. M., Majersik, J. J., Zhang, L., Sandretto, A., & Scott, P. A. (2007). Qualitative data collection and analysis methods: The INSTINCT trial. *Academic Emergency Medicine*, 14(11), 1064–1071. DOI: 10.1197/j.aem.2007.05.005 PMID: 17652268

Meyer, D., Wypych, T., Petrovic, V., Strawson, J., Kamat, S., & Kuester, F. (2018). *An air traffic control simulator for test and development of airspace management schemes. In 2018 IEEE Aerospace Conference, 1-8.*, DOI: 10.1109/AERO.2018.8396575

Michalski, D. J., & Bearman, C. (2014). Factors affecting the decision making of pilots who fly in outback Australia. *Safety Science*, 68, 288–293. DOI: 10.1016/j.ssci.2014.03.005

Mielinger, E., & Weinrich, R. (2024). Insights into plastic food packaging waste sorting behaviour: A focus group study among consumers in Germany. *Waste Management. Waste Management (New York, N.Y.)*, 178, 362–370. DOI: 10.1016/j.wasman.2024.02.038 PMID: 38430750

Miles, M. B., & Huberman, A. M. (1994). *Qualitative data analysis: An expanded source book* (2nd ed.). Sage.

Miles, M., Huberman, M., & Saldana, J. (2014). *Qualitative Data Analysis: A Methods Sourcebook* (3rd Editio). *Sage (Atlanta, Ga.)*.

Miller, C. (2020). Adult learning theory and its importance in employee training. *BizLibrary*. https://www.bizlibrary.com

Milne, C. (2005, January). Overseeing research: Ethics and the institutional review board. In *Forum Qualitative Sozialforschung/Forum: Qualitative. Social Research*, 6(1), •••.

Mishra, L. (2016). Focus group discussion in qualitative research. *TechnoLearn: An International Journal of Educational Technology*, 6(1), 1–5. DOI: 10.5958/2249-5223.2016.00001.2

Mızrak, F. (2023). Strategies for effective human resource management in the aviation industry: A case-based analysis. *Beykoz Akademi Dergisi*, 11(2), 82–109. DOI: 10.14514/beykozad.1346605

Modarress Fathi, B., Ansari, A., & Ansari, A. (2023). Green commercial aviation supply chain—A European path to environmental sustainability. *Sustainability (Basel)*, 15(8), 6574. DOI: 10.3390/su15086574

Moder, C. L. (2013). Aviation English. In Paltridge, B., & Starfield, S. (Eds.), *The handbook of English for specific purposes* (pp. 227–242). John Wiley & Sons.

Moisander, J., & Valtonen, A. (2006). *Qualitative marketing research: A cultural approach*. SAGE Publications. DOI: 10.4135/9781849209632

Molesworth, B. R., & Estival, D. (2015). Miscommunication in general aviation: The influence of external factors on communication errors. *Safety Science*, 73, 73–79. DOI: 10.1016/j.ssci.2014.11.004

Monge-Zamorano, M. (2023). Cultural heritage experiences at the smart airport. *Journal of Tourism and Heritage Research*, 6(1), 281–299.

Monteiro, A. L., & Bullock, N. (2020). A broader view of communicative competence for aeronautical communications: Implications for teaching and high-stakes testing. *The Especialist*, 41(3). Advance online publication. DOI: 10.23925/2318-7115.2020v41i3a4

Morgan, D. (1996). Focus groups. *Annual Review of Sociology. Cilt*, 22, 129–152.

Morgan, D. L. (1996). Focus groups. *Annual Review of Sociology*, 22(1), 129–152. DOI: 10.1146/annurev.soc.22.1.129

Morgan, D. L. (1997). *Focus Groups as Qualitative Research* (2nd ed.). SAGE Publication. DOI: 10.4135/9781412984287

Morgan, H. (2022). Conducting a qualitative document analysis. *The Qualitative Report*, 27(1), 64–77.

Morgan, P. B., Fletcher, D., & Sarkar, M. (2013). Defining and characterizing team resilience in elite sport. *Psychology of Sport and Exercise*, 14(4), 549–559. DOI: 10.1016/j.psychsport.2013.01.004

Morris, C. (2005). Midair collisions: Limitations of the see-and-avoid concept in civil aviation. *Aviation, Space, and Environmental Medicine*, 76(4), 357–365. PMID: 15828635

Morrow, S. L. (2005). Quality and trustworthiness in qualitative research in counseling psychology. *Journal of Counseling Psychology*, 52(2), 250–260. DOI: 10.1037/0022-0167.52.2.250

Morse, J. M. (2015). Critical analysis of strategies for determining rigor in qualitative inquiry. *Qualitative Health Research*, 25(9), 1212–1222. DOI: 10.1177/1049732315588501 PMID: 26184336

Morse, J., Barrett, M., Mayan, M., Olson, K., & Spiers, J. (2002). Verification Strategies for Establishing Reliability and Validity in Qualitative Research. *International Journal of Qualitative Methods*, 1(2), 13–22. DOI: 10.1177/160940690200100202

Moser, A., & Korstjens, I. (2018). Series: Practical guidance to qualitative research. Part 3: Sampling, data collection and analysis. *The European Journal of General Practice*, 24(1), 9–18. DOI: 10.1080/13814788.2017.1375091 PMID: 29199486

Moss, S., & Edmonds, B. (2005). Sociology and simulation: Statistical and qualitative cross-validation. *American Journal of Sociology*, 110(4), 1095–1131. DOI: 10.1086/427320

Mota, M. M., Scala, P., Schultz, M., Lubig, D., Luo, M., & Perez, E. J. (2021). The rise of the smart passenger i: analysis of impact on departing passenger flow in airports. *SESAR Innovation Days, 2021.*

Mousavi, S., & Bossink, B. A. (2017). Firms' capabilities for sustainable innovation: The case of biofuel for aviation. *Journal of Cleaner Production*, 167, 1263–1275. DOI: 10.1016/j.jclepro.2017.07.146

Mrazova, M. (2014). Sustainable development-the key for green aviation. *INCAS bulletin, 6*(1), 109.

Mueller, E. R., Kopardekar, P. H., & Goodrich, K. H. (2017). Enabling airspace integration for high-density on-demand mobility operations. In *17th AIAA Aviation Technology, Integration, and Operations Conference* (p. 3086). DOI: 10.2514/6.2017-3086

Mulhall, A. (2003). In the field: Notes on observation in qualitative research. *Journal of Advanced Nursing*, 41(3), 306–313. DOI: 10.1046/j.1365-2648.2003.02514.x PMID: 12581118

Munoz, C. & Laniado, H., (2021). Airline choice model for international round-trip flights: The role of travelers' satisfaction and personality traits. *Research in Transportation Economics,* Cilt 90.

Musa, M., & Isha, A. S. N. (2021). Holistic view of safety culture in aircraft ground handling: Integrating qualitative and quantitative methods with data triangulation. *Journal of Air Transport Management, 92*(May 2020), 102019. DOI: 10.1016/j.jairtraman.2021.102019

Nachmias, C., Nachmias, D., & Dewaard, J. (2015). *Research methods in the social sciences*. Worth Publishers.

Nakamura, H., & Kajikawa, Y. (2017). Regulation and innovation: How should small unmanned aerial vehicles be regulated? *Technological Forecasting and Social Change*, 128, 262–274. DOI: 10.1016/j.techfore.2017.06.015

Neuman, W. L. (2007). *Social Research Methods: Qualitative and Quantitative Approaches*. (6 th Ed.) Boston: Pearson.

Newington, L., & Metcalfe, A. (2014). Factors influencing recruitment to research: Qualitative study of the experiences and perceptions of research teams. *BMC Medical Research Methodology*, 14(1), 10. Advance online publication. DOI: 10.1186/1471-2288-14-10 PMID: 24456229

Nguyen, T. L. P., Nguyen, T. T. H., & Ključnikov, A. (2023). The influence of socially responsible human resource management on green behaviours in the aviation industry. *Journal of Competitiveness*.

Niemeier, H.-M., (2021). Is economics good for aviation policy? Some problems in bridging the gap between theory and policy. *Journal of Air Transport Management*, Cilt 96.

Nordqvist, M. (2022). *The Future Digital Airport Experience*. Uppsala University Press.

Nowell, L. S., Norris, J. M., White, D. E., & Moules, N. J. (2017). Thematic analysis: Striving to meet the trustworthiness criteria. *International Journal of Qualitative Methods*, 16(1), 1609406917733847. Advance online publication. DOI: 10.1177/1609406917733847

Nyboe, F., Malle, N., Hoang, V., & Ebeid, E. (2023). Open-Source Hardware/Software Architecture for Autonomous Powerline-Aware Drone Navigation and Recharging. *2023 International Conference on Unmanned Aircraft Systems (ICUAS)*, 1233-1240. https://doi.org/DOI: 10.1109/ICUAS57906.2023.10155937

Nyimbili, F., & Nyimbili, L. (2024). Types of Purposive Sampling Techniques with Their Examples and Application in Qualitative Research Studies. *British Journal of Multidisciplinary and Advanced Studies*, 5(1), 90–99. DOI: 10.37745/bjmas.2022.0419

O'Leary, S., Lieberman, S., Gulyas, A., Ogilvie, M., Bates, D., Heath, T., Pelz, C., Williams, S., & Shalet, D. (2023). Management actions to address the climate emergency: Motivations and barriers for SMEs and other societal micro/meso-level groups. *International Journal of Management Education*, 21(3), 100831. DOI: 10.1016/j.ijme.2023.100831

Ohman, A. (2005). Qualitative methodology for rehabilitation research. *Journal of Rehabilitation Medicine*, 37(5), 273–280. DOI: 10.1080/16501970510040056 PMID: 16208859

Oliver-Hoyo, M., & Allen, D. (2006). The use of triangulation methods in qualitative educational research. *Journal of College Science Teaching*, 35(4).

Onwuegbuzie, A. J., Leech, N. L., & Collins, K. M. (2010). Innovative data collection strategies in qualitative research. *Qualitative Report*, 15(3), 696–726.

Osterwalder, A., & Pigneur, Y. (2009). *Business Model Generation: A Handbook for Visionaries, Game Changers, and Challengers*. Self Published.

Ostrander, S. A. (1993). "Surely you're not in this just to be helpful" Access, Rapport, and Interviews in Three Studies of Elites. *Journal of Contemporary Ethnography*, 22(1), 7–27. DOI: 10.1177/089124193022001002

Oto, N. (2011). Environmental sustainability and airports: The example of Esenboğa Airport. (Master's thesis, Ankara University), Turkey.

Oum, T., & Lee, Y. (2002). The Northeast Asian air transport network: Is there a possibility of creating open skies in the region? *Journal of Air Transport Management. Cilt*, 8, 325–337.

Överlien, C., Aronsson, K., & Hydén, M. (2005). The focus group interview as an in-depth method? Young women talking about sexuality. *International Journal of Social Research Methodology*, 8(4), 331–344. DOI: 10.1080/1364557042000119607

Owen, G. T. (2014). Qualitative methods in higher education policy analysis: Using interviews and document analysis. *The Qualitative Report*, 19(26), 1–19. DOI: 10.46743/2160-3715/2014.1211

Özdemir, M. (2010). Qualitative Data Analysis: A Study on the Methodological Problem in Social Sciences. *Eskişehir Osmangazi University Journal of Social Sciences*, 11(1), 323–343.

Pabla, H., & Soch, H. (2023). Up in the air! Airline passenger's brand experience and its impact on brand satisfaction mediated by brand love. *Journal of Air Transport Management*, 107, 102345. DOI: 10.1016/j.jairtraman.2022.102345

Paker, T. (2015). *Qualitative Research Methods, Techniques, Analysis, and Approaches* (Seggie, F. N., & Bayyurt, Y., Eds.). Anı Publishing.

Palinkas, L. A., Horwitz, S. M., Green, C. A., Wisdom, J. P., Duan, N., & Hoagwood, K. (2015). Purposeful sampling for qualitative data collection and analysis in mixed method implementation research. *Administration and Policy in Mental Health*, 42(5), 533–544. DOI: 10.1007/s10488-013-0528-y PMID: 24193818

Pant, P. (2022). Technology social practices by Millennials and Gen Z at airport departure terminals. *Tourism Management Perspectives*, 43, 100978. DOI: 10.1016/j.tmp.2022.100978

Paraschi, E. P., Poulaki, L., & Papageorgiou, A. (2024). Sustainability challenges in airlines contemporary environmental management. *Journal of Air Transport Management*, 118, 102616. DOI: 10.1016/j.jairtraman.2024.102616

Park, W., Lee, K., & Im, K. (2014). A Proposal on the Aviation Rules of the Military UAV in the National Airspace System. *Journal of the Korean Society for Aviation and Aeronautics*, 22(1), 22–31. DOI: 10.12985/ksaa.2014.22.1.022

Patton, M. Q. (2015). *Qualitative research & evaluation methods: Integrating theory and practice* (4th Editio). Sage publications.

Patton, M. Q. (1990). *Qualitative evaluation and research methods.* SAGE Publications.

Patton, M. Q. (1999). Enhancing the quality and credibility of qualitative analysis. *Health Services Research*, 34(5 Pt 2), 1189. PMID: 10591279

Patton, M. Q. (2002). *Qualitative research and evaluation methods* (3rd ed.). Sage.

Patton, M. Q. (2015). *Qualitative research & evaluation methods* (4th ed.). Sage Publications.

Pauling, J. D., Domsic, R. T., Saketkoo, L. A., Almeida, C., Withey, J., Jay, H., & Hewlett, S. (2018). Multinational qualitative research study exploring the patient experience of Raynaud's phenomenon in systemic sclerosis. *Arthritis Care and Research*, 70(9), 1373–1384. DOI: 10.1002/acr.23475 PMID: 29473715

Pauwels, J., Buyle, S. & Dewulf, W., (2024). Regional airports revisited: Unveiling pressing research gaps and proposing a uniform definition. *Journal of the Air Transport Research Society*, Volume 2.

Pavlinovic, M., Juricic, B., & Antulov-Fantulin, B. (2017). Air traffic controllers' practical part of basic training on computer based simulation device. *2017 40th International Convention on Information and Communication Technology, Electronics and Microelectronics (MIPRO)*, 920-925. https://doi.org/DOI: 10.23919/MIPRO.2017.7973554

Payán-Sánchez, B., Plaza-Úbeda, J. A., Pérez-Valls, M., & Carmona-Moreno, E. (2018). Social embeddedness for sustainability in the aviation sector. *Corporate Social Responsibility and Environmental Management*, 25(4), 537–553. DOI: 10.1002/csr.1477

Payne, S. (2007). Qualitative methods of data collection and analysis. *Research methods in palliative care*, 139-161.

Pearce, D., Markandya, A., & Barbier, E. (2013). *Blueprint 1: For a green economy*. Routledge. [REMOVED HYPERLINK FIELD] DOI: 10.4324/9781315070223

Peasgood, T., Bourke, M., Devlin, N., Rowen, D., Yang, Y., & Dalziel, K. (2023). Randomised comparison of online interviews versus face-to-face interviews to value health states. *Social Science & Medicine*, 323, 115818. DOI: 10.1016/j.socscimed.2023.115818 PMID: 36940582

Pech, C., Cob, N., & Cejka, J. T. (2007). Understanding institutional review boards: Practical guidance to the IRB review process. *Nutrition in Clinical Practice*, 22(6), 618–628. DOI: 10.1177/0115426507022006618 PMID: 18042949

Peeters, P. ve diğerleri. (2016). Are technology myths stalling aviation climate policy? *Transportation Research Part D: Transport and Environment. Cilt*, 44, 30–42.

Pegasus Hava Taşımacılığı, A. Ş. (2023). 2023 sürdürülebilirlik raporu. https://www.pegasusyatirimciiliskileri.com/medium/image/pegasus-2023-surdurulebilirlik-raporu_1442/view.aspx

Peksatici, Ö., & Ergun, H. S. (2019). The gap between academy and industry - A qualitative study in Turkish aviation context. *Journal of Air Transport Management*, 79(March), 101687. DOI: 10.1016/j.jairtraman.2019.101687

Persiani, C., & Bagassi, S. (2013). Route planner for unmanned aerial system insertion in civil non-segregated airspace. *Proceedings of the Institution of Mechanical Engineers. Part G, Journal of Aerospace Engineering*, 227(4), 687–702. DOI: 10.1177/0954410012439975

Pfister, T., Schweighofer, M., & Reichel, A. (2016). *Sustainability*. Routledge. DOI: 10.4324/9781315882888

Pinheiro Melo, S., Barke, A., Cerdas, F., Thies, C., Mennenga, M., Spengler, T. S., & Herrmann, C. (2020). Sustainability assessment and engineering of emerging aircraft technologies—Challenges, methods and tools. *Sustainability (Basel)*, 12(14), 5663. DOI: 10.3390/su12145663

Plummer, K. (1990). *Documents of life: An introduction to the problems and literature of a humanistic method*. Unwin Hyman.

Polat, A. (2022). Nitel araştırmalarda yarı-yapılandırılmış görüşme soruları: Soru form ve türleri, nitelikler ve sıralama. *Anadolu Üniversitesi Sosyal Bilimler Dergisi*, 22(Özel Sayı 2), 161-182.

Polat, İ., Ozturk, A., Koc, R., & Inel, Y. (2024). Designing experiences: Unveiling passenger perspectives through service design at an airport. *Journal of Global Hospitality and Tourism*, 3(1), 254–268.

Polit, D. F., & Tatano Beck, C. (2004). *Nursing research: Principles and methods* (7th ed.). Lippincott Williams & Wilkins.

Powell, R. A., & Single, H. M. (1996). Focus groups. *International Journal for Quality in Health Care : Journal of the International Society for Quality in Health Care*, 8(5), 499–504. DOI: 10.1093/intqhc/8.5.499 PMID: 9117204

Puls, R., & Wittmer, A. (2021). Managing Airports. *Aviation Systems: Management of the Integrated Aviation Value Chain*, 225-249.

Puls, R., & Wittmer, A. (2021). Managing airports. In Wittmer, A., Bieger, T., & Müller, R. (Eds.), *Aviation Systems*. Springer. DOI: 10.1007/978-3-030-79549-8_7

Qing, Z., Sun, C., & Reneker, J. (2021). Evaluation of airport wayfinding accessibility with the use of a wheelchair simulator. *Transportation Research Record: Journal of the Transportation Research Board*, 2675(4), 52–60. DOI: 10.1177/0361198120980445

Qiu, R., Hou, S., Chen, X., & Meng, Z. (2021). Green aviation industry sustainable development towards an integrated support system. *Business Strategy and the Environment*, 30(5), 2441–2452. DOI: 10.1002/bse.2756

Qu, S. Q., & Dumay, J. (2011). The qualitative research interview. *Qualitative Research in Accounting & Management*, 8(3), 238–264. DOI: 10.1108/11766091111162070

Ragab, H., Polo-Peña, A. I., & Mahrous, A. A. (2024). The effect of airline service quality, perceived value, emotional attachment, and brand loyalty on passengers' willingness to pay: The moderating role of airline origin. *Case Studies on Transport Policy*, 18, 101313. DOI: 10.1016/j.cstp.2024.101313

Rahman, F. & Rahman, M. M., (2023). Analyzing service quality of domestic airlines in an emerging country- Bangladesh by structural equation models. *Journal of Air Transport Management,* Cilt 107.

Rahman, S. M. A. (2023). Ethnographic Study in Air Travel Research. In A. Hassan, N. A. A. Rahman, & N. M. Nur (Eds.), *Air Travel Industry* (1st Editio, pp. 111–127). Routledge.

Raimundo, R. J., Baltazar, M. E., & Cruz, S. P. (2023). Sustainability in the airports ecosystem: A literature review. *Sustainability (Basel)*, 15(16), 12325. DOI: 10.3390/su151612325

Ramsey, S. R., Thompson, K. L., McKenzie, M., & Rosenbaum, A. (2016). Psychological research in the internet age: The quality of web-based data. *Computers in Human Behavior*, 58, 354–360. DOI: 10.1016/j.chb.2015.12.049

Ranieri, V., Kennedy, E., Walmsley, M., Thorburn, D., & McKay, K. (2019). Rare but heard: Using asynchronous virtual focus groups, interviews and roundtable discussions to create a personalised psychological intervention for primary sclerosing cholangitis: A protocol. *BMJ Open*, 9(10), e031417. DOI: 10.1136/bmjopen-2019-031417 PMID: 31578198

Ratnasingam, P. (2005). Trust in inter-organizational exchanges: A case study in business to business electronic commerce. *Decision Support Systems*, 39(3), 525–544. DOI: 10.1016/j.dss.2003.12.005

Rawson, R., & Hooper, P. D. (2012). The importance of stakeholder participation to sustainable airport master planning in the UK. *Environmental Development*, 2, 36–47. DOI: 10.1016/j.envdev.2012.03.013

Reason, J., & Hobbs, A. (2017). *Managing maintenance error: A practical guide*. CRC Press. DOI: 10.1201/9781315249926

Rekiel, J., & de Wit, J. (2013). The security system at European airports-Tour d'Horizon. *Journal of Transportation Security*, 6(2), 89–102. DOI: 10.1007/s12198-013-0105-3

Rhoden, S., Ralston, R., & Ineson, E. M. (2008). Cabin crew training to control disruptive airline passenger behavior: A cause for tourism concern? *Tourism Management. Tourism Management*, 29(3), 538–547. DOI: 10.1016/j.tourman.2007.06.002

Ribeiro, L., Giles, S., Katkin, R., Topiwala, T., & Minnix, M. (2017). Challenges and opportunities to integrate UAS in the National Airspace System. 2017 Integrated Communications,*Navigation and Surveillance Conference (ICNS)*.https://doi.org/ DOI: 10.1109/ICNSURV.2017.8011942

Richardson, H. (1979). The FAA Approach to Air Traffic Controller Training. *Journal of Educational Technology Systems*, 7(4), 373–381. DOI: 10.2190/BVNB-23HF-KQ8C-D6GL

Rijaluddin, K., & Safitri, E. M. (2023). Analysis and Design of the Mentor On-Demand Service Interface Using the Design Thinking Method. *IJCONSIST JOURNALS*, 4(2), 11–16.

Risby, J., Guest, S. & Warnock-Smith, D., (2022). A critical analysis of Bristol Airport's employee surface access habits: Developing strategic recommendations for reducing private vehicle usage. *Research in Transportation Business & Management,* Cilt 43.

Ritchie, H., Rosado, P., & Roser, M. (2024). Breakdown of carbon dioxide, methane and nitrous oxide emissions by industry. *Our World in Data*. Erişim adresi: https://ourworldindata.org/emissions-by-industry

Robinson, O. C. (2014). Sampling in interview-based qualitative research: A theoretical and practical guide. *Qualitative Research in Psychology*, 11(1), 25–41. DOI: 10.1080/14780887.2013.801543

Rodoplu, H., & Erdoğan, T. (2022). Examining environmental sustainability practices on energy and climate change at airports. Detay Publishing, Editor: Kasım KİRACI, Number of editions: 1, Number of pages: 216, ISBN: 978-605-254-595-9, Section Pages: 41-55.4.

Rossman, G. B., & Rallis, S. F. (2017). An introduction to qualitative research: Learning in the field. *Sage (Atlanta, Ga.)*. Advance online publication. DOI: 10.4135/9781071802694

Roulston, K. (2010). Reflective interviewing: A guide to theory and practice. *Sage (Atlanta, Ga.)*. Advance online publication. DOI: 10.4135/9781446288009

Rubio-Andrada, L., Celemín-Pedroche, M. S., Escat-Cortés, M. D., & Jiménez-Crisóstomo, A. (2023). Passengers satisfaction with the technologies used in smart airports: An empirical study from a gender perspective. *Journal of Air Transport Management*, 107, 102347. DOI: 10.1016/j.jairtraman.2022.102347

Rupcic, L., Pierrat, E., Saavedra-Rubio, K., Thonemann, N., Ogugua, C., & Laurent, A. (2023). Environmental impacts in the civil aviation sector: Current state and guidance. *Transportation Research Part D, Transport and Environment*, 119, 103717. DOI: 10.1016/j.trd.2023.103717

Ruslin, R., Mashuri, S., Rasak, M. S. A., Alhabsyi, F., & Syam, H. (2022). Semi-structured interview: A methodological reflection on the development of a qualitative research instrument in educational studies. *IOSR Journal of Research & Method in Education*, 12(1), 22–29. DOI: 10.9790/7388-1201052229

Rye, L., Blakey, S., & Wilson, C. W. (2010). Sustainability of supply or the planet: A review of potential drop-in alternative aviation fuels. *Energy & Environmental Science*, 3(1), 17–27. DOI: 10.1039/B918197K

Ryen, A. (2016). Research ethics and qualitative research. In Silverman, D. (Ed.), *Qualitative Research* (pp. 31–46). Sage Publications.

Ryley, T., Davison, L., Bristow, A., & Pridmore, A. (2010). Public engagement on aviation taxes in the United Kingdom. *International Journal of Sustainable Transportation*, 4(2), 112–128. DOI: 10.1080/15568310802471735

Salas, E., Maurino, D., & Curtis, M. (2010). Human factors in aviation: An overview. In Salas, E., & Maurino, D. (Eds.), *Human factors in aviation* (2nd ed., pp. 3–19). Elsevier., DOI: 10.1016/B978-0-12-374518-7.00001-8

Saldana, J. (2022). *The Coding Manual for Qualitative Researchers* (A. Tüfekci & S. N. Şad (eds.); 2.). Pegem Publishing.

Saldaña, J. (2013). *The coding manual for qualitative researchers* (2nd ed.). Sage.

Salehi, V., Veitch, B., & Smith, D. (2021). Modeling complex socio-technical systems using the FRAM: A literature review. *Human Factors and Ergonomics in Manufacturing*, 31(1), 118–142. DOI: 10.1002/hfm.20874

Salmons, J. (2017). Using social media in data collection: Designing studies with the qualitative e-research framework. In Sloan, L., & Quan-Haase, A. (Eds.), *The SAGE handbook of social media research methods* (pp. 177–196). Sage., DOI: 10.4135/9781473983847

Sandor, Z. (2017). Challenges Caused by the Unmanned Aerial Vehicle in the Air Traffic Management. *Periodica Polytechnica Transportation Engineering*, 47(2), 96–105. DOI: 10.3311/PPtr.11204

Sant'Anna, D. A. L. M., & Hilal, A. V. G. (2021). The impact of human factors on pilots' safety behavior in offshore aviation companies: A brazilian case. *Safety Science*, 140, 105272. DOI: 10.1016/j.ssci.2021.105272

Santa, S. L. B., Ribeiro, J. M. P., Mazon, G., Schneider, J., Barcelos, R. L., & de Andrade, J. B. S. O. (2020). A Green Airport model: Proposition based on social and environmental management systems. *Sustainable Cities and Society*, 59, 102160. DOI: 10.1016/j.scs.2020.102160

Sargeant, J. (2012). Qualitative Research Part II: Participants, Analysis, and Quality Assurance. *Journal of Graduate Medical Education*, 4(1), 1–3. DOI: 10.4300/JGME-D-11-00307.1 PMID: 23451297

Sarim, M., Radmanesh, M., Dechering, M., Kumar, M., Pragada, R., & Cohen, K. (2019). Distributed Detect-and-Avoid for Multiple Unmanned Aerial Vehicles in National Air Space. *Journal of Dynamic Systems, Measurement, and Control*, 141(7), 071014. Advance online publication. DOI: 10.1115/1.4043190

Sarrah, M., Ajmal, M. M., & Mertzanis, C. (2021). Identification of sustainability indicators in the civil aviation sector in Dubai: A stakeholders' perspective. *Social Responsibility Journal*, 17(5), 648–668. DOI: 10.1108/SRJ-06-2019-0203

Saunders, M. N. (2012). Choosing research participants. In Symon, G., & Cassell, C. (Eds.), *Qualitative organizational research: Core methods and current challenges* (pp. 35–52). Sage., DOI: 10.4135/9781526435620.n3

Scheelhaase, J., Grimme, W. & Maertens, S., (2024). EU trilogue results for the aviation sector – key issues and expected impacts. *Transportation Research Procedia,* Cilt 78.

Schepers, J., & Streukens, S. (2022). To serve and protect: A typology of service robots and their role in physically safe services. *Journal of Service Management*, 33(2), 197–209. DOI: 10.1108/JOSM-11-2021-0409

Schultheis, S. (2014). Integrating advanced technology into air traffic controller training. In *14th AIAA Aviation Technology, Integration, and Operations Conference* (p. 2421). DOI: 10.2514/6.2014-2421

Schultz, M. ve diğerleri, (2020). Future aircraft turnaround operations considering post-pandemic requirements. *Journal of Air Transport Management,* Cilt 89.

Scoones, I. (2007). Sustainability. *Development in Practice*, 17(4-5), 589–596. DOI: 10.1080/09614520701469609

Scotland, J. (2012). Exploring the philosophical underpinnings of research: Relating ontology and epistemology to the methodology and methods of the scientific, interpretive, and critical research paradigms. *English Language Teaching*, 5(9), 9–16. DOI: 10.5539/elt.v5n9p9

Selimoglu, S. K., & Caliskan, A. O. (2018). Sustainability reporting in airlines industry: Evidence from Turkey. *International Journal of Sustainable Aviation*, 4(2), 114–132. DOI: 10.1504/IJSA.2018.094224

Sellner, R., & Nagl, P. (2010). Air accessibility and growth: The economic effects of a capacity expansion at Vienna International Airport. *Journal of Air Transport Management*, 16(6), 325–329. DOI: 10.1016/j.jairtraman.2010.04.003

Semtrio (2024). Havacılıkta sürdürülebilirlik: Sektörde yeşil dönüşüm. https://www.semtrio.com/blog/havacilikta-surdurulebilirlik-sektorde-yesil-donusum

Senge, P. M. (1990). *The fifth discipline: The art and practice of the learning organization*. Doubleday.

Seppänen, R., Blomqvist, K., & Sundqvist, S. (2007). Measuring inter-organizational trust-a critical review of the empirical research in 1990-2003. *Industrial Marketing Management*, 36(2), 249–265. DOI: 10.1016/j.indmarman.2005.09.003

Seymour, D. T. (1987). Focus groups and the development of new products by technologically driven companies: A comment. *Journal of Product Innovation Management*, 4(1), 50–54. DOI: 10.1111/1540-5885.410050

Sharno, M. A., & Hiloidhari, M. (2024). Social sustainability of biojet fuel for net zero aviation. *Energy for Sustainable Development*, 79, 101419. DOI: 10.1016/j.esd.2024.101419

Sharples, S., Stedmon, A., Cox, G., Nicholls, A., Shuttleworth, T., & Wilson, J. (2007). Flightdeck and air traffic control collaboration evaluation (FACE): Evaluating aviation communication in the laboratory and field. *Applied Ergonomics*, 38(4), 399–407. DOI: 10.1016/j.apergo.2007.01.012 PMID: 17499573

Shenton, A. K. (2004). Strategies for ensuring trustworthiness in qualitative research projects. *Education for Information*, 22(2), 63–75. DOI: 10.3233/EFI-2004-22201

Sheth, J. N., Jain, V., & Ambika, A. (2023). The growing importance of customer-centric support services for improving customer experience. *Journal of Business Research*, 164, 113943. DOI: 10.1016/j.jbusres.2023.113943

Shrestha, R., Oh, I., & Kim, S. (2021). A survey on operation concept, advancements, and challenging issues of urban air traffic management. *Frontiers in Future Transportation*, 2, 1. DOI: 10.3389/ffutr.2021.626935

Sia, P. Y. H., Saidin, S. S., & Iskandar, Y. H. P. (2023). Systematic review of mobile travel apps and their smart features and challenges. *Journal of Hospitality and Tourism Insights*, 6(5), 2115–2138. DOI: 10.1108/JHTI-02-2022-0087

Sikdar, S. K. (2004). Science of sustainability. *Clean Technologies and Environmental Policy*, 7(1), 1–2. DOI: 10.1007/s10098-004-0261-z

Silverman, D. (2008). Doing qualitative research: A comprehensive guide. *Sage (Atlanta, Ga.)*.

Silverman, D. (2014). *Interpreting Qualitative Data*. SAGE Publications Ltd.

Silverman, D. (2017). *Doing qualitative research* (5th ed.). SAGE.

Silverman, R. M., & Patterson, K. (2021). *Qualitative research methods for community development*. Routledge. DOI: 10.4324/9781003172925

Sim, J., & Waterfield, J. (2019). Focus group methodology: Some ethical challenges. *Quality & Quantity*, 53(6), 3003–3022. DOI: 10.1007/s11135-019-00914-5

Simpson, A. J., & Stoker, J. (2006). Safety challenges in flying UAVS (unmanned aerial vehicles) in non-segregated airspace. In *The First Institution of Engineering and Technology International Conference on System Safety, 2006.* (pp. 8-pp). IET.

Simpson, A., Stoker, J., & Brennan, V. (2009). *UAS Safety in Non-segregated Airspace*. INTECH Open Access Publisher. DOI: 10.5772/6492

Singh, H., Khalajzadeh, H., Paktinat, S., Graetsch, U. M., & Grundy, J. (2022). Modelling human-centric aspects of end-users with iStar. *Journal of Computer Languages*, 68, 101091. DOI: 10.1016/j.cola.2022.101091

Singh, J., Rana, S., Abdul Hamid, A. B., & Gupta, P. (2023). Who should hold the baton of aviation sustainability? *Social Responsibility Journal*, 19(7), 1161–1177. DOI: 10.1108/SRJ-05-2021-0181

Skouloudis, A., Evangelinos, K., & Moraitis, S. (2012). Accountability and stakeholder engagement in the airport industry: An assessment of airports' CSR reports. *Journal of Air Transport Management*, 18(1), 16–20. DOI: 10.1016/j.jairtraman.2011.06.001

Smith, A., & Douglas, A. (1998). Introducing new technology to the air traffic controller: Implications for skill acquisition and training. *17th DASC. AIAA/IEEE/SAE. Digital Avionics Systems Conference. Proceedings (Cat. No.98CH36267)*, 2, F54/1-F54/8 vol.2. https://doi.org/DOI: 10.1109/DASC.1998.739833

Smith, M., & Bowers-Brown, T. (2010). Different kinds of qualitative data collection methods. *Practical research and evaluation: A start-to-finish guide for practitioners*, 111-125.

Smith, B. (2018). Generalizability in qualitative research: Misunderstandings, opportunities and recommendations for the sport and exercise sciences. *Qualitative Research in Sport, Exercise and Health*, 10(1), 137–149. DOI: 10.1080/2159676X.2017.1393221

Smith, J. (1996). Qualitative methodology: Analyzing participants' perspectives. *Current Opinion in Psychiatry*, 9(6), 417–421. DOI: 10.1097/00001504-199611000-00010

Smith, P., Gaffney, M. J., Shi, W., Hoard, S., Armendariz, I. I., & Mueller, D. W. (2017). Drivers and barriers to the adoption and diffusion of Sustainable Jet Fuel (SJF) in the U.S. Pacific Northwest. *Journal of Air Transport Management*, 58, 113–124. DOI: 10.1016/j.jairtraman.2016.10.004

Snelson, C. L. (2016). Qualitative and Mixed Methods Social Media Research: A Review of the Literature. *International Journal of Qualitative Methods*, 15(1), 1609406915624574. DOI: 10.1177/1609406915624574

Sobieralski, J. B. (2020). COVID-19 and airline employment: Insights from historical uncertainty shocks to the industry. *Transportation Research Interdisciplinary Perspectives*, 5, 100123. DOI: 10.1016/j.trip.2020.100123 PMID: 34173453

Spencer, K., Sanders, C., Whitley, E. A., Lund, D., Kaye, J., & Dixon, W. G. (2016). Patient perspectives on sharing anonymized personal health data using a digital system for dynamic consent and research feedback: A qualitative study. *Journal of Medical Internet Research*, 18(4), e5011. DOI: 10.2196/jmir.5011 PMID: 27083521

Spratling, R. (2013). Recruitment of medically fragile children and adolescents: Lessons learned from qualitative research. *Journal of Pediatric Health Care*, 27(1), 62–65. DOI: 10.1016/j.pedhc.2012.08.001 PMID: 23084469

Sreenath, S., Sudhakar, K., & Yusop, A. F. (2021). Sustainability at airports: Technologies and best practices from ASEAN countries. *Journal of Environmental Management*, 299, 113639. DOI: 10.1016/j.jenvman.2021.113639 PMID: 34479146

Sridhar, B., & Chatterji, G. (1997). *Computationally efficient conflict detection methods for air traffic management*. Proceedings of the 1997 American Control Conference (Cat. No.97CH36041), 2, 1126-1130 vol.2. https://doi.org/DOI: 10.1109/ACC.1997.609708

Srivastava, S., Gupta, S., Dikshit, O., & Nair, S. (2020). A review of UAV regulations and policies in India. *Proceedings of UASG 2019: Unmanned Aerial System in Geomatics 1*, 315-325. DOI: 10.1007/978-3-030-37393-1_27

Stamper, B. (2021). Co-creating sustainable aviation solutions: A participatory approach. *Journal of Air Transport Management*, 91, 102022. DOI: 10.1016/j.jairtraman.2021.102022

Stamper, D. (2021). Exploring stakeholder perceptions of innovation in aviation: A qualitative perspective. *Journal of Transportation Innovation*, 12(3), 45–62.

Stamper, D. (2021). Sustainability in aviation: Challenges and opportunities. *Aviation Studies Journal*, 12(2), 45–63. DOI: 10.1080/ASJ-2021-02-0067

Steen, R., Norman, J. E., Bergström, J., & Damm, G. F. (2024). Dark knights: Exploring resilience and hidden workarounds in commercial aviation through mixed methods. *Safety Science*, 175, 106498. DOI: 10.1016/j.ssci.2024.106498

Stemler, S.E. (2015). Content Analysis. *Emerging Trends in the Social and Behavioral Sciences,* 1-14.

Stephenson, C., Lohmann, G., & Spasojevic, B. (2018). Stakeholder engagement in the development of international air services: A case study on Adelaide Airport. *Journal of Air Transport Management*, 71(March), 45–54. DOI: 10.1016/j.jairtraman.2018.06.006

Stiebe, M. (2023). Stakeholder Perceptions on Sustainability Challenges and Innovations in General Aviation. *Sustainability (Basel)*, 15(23), 16505. DOI: 10.3390/su152316505

Stiles, W. (1993). Quality control in qualitative research. *Clinical Psychology Review*, 13(6), 593–618. DOI: 10.1016/0272-7358(93)90048-Q

Stoll-Kleemann, S., O'Riordan, T., & Jaeger, C. C. (2001). The psychology of denial concerning climate mitigation measures: Evidence from Swiss focus groups. *Global Environmental Change*, 11(2), 107–117. DOI: 10.1016/S0959-3780(00)00061-3

Stolzer, A. J., Sumwalt, R. L., & Goglia, J. J. (2023). *Safety management systems in aviation*. CRC Press. DOI: 10.1201/9781003286127

Story, V., Zolkiewski, J., Verleye, K., Nazifi, A., Hannibal, C., Grimes, A., & Abboud, L. (2020). Stepping out of the shadows: Supporting actors' strategies for managing end-user experiences in service ecosystems. *Journal of Business Research*, 116, 401–411. DOI: 10.1016/j.jbusres.2020.04.029

Strokosch, K., & Osborne, S. P. (2023). Design of services or designing for service? The application of design methodology in public service settings. *Policy and Politics*, 51(2), 231–249. DOI: 10.1332/030557321X16750746455167

Subaşı, M., & Okumuş, K. (2017). Bir araştırma yöntemi olarak durum çalışması. *Atatürk Üniversitesi Sosyal Bilimler Enstitüsü Dergisi*, 21(2), 419–426.

Subaşı, M., & Okumuş, K. (2017). Case Study as a Research Method. *Atatürk University Journal of Social Sciences*, 21(2), 419–426.

Sullivan, J. R. (2012). Skype: An appropriate method of data collection for qualitative interviews? *The Hilltop Review*, 6(1), 10.

Sullivan, P., & Girginer, H. (2002). The use of discourse analysis to enhance ESP teacher knowledge: An example of Aviation English. *English for Specific Purposes*, 21(4), 97–404. DOI: 10.1016/S0889-4906(01)00042-4

Sun, C., Li, S., Lin, Y., & Hu, W. (2022). From Visual Behavior to Signage Design: A Wayfinding Experiment with Eye-Tracking in Satellite Terminal of PVG Airport. In *Proceedings of the 2021 DigitalFUTURES: The 3rd International Conference on Computational Design and Robotic Fabrication (CDRF 2021) 3* (pp. 252-262). Springer Singapore. DOI: 10.1007/978-981-16-5983-6_24

Sun, Q., Phoebe, J., & Ziwei, L. (2022). Service design practice and its future relevance. In *The 23rd DMI: Academic design management conference proceedings* (pp. 359-368). DMI.

Suri, H. (2011). Purposeful sampling in qualitative research synthesis. *Qualitative Research Journal*, 11(2), 63–75. DOI: 10.3316/QRJ1102063

Sustainability Practices, P. W. C. (2011). *Turkish Business World Assessment Report*. PWC.

Sutton, J., & Austin, Z. (2015). Qualitative research: Data collection, analysis, and management. *The Canadian Journal of Hospital Pharmacy*, 68(3), 226. DOI: 10.4212/cjhp.v68i3.1456 PMID: 26157184

Taherdoost, H. (2021). Data collection methods and tools for research; a step-by-step guide to choose data collection technique for academic and business research projects. [IJARM]. *International Journal of Academic Research in Management*, 10(1), 10–38.

Tan, L. K. L., Lim, B. C., Park, G., Low, K. H., & Yeo, V. C. S. (2021). Public acceptance of drone applications in a highly urbanized environment. *Technology in Society*, 64, 101462. DOI: 10.1016/j.techsoc.2020.101462

Taraglio, S., Nanni, V., & Taurino, D. (2012). Hardware on-Board Implementation of a Possible Solution to Introduce UAVs into Non Segregated Areas. In *ICINCO (1)* (pp. 177-182).

Tarnavska, T., Glushanytsia, N., Bilous, N., Lyashenko, L., & Palamarchuk, K. (2021). English for aviation safety: Language training methodology. *The Asian ESP Journal*, 17(1), 5.

Teperi, A. M., Paajanen, T., Asikainen, I., & Lantto, E. (2023). From must to mindset: Outcomes of human factor practices in aviation and railway companies. *Safety Science*, 158, 105968. DOI: 10.1016/j.ssci.2022.105968

Tepylo, N., Straubinger, A., & Laliberte, J. (2023). Public perception of advanced aviation technologies: A review and roadmap to acceptance. *Progress in Aerospace Sciences*, 138, 100899. DOI: 10.1016/j.paerosci.2023.100899

Thelwall, M., & Nevill, T. (2021). Is research with qualitative data more prevalent and impactful now? Interviews, case studies, focus groups and ethnographies. *Library & Information Science Research*, 43(2), 101094. DOI: 10.1016/j.lisr.2021.101094

Then, K. L., Rankin, J. A., & Ali, E. (2014). Focus group research: What is it and how can it be used? *Canadian Journal of Cardiovascular Nursing*, 24(1). PMID: 24660275

Theunissen, E., Goossens, A., Bleeker, O., & Koeners, J. (2005). UAV mission management functions to support integration in a strategic and tactical ATC and C2 environment. In *AIAA modeling and simulation technologies conference and exhibit* (p. 6310). DOI: 10.2514/6.2005-6310

Tian, W., & Caponecchia, C. (2020). Using the Functional Resonance Analysis Method (FRAM) in Aviation Safety: A Systematic Review. *Journal of Advanced Transportation*, 1, 8898903. DOI: 10.1155/2020/8898903

Tiewtrakul, T., & Fletcher, S. R. (2010). The challenge of regional accents for aviation English language proficiency standards: A study of difficulties in understanding in air traffic control–pilot communications. *Ergonomics*, 53(2), 229–239. DOI: 10.1080/00140130903470033 PMID: 20099176

Tisdall, L., & Zhang, Y. (2020). Preparing for 'COVID-27': Lessons in management focus – An Australian general aviation perspective. *Journal of Air Transport Management*, 89, 101922. DOI: 10.1016/j.jairtraman.2020.101922 PMID: 32901185

Tisdall, L., Zhang, Y., & Zhang, A. (2021). COVID-19 impacts on general aviation – Comparative experiences, governmental responses and policy imperatives. *Transport Policy*, 110, 273–280. DOI: 10.1016/j.tranpol.2021.06.009 PMID: 36567696

Tomlinson, B. (2013). *Developing materials for language teaching*. Bloomsbury.

Topçu, G. (2023). Bibliometric analysis on the place of sustainability in aviation in international literature. [JSHSR]. *International Journal of Social and Humanities Sciences Research*, 10(99), 2419–2428.

Torum, O., & Yılmaz, A. K. (2009). Sustainability Management in Aviation: Research on Sustainability Practices for Airports in Turkey. *Journal of Aeronautics and Space Technologies*, 4(2), 47–58.

Tracy, S. J. (2024). *Qualitative research methods: Collecting evidence, crafting analysis, communicating impact*. John Wiley & Sons.

Tremblay, M. C., Hevner, A. R., & Berndt, D. J. (2010). *The Use of Focus Groups in Design Science Research. %1 içindeA. Hevner & S. Chatterjee, düz. Design Research in Information Systems: Theory and Practice. basım yeri bilinmiyor*. Springer.

Trimmer, R. E., Mandy, W. P., Muntoni, F., & Maresh, K. E. (2024). Understanding anxiety experienced by young males with Duchenne muscular dystrophy: A qualitative focus group study. *Neuromuscular Disorders*, 34, 95–104. DOI: 10.1016/j.nmd.2023.12.002 PMID: 38159461

Trotter, R. T.II. (2012). Qualitative research sample design and sample size: Resolving and unresolved ıssues and inferential İmperatives. *Preventive Medicine*, 55(5), 398–400. DOI: 10.1016/j.ypmed.2012.07.003 PMID: 22800684

Tueanrat, Y., Papagiannidis, S., & Alamanos, E. (2021). Going on a journey: A review of the customer journey literature. *Journal of Business Research*, 125, 336–353. DOI: 10.1016/j.jbusres.2020.12.028

Tuna, Ö., & Besler, S. (2015). Kurumsal sürdürülebilirlik anlayışı ve uygulamaları: Örnek olay çalışması. *Journal of Faculty of Political Science*, 52, 171–200.

Tuncal, A., & Uslu, S. (2021). Kentsel Hava Hareketliliği Kavramının Gelişiminde İki Önemli Faktör: ATM ve Toplum. *Karamanoğlu Mehmetbey Üniversitesi Sosyal ve Ekonomik Araştırmalar Dergisi*, 23(41), 564–577.

Turan, G. (2018). *Çok Kriterli Karar Verme. B. F. Yıldırım, & E. Önder içinde, Çok Kriterli Karar Verme Yöntemleri*. Dora Basım Yayınları.

Türk Hava Yolları (THY). (2024). 2023 Sürdürülebilirlik raporu. https://cdn.turkishairlines.com/m/38a092fe993bc5a0/original/2023-Surdurulebilirlik-Raporu.pdff

Türk Hava Yolları (THY). (2024). Sürdürülebilirlik yolcuğumuz. https://cdn.turkishairlines.com/m/3fe0ae64a7afc39e/original/PowerPoint-Presentation.pdf

Tüyen, Z. (2020). İşletmelerde sürdürülebilirlik kavramı ve sürdürebilirliği etkileyen etmenler. İstanbul Ticaret Üniversitesi Sosyal Bilimler Dergisi, 19(37), 91-117.

UNESCO. (1978). *Intergovernmental Conference on Environmental Education: Final Report*. UNESCO and UNEP. https://unesdoc.unesco.org/images/0003/000327/032763eo.pdf

Usman, A., Azis, Y., Harsanto, B., & Azis, A. M. (2022). Airport service quality dimension and measurement: A systematic literature review and future research agenda. *International Journal of Quality & Reliability Management*, 39(10), 2302–2322. DOI: 10.1108/IJQRM-07-2021-0198

Uysal, M. P., & Sogut, M. Z. (2017). An integrated research for architecture-based energy management in sustainable airports. *Energy*, 140, 1387–1397. DOI: 10.1016/j.energy.2017.05.199

Uzgör, M. (2024). *Evaluation of airport environmental management performance with multi criteria decision making methods: An application at five major airports in Türkiye*. Anadolu University.

Uzgör, M., & Şengür, F. (2022). Investigating an underutilized subsidized routes scheme: Underlying reasons and policy recommendations. *Case Studies on Transport Policy*, 10(1), 287–299. DOI: 10.1016/j.cstp.2021.12.010

Van Manen, M. (2016). *Researching lived experience: Human science for an action sensitive pedagogy*. Routledge.

van Oel, C. J., & van den Berkhof, F. (2013). Consumer preferences in the design of airport passenger areas. *Journal of Environmental Psychology. Cilt*, 36, 280–290.

Vannucci, V., & Pantano, E. (2020). Digital or human touchpoints? Insights from consumer-facing in-store services. *Information Technology & People*, 33(1), 296–310. DOI: 10.1108/ITP-02-2018-0113

Varhol, R. J., Man Ying Lee, C., Hindmarsh, S., Boyd, J. H., Robinson, S., & Randall, S. (2024). Consumer attitudes, barriers and facilitators to sharing clinical data for research purposes: Results from a focus group synthesis. *Heliyon*, 10(14), e34431. DOI: 10.1016/j.heliyon.2024.e34431

Vaughn, S., Schumm, J. S., & Sinagub, J. (1996). *Focus group interviews in education and psychology*. SAGE. DOI: 10.4135/9781452243641

Vaz, C. A., Junior. (2018). Remembering Tenerife: 40 years later. *SF Journal of Aviation and Aeronaut Science, 1*(2). https://scienceforecastoa.com/Articles/SJAAS-V1-E2-1012.pdf

Vink, J., Koskela-Huotari, K., Tronvoll, B., Edvardsson, B., & Wetter-Edman, K. (2021). Service ecosystem design: Propositions, process model, and future research agenda. *Journal of Service Research*, 24(2), 168–186. DOI: 10.1177/1094670520952537

Vishnevsky, T., & Beanlands, H. (2004). Qualitative research. *Nephrology Nursing Journal : Journal of the American Nephrology Nurses' Association*, 31-2, 234–238. DOI: 10.4324/9780203299975-17 PMID: 15114810

Volz, U., Böhnke, J., Knierim, L., Richert, K., Roeber, G. M., & Eidt, V. (2015). *Financing the green transformation: How to make green finance work in Indonesia*. Springer. DOI: 10.1057/9781137486127

Voyer, M., Gollan, N., Barclay, K., & Gladstone, W. (2015). 'It s part of me'; understanding the values, images and principles of coastal users and their influence on the social acceptability of MPAs. *Marine Policy*, 52, 93–102. DOI: 10.1016/j.marpol.2014.10.027

Wach, E., & Ward, R. (2013). Learning about qualitative document analysis.

Walby, K. (2015). Marshall, Catherine, and Gretchen B. Rossman, Designing Qualitative Research. *Canadian Journal of Sociology*, 40(3), 399–402. DOI: 10.29173/cjs25398

Walker, D. M. (2013). The internet as a medium for health service research. Part 1. *Nurse Researcher*, 20(4), 18–21. DOI: 10.7748/nr2013.03.20.4.18.e294 PMID: 23520708

Wang, D., Wang, X., & Xia, N. (2018). How safety-related stress affects workers' safety behavior: The moderating role of psychological capital. *Safety Science. Safety Science*, 103, 247–259. DOI: 10.1016/j.ssci.2017.11.020

Wang, H., Dattel, A., Mummert, E., & Haris, S. (2022). Assessing Air Traffic Controllers' Stress and Performance with Uav Integration in Future Air Traffic Management. *Proceedings of the Human Factors and Ergonomics Society Annual Meeting*, 66(1), 38–38. DOI: 10.1177/1071181322661332

Wang, R., Xie, X., & Ma, H. (2021). Evolution of omni-channel business models: A new community-based omni-channel and data-enabled ecosystem. *Journal of Contemporary Marketing Science*, 4(3), 385–396. DOI: 10.1108/JCMARS-11-2020-0045

WCED. (1987). *Our common future*. Oxford University Press.

Wegler, P. (Ed.). (2015). Training needs beyond the ICAO language proficiency requirements. In A. Borowska & A. Enright (Eds.), *Changing perspectives on aviation English training* (pp. 53–60). Uniwersytet Warszawski.

Weiber, I., Eklund, M., & Tengland, P. A. (2016). The characteristics of local support systems, and the roles of professionals, in supporting families where a mother has an intellectual disability. *Journal of Applied Research in Intellectual Disabilities*, 29(3), 197–210. DOI: 10.1111/jar.12169 PMID: 25754531

Weidig, J., Weippert, M., & Kuehnl, C. (2024). Personalized touchpoints and customer experience: A conceptual synthesis. *Journal of Business Research*, 177, 114641. DOI: 10.1016/j.jbusres.2024.114641

Welker, A., & Kamberelis, G. (2023). *Imagining and reimagining focus group research in education* (4th ed.). International Encyclopedia of Education. DOI: 10.1016/B978-0-12-818630-5.11023-1

Wellings, K., Branigan, P., & Mitchell, K. (2000). Discomfort, discord and discontinuity as data: Using focus groups to research sensitive topics. *Culture, Health & Sexuality*, 2(3), 255–267. DOI: 10.1080/136910500422241

Wennersten, R., & Qie, S. (2018). United Nations sustainable development goals for 2030 and resource use. *Handbook of sustainability science and research*, 317-339.

Wensveen, J. G. (2011). *Air Transportation: A Management Perspective* (7th ed.). Ashgate.

Werbach, A. (2011). Strategy for sustainability. *Strategic Direction, 27*(10).

Westrum, R., & Adamski, A. J. (2016). Organizational factors associated with safety and mission success in aviation environments. In Dismukes, R. K. (Ed.), *Human error in aviation* (pp. 475–512). Routledge.

West, S., Gaiardelli, P., & Saccani, N. (2022). *Modern industrial services: a cookbook for design, delivery, and management.* Springer Nature. DOI: 10.1007/978-3-030-80511-1

West, S., Stoll, O., Østerlund, M., Müller-Csernetzky, P., Keiderling, F., & Kowalkowski, C. (2020). Adjusting customer journey mapping for application in industrial product-service systems. *International Journal of Business Environment*, 11(3), 275–297. DOI: 10.1504/IJBE.2020.110911

Westwood, S., Pritchard, A., & Morgan, N. J. (2000). Gender-blind marketing: Businesswomen's perceptions of airline services. *Tourism Management*, 21(4), 353–362. DOI: 10.1016/S0261-5177(99)00069-2

Weyrich, P., Schrauth, M., Kraus, B., Habermehl, D., Netzhammer, N., Zipfel, S., & Nikendei, C. (2008). Undergraduate technical skills training guided by student tutors–analysis of tutors' attitudes, tutees' acceptance and learning progress in an innovative teaching model. *BMC Medical Education*, 8(1), 1–9. DOI: 10.1186/1472-6920-8-18 PMID: 18400106

White, G. E., & Thomson, A. N. (1995). Anonymized focus groups as a research tool for health professionals. *Qualitative Health Research*, 5(2), 256–261. DOI: 10.1177/104973239500500209

Wiegmann, D. A., & Shappell, S. A. (2003). *A Human Error Approach to Aviation Accident Analysis: The Human Factors Analysis and Classification System* (1st ed.). Routledge.

Wikipedia. (2024). List of unmanned aerial vehicle-related incidents. In *Wikipedia, The Free Encyclopedia*. Retrieved May 18, 2024, from https://en.wikipedia.org/wiki/List_of_unmanned_aerial_vehicle-related_incidents

Wilkinson, S. (1998). Focus group methodology: A review. *International Journal of Social Research Methodology*, 1(3), 181–203. DOI: 10.1080/13645579.1998.10846874

Winter, S. R., Thropp, J. E., & Rice, S. (2019). What factors predict a consumer's support of environmental sustainability in aviation? A multi-model analysis. *International Journal of Sustainable Aviation*, 5(3), 190–204. DOI: 10.1504/IJSA.2019.103502

Wittenbaum, G., & Bowman, J. (2004). A social validation explanation for mutual enhancement. *Journal of Experimental Social Psychology*, 40(2), 169–184. DOI: 10.1016/S0022-1031(03)00091-X

Wolcott, H. F. (1994). *Transforming qualitative data: Description, analysis, and interpretation*. Sage.

Wolcott, H. F. (2001). *Writing Up Qualitative Research* (2nd ed.). SAGE Publications.

Wong, L. P. (2008). Focus group discussion: A tool for health and medical research. *Singapore Medical Journal*, 49(3), 256–260. PMID: 18363011

Wongyai, P. H., Ngo, T., Wu, H., Tsui, K., & Nguyen, T. H. (2024). Self-Service Technology in Aviation: A Systematic Literature Review. *Journal of the Air Transport Research Society*, 100016.

Wood, L. M., Sebar, B., & Vecchio, N. (2020). Application of rigour and credibility in qualitative document analysis: Lessons learnt from a case study. *The Qualitative Report*, 25(2), 456–470. DOI: 10.46743/2160-3715/2020.4240

Woodrow, L. (2022). *Introducing researching english for specific purposes*. Routledge. DOI: 10.4324/9781003049678

Wu, C., Jiang, Q., & Yang, H. (2018). Changes in cross-strait aviation policies and their impact on tourism flows since 2009. *Transport Policy*, 63, 61–72. DOI: 10.1016/j.tranpol.2017.11.008

Wu, P. J., & Yang, C. K. (2021). Sustainable development in aviation logistics: Successful drivers and business strategies. *Business Strategy and the Environment*, 30(8), 3763–3771. DOI: 10.1002/bse.2838

Xia, L., Xu, Y., Zhang, Y., Jiang, H., & Cui, B. (2024). Impact of airline social media marketing on purchase intention:Evidence from China using PLS-SEM. *Transport Economics and Management*, 2, 249–262. DOI: 10.1016/j.team.2024.09.003

Xu, Y. (2013). TCAS/ADS-B integrated surveillance and collision avoidance system. In *Conference of the 2nd International Conference on Computer Science and Electronics Engineering (ICCSEE 2013)* (pp. 666-669). Atlantis Press. DOI: 10.2991/iccsee.2013.169

Yalçınkaya, A., & Yalçınkaya, L. A. (2020). The Emergence, Rise, and Fall of Nuri Demirağ as an Entrepreneur: A Historical Analysis within Macro Institutional Theory. *The Journal of American Science*, 18(38), 839–864.

Yamamoto, H., Sattar, R. A., Wang, J., & Mendez, L. F. (2016). *Drone-enabled Foreign Object Debris (FOD) Removal System in ad hoc Situations*.

Yapı Kredi Bankası (YKB). (2022). Finansal sürdürülebilirlik nedir? https://www.yapikredi.com.tr/blog/surdurulebilirlik/finansal-okuryazarlik/detay/finansal-surdurulebilirlik-nedir

Yazgan, E. (2018). Development taxonomy of human risk factors for corporate sustainability in aviation sector. *Aircraft Engineering and Aerospace Technology*, 90(6), 1012–1022. DOI: 10.1108/AEAT-04-2018-0133

Yıldırım, A., & Şimşek, H. (2003). *Sosyal bilimlerde nitel araştırma yöntemleri*. Seçkin.

Yıldırım, A., & Şimşek, H. (2013). *Qualitative Research Methods in Social Sciences* (9th ed.). Seçkin Yayıncılık.

Yıldırım, A., & Şimşek, H. (2016). *Sosyal Bilimlerde Nitel Araştırma Yöntemleri* (10th ed.). Seçkin Yayıncılık.

Yin, R. (2017). *Case Study Research and Applications Design and Methods*. Sage Publishing.

Yin, R. K. (1981). The Case Study Crisis: Some Answers. *Administrative Science Quarterly*, 26(1), 58–65. http://search.ebscohost.com/login.aspx?direct=true&db=ufh&AN=5007419&site=ehost-live%5Cnhttp://content.ebscohost.com/ContentServer.asp?T=P&P=AN&K=5007419&S=R&D=ufh&EbscoContent=dGJyMMTo50Seqa44y9f3OLCmr0yeprdSrqq4SLCWxWXS&ContentCustomer=dGJyMOzpr1GwqLRIu. DOI: 10.2307/2392599

Yin, R. K. (2009). *Case study research: Design and method* (Vol. 4). Sage Publishing.

Yin, R. K. (2011). *Qualitative research from start to finish.* Guilford Press.

Yin, R. K. (2018). *Case study research and applications: Design and methods* (6th ed.). Sage Publications.

Yum, K. K., Choi, J. O., & Hyun, S. S. (2024). A study on the effect of job stress factors on mental health and service sabotage: Focusing on flight attendants in foreign airlines. *Journal of Air Transport Management*, 115, 102541. DOI: 10.1016/j.jairtraman.2024.102541

Yurdakul, M. (2024). Şirketlerde sürdürülebilirliği stratejik olarak gündeme getiren dinamikler. *Yönetim Bilimleri Dergisi*, 22(52), 282–307. DOI: 10.35408/comuybd.1351915

Zafiharimalala, H., Robin, D., & Tricot, A. (2014). Why Aircraft Maintenance Technicians Sometimes Do Not Use Their Maintenance Documents: Towards a New Qualitative Perspective. *The International Journal of Aviation Psychology*, 24(3), 190–209. DOI: 10.1080/10508414.2014.918444

Zahari, H. M., Zain, R. M., & Azhar, H. B. (2023). Summary of Qualitative Methods in Air Travel Research. In A. Hassan, N. A. A. Rahman, & N. M. Nur (Eds.), *Air Travel Industry: Theories, Methods and Recent Issues* (1st Editio, pp. 65–74). Routledge.

Zdanowicz, K., Herriotts, P., Payre, W., Mangurenje, D., & Birrell, S. (2022). The World's First 'Pop-Up'Urban Airport: A User-Centred Design Approach to Understand the Customer Journey. In *Human Interaction, Emerging Technologies and Future Systems V: Proceedings of the 5th International Virtual Conference on Human Interaction and Emerging Technologies, IHIET 2021, August 27-29, 2021 and the 6th IHIET: Future Systems (IHIET-FS 2021), October 28-30, 2021, France* (pp. 483-490). Springer International Publishing.

Zeng, Y., Zhang, R., & Lim, T. (2016). Wireless communications with unmanned aerial vehicles: Opportunities and challenges. *IEEE Communications Magazine*, 54(5), 36–42. DOI: 10.1109/MCOM.2016.7470933

Zhang, Z., Zhang, J., Wang, P., & Chen, L. (2018). Research on Operation of UAVs in Non-isolated Airspace. *Computers, Materials & Continua*, 57(1), 151–166. DOI: 10.32604/cmc.2018.02890

Zieba, M., & Johansson, E. (2022). Sustainability reporting in the airline industry: Current literature and future research avenues. *Transportation Research Part D, Transport and Environment*, 102, 103133. DOI: 10.1016/j.trd.2021.103133

Zsedrovits, T., Zarándy, Á., Vanek, B., Péni, T., Bokor, J., & Roska, T. (2011). Visual Detection and Implementation Aspects of a UAV See and Avoid System. *2011 20th European Conference on Circuit Theory and Design (ECCTD)*, 472-475. https://doi.org/DOI: 10.1109/ECCTD.2011.6043389

About the Contributors

Yeşim Kurt received her bachelor's, master's and doctoral degrees from Anadolu University Aviation Management Department. Later, she served as vice dean and department head at Kırklareli University Lüleburgaz Faculty of Aviation and Astronautics for 4 years. Today, she works as an academician in the Aviation Management department at Balıkesir University School of Edremit Civil Aviation.

***

Ozlem Colak is an Asst. Prof. at the Aviation Management Department, Faculty of Aeronautics and Aerospace, Gaziantep University. Dr. Colak has an MSc in Air Transport Management from the University of Surrey (UK) and obtained her PhD from Loughborough University, United Kingdom.

Nicole Falzon has 13 years of experience in the EFL industry, teaching in the UK and Malta. She holds an MA in Applied Linguistics and TESOL, focusing on Aviation English course design. Her academic and professional interests include linguistics and specialised language training, particularly in the context of English for Specific Purposes.

Gülaçtı Şen, who has been in the professional business world since 2008, completed the Department of Business Administration at Anadolu University, Faculty of Economics and Administrative Sciences between 2007-2012. He completed his Master's degree in Business Administration at Beykent University, Institute of Graduate Education in 2015. He completed his PhD in Business Administration at Beykent University, Institute of Graduate Education in 2020. His master's and doctoral theses are related to the field of entrepreneurship and the aviation industry. Between 2010-2016, he served Turkish Airlines as a Passenger Services Officer at Turkish Ground Service A.Ş. in the aviation sector. ŞEN, who has been teaching aviation in different aviation programs at universities since 2016, has been serving as an Assistant Professor at Balıkesir University, Edremit Civil Aviation School, Aviation Department since 2022. His research covers qualitative and qualitative research techniques that aim to provide new empirical evidence in the field of management (especially for the aviation industry). His books Career in Civil Aviation were published in 2019, A Day in the Terminal - Interviews with Employees of Different Aviation Professions in 2021, and Strategic Management in Aviation: Passenger Services Management in 2022.

Elad Harison is the head of Industrial Engineering and Management Department at the Shenkar College of Engineering and Design (Israel) and an affiliated researcher at the United Nations University-MERIT (The Netherlands). He holds a PhD from the Maastricht University (The Netherlands). His research interests are in the areas of innovation, services and ICT. He has written a book on the economics and policy of software intellectual property rights and has published in international journals including: Research Policy, the International Journal of Information Management, Economics of Innovation and New Technology and the Journal of Air Transport Management. He has been a policy expert on Israeli innovation and on the patentability of computer implemented inventions in EC projects and a member of the EC's Expert Group on Intellectual Property Rights in Internet Collaborations.

Fehmina Khalique, Professor OB & HR, Lloyd Business School, Greater Noida, is a Ph.D. and MBA in human resource management and also Editor for Lloyd Business Review Journal. She is associated with academics for the past 17 years with 4 years of industry experience and enjoys being a corporate trainer, counsellor, and consultant. Her research interest lies in cross-cultural management, international management, human resource management, and human capital management. She has several high-quality research papers to her credit and has co-authored a book with a reputed publisher. In her free time, she likes to do some social work and deliver free career counselling sessions for students.

Nusrat Khan is an expert in the area of HR and OB. She has worked with national and International NGOs. She also worked as a researcher for projects with an internationally recognized Distinguished Professor of Western Kentucky University, Kentucky, USA. She has also worked for the corporates both nationally and internationally. She has taught in several well-known universities. While working in these universities she had held several administrative duties as well. She is presently working as a Professor at GD Goenka University in the School of Management. At GD Goenka University she has also been serving as Head Post Graduate Programmes, MBA Admission head, HR Area Head, Teaching various management subjects, also served as a class teacher, handling mentees, had been a member of the Internal Quality Assurance Cell (IQAC) as prescribed by National Assessment and Accreditation Council (NAAC), Member of the Departmental Research Committee for Doctor of Philosophy (Ph.D.) program of G.D. Goenka University, Member of the Examination Moderation Committee, served as an Associate editor of GD Goenka Journal. She is also a Mendeley Advisor from Mendeley (Elsevier UK).

Yael Lahav, Ph.D., is a Lecturer at the Department of Industrial Engineering and Management at the Shenkar College of Engineering and Design, and at the department of management at Bar Ilan University. She specializes in management in the public sector. Dr. Lahav obtained her Ph.D. from the Bar Ilan University. She participated in research on supply chain.

Hilal Tuğçe Lapçın has completed her bachelor's, master's, and doctoral degrees in the field of aviation management. She currently serves as a Assistant Professor at Samsun University. Her research areas include airport management, airline management, competition, and strategy. She is married and a mother of one child.

Armağan Macit completed his undergraduate, graduate and doctoral education in aviation management. After 5 years of industry experience, he continues his academic career at Ege University Aviation Vocational School. He conducts studies on strategic management in aviation, airport management, decision making with artificial intelligence in aviation and sustainability in aviation.

Deniz Macit completed her undergraduate education in finance and her master's and doctorate in economics. With her interest in the aviation sector, she also completed her undergraduate education in aviation management. She continues her academic career at Ege University Aviation Vocational School. She conducts academic studies in the fields of transportation economics, aviation economics and aviation finance.

Kartikay Saini, an ex Naval Commander of the Indian Navy, is the Chairman of Scottish High International School, India. He is the former Chairman of Special Olympics of India and Board Member of Special Olympics Worldwide, Washington. With more than 36 years of experience in the corporate, defense, and education world, Cdr Saini had been recognised worldwide and felicitated at House of Commons London, the President of Mauritius,, the Govt of Sri Lanka, the President of Dominican Republic, Education world, Forbes Magazine, among others. Presently he is a research scholar at GD Goenka University.

Arif Tuncal is an air traffic controller and researcher. He holds a Ph.D. in Air Traffic Control from Eskisehir Technical University. He continues to focus on topics such as air traffic control, air traffic management, and unmanned aerial vehicles.

Mustafa Uzgör is currently working at Muğla Sıtkı Koçman University. He has completed his bachelor's degree in aviation management at Erciyes University and his master's and doctoral degrees in aviation management at Anadolu University. He also has industry experience in the operational department at HAVAŞ Ground Handling Company. His research areas include airport and airline management, aviation policies, environmental management in aviation, and multi-criteria decision-making. His research interests focus on unmanned aerial vehicle transportation, urban air mobility, airport capacity, and operational management.

Daniel Xerri is an Associate Professor in Applied Linguistics and TESOL at the University of Malta, and the Chairperson of the ELT Council. His main research interests are professional learning, teacher research, and creativity education. www .danielxerri.com

Index

A

B

C

D

E

F

I

L

M

N

O

P

U

www.ingramcontent.com/pod-product-compliance
Lightning Source LLC
LaVergne TN
LVHW081349300926
849161LV00008B/2110